Campfires in Context: Hunter-Gatherer Fire Technology and the Archaeological Record of the Southern High Plains, USA

Paul N. Backhouse

BAR International Series 2114
2010

Published by

Archaeopress
Publishers of British Archaeological Reports
Gordon House
276 Banbury Road
Oxford OX2 7ED
England
bar@archaeopress.com
www.archaeopress.com

BAR S2114

Campfires in Context: Hunter-Gatherer Fire Technology and the Archaeological Record of the Southern High Plains, USA

ISBN 978 1 4073 0655 1

Printed in England by Blenheim Colour Ltd

All BAR titles are available from:

Hadrian Books Ltd
122 Banbury Road
Oxford
OX2 7BP
England
bar@hadrianbooks.co.uk

The current BAR catalogue with details of all titles in print, prices and means of payment is available free from Hadrian Books or may be downloaded from www.archaeopress.com

Experimental Research at the Lubbock Lake Landmark, July 2003

FOREWORD

Hearthstones are ubiquitous on the Southern High Plains, often ignored, seldom mapped in or collected, and even less frequently analyzed either individually or within the structure from which it came. Yet, as demonstrated by this monograph, these hot rocks have a tremendous tale to tell, either individually or as a constitute part of a feature. As an exception to the general rule, for almost 40 years, the Lubbock Lake Landmark regional research program has been gathering data on hearthstones and hearths (Paul's localized thermal features) throughout the Southern High Plains, conducting experimental work, and exploring ways to extract meaning and interpret human behavior. Paul, since his undergraduate days, has been steeped in that general program and mentored and encouraged to pursue his growing interest in hot rock technology and experimental work. Paul now has become the teacher, providing a studied approach and productive perspective that should open up new avenues for analysis and interpretation. This groundbreaking publication provides the opportunity for new insights into hunter-gatherer groups in terms of, among other aspects, technology, economy, and landscape use.

Landscape use, and the relationship of hunter-gatherers to the landscape, is perhaps of greatest interest as it ties hearthstones and hearths directly to the theoretical framework of a landscape approach. That approach is broad and interdisciplinary, accommodating various spatial scales and perspectives. In this approach, the landscape is formed from the relationship between environmental elements, i.e., landforms, soils, climate, and human interactions with the environment (Johnson, 2008; Heilen, 2005). The landscape is viewed as layered, each layer and its various parts and elements interacting synergistically, with the layers being physiographic, biotic, and cultural. Landscape elements, located on landforms, have specific geomorphic, topographic, and ecological characteristics. Both movement and foraging patterns are influenced by the spatial arrangement of landscape elements. Landscape elements provide potential for cultural activity and material remains, and they influence where such stopping points will be located. Hearthstones and hearths reflect all three layers and represent not only those stopping points but also places of human activity. And that activity represented by hearths and hearthstones is not always the same.

Hearths, regardless of their function and purpose, contain a wealth of information – a major point of this monograph. Fuel used is a reflection of the biotic layer as is the foodstuffs being cooked if that was the purpose of the hearth. A number of useful techniques now are available (among which are charcoal identification, phytolith analysis, residue and lipid analyses) to extract that biotic information and examine the decision-making processes that go into the construction and use of hearths. That information also provides a window into the environment in terms of native trees and plants as well as the why of preferential use. For the Southern High Plains, local folklore (including publications) has honey mesquite (*Prosopis glandulosa*) as an invader from South Texas brought in by cattle in the late 1800s and the region a starkly treeless plains. Neither is the case. Mesquite is a native tree, being used as fuel since at least Middle Archaic times. It is a favored fuel; time and again, identifications come back as mesquite regardless of time period or location. Oak (*Quercus*) also is used occasionally as was elm (*Ulmus*), acacia (*Acacia*), cottonwood (*Populus*), and willow (*Salix*). These are native trees on the Southern High Plains. Mesquite and oak also are hardwoods that burn hot and long with little smoke, favored today for these qualities. They sustain high temperatures for longer periods, critical to cooking geophytes (basically root foods) that require 24 hours or more baking time to render them digestible and nutritious. Hardwoods, however, are difficult to start and need a goodly supply of kindling such as twigs and grass.

Paul's work is complementary to and builds on the work of such US authors as Ellis (1997) and Wandsnider (1997) as well as the seminal work of Thoms (2003, 2008, 2009). These authors not only recognize the significance of fire technology and hot rock technology in their potential to inform, but also the need to view these features systematically and create standardized categories and a classification system. Their focus is on food-related hearths. Paul's effort goes beyond and is a larger umbrella encompassing a holistic approach in which to place all thermal features, not just a segment. As he notes, thermal features are not only for cooking purposes and not all thermal features were constructed with or used hearthstones. The earliest hearths known for the Southern High Plains are very shallow, small, rock-less basins with ash and charcoal or charcoal stained sediments. These may be warming hearths. Or as Thoms (2003, 2009) notes, a bed of coals where low cost foods such as lean meat or fish or a certain category of root foods could be prepared directly on the coals rather than in a baking oven.

Hearths in general, and pit hearths categorized as baking or earth ovens in particular, can be a useful way to measure land use intensification through time (Thoms, 2003, 2008, 2009). These baking ovens are used to cook geophytes, a more costly food item due to energy expenditure requirements. Time and energy is offset by abundance, nutritional value, and ability to feed increasing numbers of people from the same landscape. More people, then, can exist within that landscape through utilizing more of the food-resource potential (such as vegetal products underused or not previously used) and using them in an effective way in as efficient manner as possible. This aspect of viewing hearths as a measure of land use intensification certainly drives to the heart of hunter-gatherer optimal foraging theory and underscores Paul's point of the great explanatory information and illumination of hunter-gatherer behavior from the lowly hearths and even lowlier hearthstones. The use of hot rock technology and baking ovens also may be measures of stress and population growth. Their use can be seen as a mechanism to counteract periods of game uncertainly, such as during drought. Through time, population levels may be increasing if land use intensification continues or increases without climatic modulations. These measures harken back to Paul's point of hearths as analytical equals to lithics and bone in their potential to inform about hunter-gatherer behavior and social organization. They underscore Paul's call for systematic investigation of fire technology and hot rock technology at regional scales, while emphasizing the explanatory power at local scales.

The Southern High Plains is a vast area. The general dearth of systematic surveys has led to an uneven and arbitrary distribution of recorded sites. Intense collecting and vandalism has occurred since at least the 1920s, with site destruction through urbanization and agricultural practices since the 1880s. Erosion plays a major role in both visibility and destruction, with processes driven by drought and poor land management practices in the 20th century. The recorded information is of variable quality. These obstacles and the biases that may interject are not limited to the Southern High Plains but rather are a reflection of a more global situation. Paul is aware of these biases and his approach to the data is designed to compensate and make all data as useful as possible. It is that approach and research design that has global application – another major point made in the monograph. Nevertheless, Paul also recognizes that the data trends may be deceptive and much research still is to be done to verify and expand any model.

This monograph represents a tremendous step forward in understanding fire technology and hot rock technology and their role and relationship within hunter-gatherer societies. It should not only elevate the status of hearths and hearthstones as worthy of study within hunter-gatherer research but equally important, also present new avenues for that research.

Dr. Eileen Johnson

Senior Curator, Museum of Texas Tech University

Horn Professor of Museum Science

Director, Lubbock Lake Landmark

December 18, 2009

References:

Heilen, Michael. 2005. An Archaeological Theory of Landscapes. Unpublished doctoral dissertation, University of Arizona, Tucson.

Johnson, Eileen. 2008. Landscapes and Peoples of the Llano Estacado. In: Laura L. Scheiber and Bonnie Clark (eds.), *Archaeological Landscapes on the High Plains*, pp. 115-156. University Press of Colorado, Boulder.

Ellis, Linda W. 1997. Hot Rock Technology. In: Stephen L. Black, Linda W. Ellis, Darrell G. Creel, Glen T. Goode (eds.), Hot Rock Cooking on the Greater Edwards Plateau: Four Burned Rock Midden Sites in West Central Texas (Vol. 1). University of Texas, Texas Archeological Research Laboratory, *Studies in Archeology*, 22:43-81.

Thoms, Alston V. 2003. Cook-stone Technology in North America: Evolutionary Changes in domestic Fire Structures During the Holocene. In: M. Frere-Sautot (ed.), Le Feu Domestique et ses Structures au Neolithique et aux Ages des Metaux. Editions Monique Mergoil, Saint-Apollinaire, France, *Collections Prehistoires*, 9:87-96.

Thoms, Alston V. 2008. The Fire Stones Carry: Ethnographic Records and Archaeological Expectations for Hot-Rock Cookery in Western North America. *Journal of Anthropological Archaeology*, 27:443-460.

Thoms, Alston V. 2009. Rock of Ages: Propagation of Hot-Rock Cookery in Western North America. *Journal of Archaeological Science*, 36(3):573-591.

Wandsnider, LuAnn. 1997. The Roasted and the Boiled: Food Composition and Heat Treatment with Special Emphasis on Pit-hearth Cooking. J*ournal of Anthropological Archaeology*, 16:1-48.

LIST OF CONTENTS

		Page Number
	Title Page	i
	Foreword	v
	List of Contents	ix
	Acknowledgements	xii
	List of Tables	xiii
	List of Illustrations	xv
	Preface	xxiii
1	**Introduction**	1
	Fire Hearths, Archaeology, and Hunter-Gatherers	1
	Purpose of Research	2
	Project Description	2
	Project Context - Southern High Plains, USA	4
	Chapter Synopsis	4
2	**Background**	7
	Introduction.	7
	Archaeological Evidence	8
	Micro-scale Research	9
	Macro-scale Research	17
	Regional Level Research	21
	Thinking from Features, a Theoretical Background	21
	Portable Material Culture and the Traditional Perspective	22
	Interpretative Potential	22
	Summary	23
3	**Approach and Methodology**	25
	Research Approach	25
	Research Method	25
	Data Capture and Coding	26
	Texas Section	28
	New Mexico Section	30
	Search Criteria	30
	Geographic Information System	33
	Developing a Relational Database	34
	Applying the Classificatory Structure	36
	Heuristic Research	37
	Ethnographic Research	37
	Experimental Research	39

		Page Number
	Field Research	40
	Summary	41
4	**Geographic Focus**	43
	Physiographic Setting	43
	Great Plains	44
	High Plains	45
	Southern High Plains (Llano Estacado)	45
	Environmental	46
	Climatic Record	46
	Flora and Fauna	47
	Archaeological	49
	History of Research	51
	Key Sites	52
	Cultural Chronology	55
	The Ten-Mile Buffer Zone	57
	Conclusion	58
5	**Theoretical Framework**	61
	An Interpretative Model for the Investigation of LTFs	61
	Research Model	62
	A Hierarchical Key as a Means of Feature Classification	63
	The Five Classificatory Keys	66
	Assumptions and Limitations of Research	68
	Conclusion	69
6	**Ethnographic Context**	71
	Historic Documentary Sources	72
	Anthropological Fieldwork	73
	Domestic Facilities	75
	Cooking Facilities	76
	Symbolic Facilities	81
	Other LTF technologies	83
	Informant Interview	83
	Conclusion	85
7	**Experimental Research**	87
	Background	88
	Experimental Objectives in Light of the Regional Dataset	89
	Experimental Research Goals	92
	The Experiments	93
	Methods	93
	Hot-Rock Technology	94

Page Number

Site Formation Processes 96
Performance Characteristics 100
Summary and Conclusions 102

8 Archaeological Results and Analysis 105
Results of The Compilation and Characterisation of the Regional Dataset 105
Geospatial 106
Spatiotemporal 114
Technological 121
Synopsis of the Archaeological Results and Conclusion 138

9 Discussion and Potential for Application to Other Regions 139
Introduction 139
Drawing Together the Evidence for Domestic Fire Technology on the Southern HighPlains 139
Implications for Understanding Prehistoric Populations and Technology 140
Implications for Understanding Site Formation Processes and Regional Sources of Bias 151
Potential for the Investigation of Other Regions 152
Southern British Isles Dataset 154
Resituating Fire Technology Research in Hunter-Gatherer Studies 158
Establishing Research Values 159
Conclusions 160

10 Conclusion 161
Prehistoric Fire Technology on The Southern High Plains 165
Potential for Extension 166
Limitations 167
Suggestions for Future Research 167

References Cited 169

ACKNOWLEDGEMENTS

This manuscript is largely the result of several fortuitous events and I would like to sincerely thank the people who have spent the time and effort to help an Englishman appreciate the vast landscape of the Llano Estacado. Dr. Eileen Johnson gave me the opportunity to volunteer and later work at the Lubbock Lake site. Arriving from Las Vegas in 1997 as a volunteer, I stumbled from an ancient army issue tent into the intense heat and giant Texas sky. I thought I would give it a couple of days, make my excuses and be on the next plane to England. Nine years, a wife, two cats, and green card later I was still there, caught by the tumbleweeds, dust, and beauty of the High Plains.

Along the way, many people have taken the time to share their research and discuss the archaeological record of the region. In Bournemouth, Dr. Timothy Darvill and Jeff Chartrand have expertly and patiently led me through the research process. Their commitment to a cohesive archaeology that is not constrained by 'academic' or 'CRM' stereotypes has greatly helped along my thinking and the development of this project. The experimental research benefited from the comments of Dr. Alston Thoms (Texas A&M University), Dr. Michael Petraglia (Oxford University), and Dr. Christopher Lintz (Geo-Marine, Inc.). The Geosciences Department at Texas Tech University, particularly Dr. Thomas Lehaman, Dr. Melanie Barnes, and Dr. Harold Gurrola, all generously contributed their time, equipment, and expertise. Dr. Vance Holliday's (University of Arizona) geoarchaeological knowledge and congenial nature directly benefited my understanding of the region. His (and Dr. James Meyer's) annual Plains pilgrimage always made us remember there's a wider archaeological community out there. I sincerely appreciated the support and lively discussions with Dr. Eileen Johnson, who, has continuously supported my research and whose commitment to Lubbock Lake and the archaeology of Southern High Plains is unparalleled. My knowledge of the Southern Plains benefited from the generous advice and kind words of Dr. Paul Katz (PRIAM). The completion of the experimental research was greatly helped by the involvement of Alexander Brackenreed-Johnston, Dr. Briggs Buchanan, Ray Olachia, Dan Rafuse, Alejandra Matarrese, Blake Morris, Nicola Goodwyn, Doug Cunningham, and the lab. girls who never missed an opportunity to roast 'Smores' over an open fire. Mrs. Ardis Johnson deserves special mention for proof reading this manuscript more than once. I would also like to thank George MacDonell (Bureau of Land Management) and Andy Chopp (Museum of London Archaeological Service) for assistance with obtaining permission for images.

My sincere thanks to the staff and volunteers at Lubbock Lake with whom I had the privilege to work and who have put up with me on many long hot days in the field. The effort to bring my thesis to publication was greatly aided by the encouragement of Dr. Clive Gamble (Royal Holloway, University of London), two anonymous peer reviewers, and the excellent publication team at BAR.

To my wife Tara, thank you for the love, support, and innumerable cups of tea, without which I could not have completed this manuscript. To my parents in England, thank you for letting me know anything is possible and my new family in America, thank you for all your support.

Along the way this research benefited from the support of the South Plains Archaeological Society and their 'Llano Estacado Fellowship'.

LIST OF TABLES

Page Number

CHAPTER 2

Table 2.1. Example of the global variability in structures interpreted as evidence for hunter-gatherer localised thermal feature technology. 12

Table 2.2. Examples of typical terminology used in archaeological literature to classify localised thermal feature types. 15

CHAPTER 3

Table 3.1. Data for counties located within the Texas portion of the study area. 29

Table 3.2. Data for counties located within the New Mexico portion of the study area. 30

Table 3.3. UTM (North American Datum (NAD) 1927 datum) coordinates for the search area in the New Mexico portion of the study area. 32

Table 3.4. Example of feature descriptions contained on site-survey forms in the Texas portion of the study area. 32

Table 3.5. Standard thermal feature definitions used for New Mexico site-survey forms. 33

Table 3.6. Summary of organizations providing GIS data used in this project. 34

Table 3.7. Summary of field work undertaken between 2002 and 2007 by the LLLRRP at hunter-gatherer sites with LTFs on the Southern High Plains. 41

CHAPTER 4

Table 4.1. Edible native plants identified on the Southern High Plains and in adjacent areas. 51

CHAPTER 7

Table 7.1. Summary of the objectives of the experimental programme by experiment. 93

CHAPTER 8

Table 8.1. Frequency of sites identified in the present study versus those surveyed in south-eastern New Mexico, presented by cultural component. 115

Table 8.2. Summary of Archaic period sites identified in the regional LTF dataset. 118

Table 8.3. Summary of the physical characteristics of the Ceramic period LTFs investigated at the Lubbock Lake Landmark. 133

Table 8.4. Summary of faunal resources identified in association with LTFs on the Southern High Plains by major time period. 135

Tale 8.5. Summary of the physical characteristics of the Protohistoric period LTFs investigated at the Lubbock Lake Landmark. 136

Page Number

Table 8.6. Summary of the archaeological evidence for hunter-gatherer domestic fire technology by major cultural time period. 137

CHAPTER 9

Table 9.1. Summary of the size of features recorded by major temporal period. 150

Table 9.2. Summary results of literature search of post-glacial hunter-gatherer sites with positive evidence of LTF in the Southern Britain Investigation area. 155

LIST OF ILLUSTRATIONS

Page Number

CHAPTER 1

Figure 1.1. Location of the Southern High Plains: a) in relation to the Great Plains physiographic province; b) regional overview (diagram b) modified from Holliday 1997, fig. 1.2). 5

CHAPTER 2

Figure 2.1. Overview of the range of hunter-gatherer fire technologies identified in the ethnographic and archaeological record (data from Rolland 2004:table I; Bellomo 1994, fig.2). 8

Figure 2.2. Location of LTF sites mentioned and discussed in Chapter 2. 9

Figure 2.3. High resolution (6in/15.2cm pixel resolution) aerial photograph taken in March 2005 showing two thermal features within the experimental research area at the Lubbock Lake Landmark, Texas; top insert shows an oblique view of the corresponding feature, only 0.5m in diameter photograph also taken in March 2005; bottom insert shows an oblique view of the corresponding feature, a large 1.5m diameter earth oven, photograph taken in June 2005 (aerial photograph available from: http://maps2005.ci.lubbock.tx.us). 10

Figure 2.4. Example of some typical LTFs recorded during archaeological fieldwork: a) cluster of burned rocks exposed on the modern ground surface at 41LU35 in Texas, USA; b) discrete area of burning revealed on the modern ground surface after a blowout episode in a dune field, unnamed site, Southern High Plains, Texas. (figures Courtesy of the Anthropology Division, Museum of Texas Tech University). 13

Figure 2.5. Excavation of thermal pit feature during the 2001 field season at Tahoka Lake (41LY52) Area 14: a) field photograph of the feature exposed in the arroyo wall; b) profile of feature with the locations of the hearthstones indicated within the feature fill; c) Plan view of the excavation area. Note the hearthstones exposed in the wall of unit 2496N529W (figures modified from Backhouse 2003, fig. 6.2, 6.4, 6.3). 13

Figure 2.6. Hearthstone cluster on the modern ground surface at PLK Locality 4 identified during pedestrian survey on a ranch in west Texas. Note the dark discoloration of the rocks. 14

Figure 2.7. Overview of the range of archaeometric techniques applied to the investigation of LTFs encountered during archaeological fieldwork. 16

Figure 2.8. Intersecting LTFs excavated into the aeolian upland sediments at 41LU1, Area 10 in Lubbock County, Texas, USA. Note: field photograph taken after the complete excavation of several of the features (photograph courtesy of the Anthropology Division, Museum of Texas Tech University). 18

Figure 2.9 Results of samples sent for radiocarbon dating from 41LU1 Area 10 plotted using OxCal Viewer Version 3.9 (data from Buchanan 2002, table 7.1; Backhouse 2002, table 10.5). 18

CHAPTER 3

Figure 3.1. Conceptual timeline showing the main methodological components of the research project. 27

Figure 3.2. Simplified process chart showing the component steps of data capture. 27

CHAPTER 3 continued

Figure 3.3. Southern High Plains region showing the County boundaries. Note: The ten-mile boundary around the physiographic region indicated in green. 28

Figure 3.4. Distribution of identified archaeological sites prior to geoprocessing. 31

Figure 3.5. Distribution of archaeological sites within the investigation area following geoprocessing. 31

Figure 3.6. Example record output from the sites database. Note: the northing and easting are obscured in this example to protect the location of the archaeological site, site location data in Texas are confidential under state (§191.004 Texas Natural Resources Code (VTCA 1995)) and federal (Archeological Resources Protection Act of 1979, 16 USC §470aa et. seq.) laws. 34

Figure 3.7. Example of the decision making process in the assignment of the primary and secondary LTF components at 41LU6 based on the descriptive data compiled from the site survey record. 37

Figure 3.8. Reconstruction drawing of hunter-gatherer residential activities in relation to a centrally placed LTF (figure Sidell et al. 2002:20, fig. 15, copyright Museum of London Archaeological Service). 38

Figure 3.9. Ray Olachia instructs local high school teachers, students, and professional archaeologists in the construction of an earth oven at the Lubbock Lake research area. photograph: authors own 39

Figure 3.10. Excavation block at 41LU118 during full-scale excavation. photograph: authors own 41

CHAPTER 4

Figure 4.1. Physiographic location of the Southern High Plains on the North American Plains overlaid with the location of modern political boundaries. 44

Figure 4.2. Physiographic setting of the Southern High Plains (Llano Estacado) with neighbouring geographic areas indicated. 45

Figure 4.3. Geomorphic development of the Southern High Plains; photograph of the eastern escarpment showing the exposed upper caprock section above the low Rolling Plains. (P.N. Backhouse) 46

Figure 4.4. Modern precipitation pattern for the Southern High Plains Regions. Note that rainfall increases from southwest to northeast across the region (data National Oceanic Atmospheric Administration). 46

Figure 4.5. Generalised temporal framework for the Southern High Plains showing climatic and cultural periods, with the approximate appearance of localised thermal feature technology indicated in red (modified after Johnson 1987a, fig. 2.7; chronostratigraphic conventions follow Holliday 1997:16). 47

Figure 4.6. The modern environment on and around the Southern High Plains: a) short-grass prairie in a virtually featureless landscape; b) a prairie dog emerges from his subterranean burrow excavated into the prairie; c) the desert scrubland just off of the south eastern escarpment, note the honey mesquite to the right of the photograph and prickly pear cactus in the foreground; d) looking back at the western Mescalero escarpment across the dune fields of New Mexico; e) water flowing in a perennial stream in the eastern canyonlands; f) Stands of juniper growing along the eastern escarpment and rough canyonlands. (P.N. Backhouse) 49

CHAPTER 4 continued

Figure 4.7. Modern farming intensity by county (data U.S. Census Bureau, Census 2000). 50

Figure 4.8. Intensity of archaeological research on the Southern High Plains presented by decade. (data sampled from TARL and ARMS and Accessed September 2005). 52

Figure 4.9. Key sites identified on and in proximity to the Southern High Plains. 53

Figure 4.10. Late Quaternary valley fill exposed in the type section at Lubbock Lake (41LU1). Note the stratigraphic sequence spans the late Pleistocene through Holocene (photograph courtesy of the Anthropology Division, Museum of Texas Tech University). 54

Figure 4.11. Boyd's Late Prehistoric (c. 750 – 1450 B.P.) culture complexes and phases for the Southern High Plains region (modified after Boyd 1995, fig. 3). 56

Figure 4.12. Southern High Plains region showing the arbitrary ten-mile extension used in the current programme of research. 58

CHAPTER 5

Figure 5.1. Research model for the investigation of localised thermal features. 63

Figure 5.2. Example of the types of information typically recorded for a hypothetical site in Texas, USA. and on file at the Texas Archaeological Research Laboratory and the Texas Historical Commission Sites Database. 64

Figure 5.3. Hierarchical key for the classification of localised thermal features. 65

Figure 5.4. Examples of an intentionally excavated LTF structure encountered during archaeological research. Note the stained sediments encountered in the wall of an excavated trench (photograph courtesy of the Anthropology Department, Museum of Texas Tech University). 66

Figure 5.5. North American example of a thermally altered soil stain defining the spatial boundary of a feature recorded during excavation at 41LU1, area 10 (photograph courtesy of the Anthropology Department, Museum of Texas Tech University). 67

Figure 5.6. Example of a structurally intact feature; note the presence of fine ash around the exposed hearthstones and the obvious visual difference between the carbonized material within the feature and the surrounding sediments. 67

Figure 5.7. Example of the two measurements for the determination of diameter and depth of an LTF. In this example maximum width is greater than the maximum depth resulting in a basin-shaped structure. 67

Figure 5.8. Caliche hot-rocks within an experimental LTF at Lubbock Lake Landmark, Texas (photograph courtesy of the Anthropology Department, Museum of Texas Tech University). 68

CHAPTER 6

Figure 6.1. Locations of the major Native American ethnic groups on the Great Plains by the early 19th Century (modified from Carlson 1998:3). 74

CHAPTER 6 continued

Figure 6.2. Direct and indirect container boiling methods (after Holmes 1903, fig. 2). 78

Figure 6.3. Ethnographic spatial distribution of direct fire boiling technology in relation to the Southern Plains and the Llano Estacado Research Area (data from Driver and Massey 1957, maps 41 and 42) . 79

Figure 6.4. Ethnographic spatial distribution of indirect fire boiling technology in relation to the Southern Plains and the Llano Estacado Research Area (data from Driver and Massey 1957, maps 43 and 44) . 79

Figure 6.5. Ethnographic spatial distribution of earth oven technology in relation to the Southern Plains and the Llano Estacado Research Area (data from Driver and Massey 1957, map 45). 81

Figure 6.6. Ethnographic spatial distribution of meat drying technology in relation to the Southern Plains and the Llano Estacado Research Area (data from Driver and Massey 1957, map 53) . 82

Figure 6.7. Men of European American descent dominate the race to light an educational earth oven, an event in direct contrast to Ray Olachia's stated Mescalero Apache sexual task divisions (Note: the peripheral positions of European American females in this activity). 84

Figure 6.8. Women of European American descent dominate the preparation of food for placement in the earth oven (Note: the peripheral positions of the older European American men in this activity). 84

CHAPTER 7

Figure 7.1. Frequency of prehistoric archaeological sites identified with evidence for hot-rock use. 90

Figure 7.2. Temporal distribution of evidence for hot-rock use at prehistoric sites with LTF components within the study area. Note (TY) = age of site estimated from typological analysis of artefacts, (DA) = age of site secured from radiometric assay e.g. radiocarbon dated. 90

Figure 7.3. Geographic distribution of prehistoric archaeological sites within the LTF database with evidence for hot-rock use. 91

Figure 7.4. Classification of the primary LTF types at prehistoric archaeological sites within the study area. 92

Figure 7.5. Experimental research area at Lubbock Lake Landmark (P.N. Backhouse) 94

Figure 7.6. Alejandra Matarrese mapping the displacement of hearthstones during a bitterly cold winter's day on the Southern High Plains (photograph courtesy of the Anthropology Division, Museum of Texas Tech University). 94

Figure 7.7. Experiment 4 facilitated the field identification of Ogallala Fm. gravels as indicators for hunter-gatherer domestic fire technologies; a) large structurally intact LTF with Potter member hearthstones identified during pedestrian survey at PLK-Locality 4; b) isolated Ogallala Formation hearthstone (indicated by black arrow) from an eroded hearth identified as 41GR146 (P.N. Backhouse). 95

Figure 7.8. Spatial displacement of simulated cultural material around a hearth after one year; note the extreme horizontal displacement of specimens in the north of the experimental area along the route of the tour groups. 97

CHAPTER 7 continued

Figure 7.9. Basin hearth taphonomy experiment after two years of experimentation: a) overall view of the feature; b) close-up of an internal hearthstone that has fractured. 99

Figure 7.10. Heat dissipation from an experimental basin hearth plotted for a one-hour period: a) 0.5 meters from the hearth; b) 1.5 meters from the hearth; c) 2.5 meters from the hearth; and d) 3.5 meters from the hearth. 100

Figure 7.11. Light dissipation from an experimental basin hearth plotted for a one-hour period: a) 0.5 meters from the hearth; b) 1.5 meters from the hearth; c) 2.5 meters from the hearth; and d) 3.5 meters from the hearth. 101

CHAPTER 8

Figure 8.1. Archaeological sites with LTFs by state within the study area. 106

Figure 8.2. Spatial distribution of sites with LTFs identified within the research area. 107

Figure 8.3. Number of sites with evidence for domestic fire technology by State and County. 108

Figure 8.4. Spatial distribution of sites with LTFs identified within the research area overlaid on the modern vegetation communities along the Texas potion of the area (data: 1:100,000 scale vegetation community dataset, Texas Park and Wildlife). 109

Figure 8.5. Spatial distribution of sites overlaid by modern farming intensity by county (data U.S. Census Bureau, Census 2000 Data). 110

Figure 8.6. Bivariate scatterplot showing percentage of County classified as farmland versus the number of LTF sites identified per square mile, excluding Armstrong, Bailey, Borden, Chochran, Dawson, Glasscock, Hall, Hemphill, Hockley, Mitchell, Upton, Wheeler, and Yokam Counties due to the lack of identified sites (agricultural data courtesy of the U.S. Census Bureau, Census 2000). 111

Figure 8.7. Distribution and size of sites identified along the Mescalero Escarpment and adjacent dunes relative to Bureau of Land Management (BLM) land. 112

Figure 8.8. Distribution and size of sites identified along the Mescalero Escarpment and adjacent dunes relative to modern land cover (vegetation data from the Natural Resources Conservation Service, www.nrcs.usda.gov). 113

Figure 8.9. LTF sites identified as large-scale survey projects in the Texas portion of the study area. 113

Figure 8.10. Temporal distribution of prehistoric sites with LTF components within the study area. 116

Figure 8.11. Spatiotemporal distribution of archaeological sites with LTF components within the study area. 117

Figure 8.12. Distribution of Paleoindian archaeological sites with LTFs within the study area (BWD Locality 1 added). 118

Figure 8.13. Distribution of Archaic archaeological sites with LTFs within the study area. 119

Figure 8.14. Distribution of Ceramic/Mogollon archaeological sites with LTFs within the study area. 120

Figure 8.15. Distribution of Protohistoric archaeological sites with LTFs within the study area. 120

Figure 8.16. Mode of archaeological investigation undertaken at sites with LTFs by state. Note: where more than one mode of investigation occurs then the most intensive is selected regardless of date. 121

Figure 8.17. Frequency of LTFs at prehistoric archaeological sites within the study area. 122

Figure 8.18. Frequency of LTFs by geographic location at prehistoric archaeological sites within the study area. 123

Figure 8.19. Frequency of different LTF components at prehistoric archaeological sites within the study area. Note: the lack of variation in identified components indicating that in most cases only one type of LTF was discernable when applying the hierarchical key. 123

Figure 8.20. Classification of the primary LTF components at prehistoric archaeological sites within the study area. 124

Figure 8.21. Distribution of the primary LTF components identified at sites and presented by State (note: including BWD, Locality 1). 125

Figure 8.22. Classification of the secondary LTF components at prehistoric archaeological sites within the study area. 126

Figure 8.23. Classification of the tertiary LTF components at prehistoric archaeological sites within the study area. 127

Figure 8.24. Temporal distribution of primary LTF components at prehistoric sites within the study area (note: including BWD, Locality 1). 127

Figure 8.25. Photograph of the fire basin at the Milnesand site. Text caption hand written by James Warnica states 'Ted Williamson, Sr. Standing at hearth at the Milnesand site Feb. 21, 1965. Photo was Taken from Sellards 1954 Test Pit, Looking to the N.E.' (photograph: James Warnica, courtesy of the Anthropology Division Museum of Texas Tech University). 128

Figure 8.26. Photograph of the fire basin at the Milnesand site. Text caption hand written by James Warnica states 'Overhead view of hearth exposed at the Milnesand Site Feb. 21, 1965. Finger of Ted Williamson, Sr. points to charred bone' (photograph: James Warnica, courtesy of the Anthropology Division Museum of Texas Tech University). 129

Figure 8.27. Photograph of the fire basin at the Milnesand site. Text caption hand written by James Warnica states 'Bottom of hearth exposed at the Milnesand site. Finger of Ted Williamson, Sr. points to charred bone. Feb. 21, 1965' (photograph: James Warnica, courtesy of the Anthropology Division Museum of Texas Tech University). 129

CHAPTER 8 continued

Figure 8.28. Photograph of the fire basin, most probably at the Williamson site. Text caption hand written by James Warnica states 'View of hearth at a Folsom site on the former game refuge S.E. of Milnesand Feb. 23, 1965. The sand was burned but it was not collected. There were no burned bones in the hearth' (Photograph: James Warnica, courtesy of the Anthropology Division Museum of Texas Tech University). 129

Figure 8.29. Fuel rich Juniper landscape found off the caprock along the eastern edge of the Llano Estacado. (P.N. Backhouse) 131

Figure 8.30. Excavation of a Late Archaic LTF (fire basin with hot-rocks) at PLK-Locality 19: a) field crew recording and mapping the LTF; b) half-section profile revealing the blackened sediments and largely intact structure at the base of the feature. 131

Figure 8.31. Early Ceramic LTF (hearth with hot-rocks) identified along an eroded intermittent drainage channel at PLK-Locality 4. 132

Figure 8.32. 41LU1, Area 10 plan drawing showing the relationship of the overlapping LTFS (fire basins with hot-rocks) to the east of the area and the more domestic focus to the west of the site; note radiocarbon assays are listed for individual features from oldest to youngest (data: Backhouse 2002:table 10.5; Buchanan 2002:table 7.1). 134

Figure 8.33. Summary of the archaeological evidence for localised thermal features identified within the Southern High Plains study area. 138

CHAPTER 9

Figure 9.1. Hunter-gatherer campsite distribution and modern biomass index for the Southern High Plains Study area; Note that the darker shaded areas indicate areas of higher biomass(data: Normalized Differential Vegetation Index (NDVI) acquired from NASA's Moderate-resolution Imaging Spectroradiometer (MODIS), averaged data acquired between 1990 and 2005 and available from the USGS Seamless Data Server (www.seamless.usgs.gov)). 141

Figure 9.2. Hunter-gatherer campsite distribution and modern ranges of major tree species that may be suitable as fuel in domestic fire technologies: a) Netleaf Hackberry (Celtis reticulata); b) Eastern Cottonwood (Populus deltoids); c) One Seed Juniper (Juniperus monosperma); d) Mesquite (Prospis prospubescens); e) Harvard Oak (Quercus havardii); f) Pinchot Juniper (Juniperus pinchotti) (data: available from the United States Geologic Service and modified from Digital Representations of Tree Species Range Maps from "Atlas of United States Trees" by Elbert L. Little, Jr. (and other publications) and freely available from http://esp.cr.usgs.gov/data/atlas/little/). 142

Figure 9.3. General population model based on the localised thermal feature dataset for the Southern High Plains study area. 143

Figure 9.4. Modes of engaging the archaeological record on the Southern High Plains by major temporal period. 147

Figure 9.5. Spatial distribution of Ceramic period sites with evidence for LTFs and relationship to previously developed models for the spatial extent of the Jornada Mogollon culture (data Corley 1965). 148

Figure 9.6. Examples of a 'Ring Midden' feature recorded along the western margin of the Pecos river. Note both features were located directly to the west of the present study area (photograph courtesy of the Bureau of Land Management, Carlsbad Office). 149

CHAPTER 9 continued

Figure 9.7. Frequency of radiocarbon determinations obtained on Ring Midden features identified in southeastern New Mexico (data: Katz and Katz 2000, table 35). 149

Figure 9.8. Summary of the regional record for hunter-gatherer fire technology on the Southern High Plains and surrounding area (climate and Cultural Periods after Johnson and Holliday 1995, 2004). 153

Figure 9.9. Distribution of hunter-gatherer sites with evidence for localised thermal features identified within the southern British study area. 157

Figure 9.10. Range of types of LTF in the southern British study area after classification. 158

PREFACE

Research reported herein represents an original project devised for the purposes of exploring hunter-gatherer domestic fire technology as represented in the archaeological record of the Southern High Plains of North America. The project, was conducted within the mutually complementary objectives of the Lubbock Lake Landmark Regional Research Programme which seeks to examine grasslands hunter-gatherers and adaptation to ecological change on the Southern Plains.

Vandalism and looting of archaeological sites in North America is a major heritage management problem and the geographic locations of archaeological sites is restricted under state (§191.004 Texas Natural Resources Code (VTCA 1995)) and federal (Archeological Resources Protection Act of 1979, 16 USC §470aa et. seq.) laws. Throughout the following manuscript site distribution maps are deliberately scaled to a regional perspective and omit elements of the modern built environment that might otherwise give more detailed information as to the location of a site. Qualified readers interested in accessing specific site data should contact the relevant State Historic Preservation Office to obtain access to these restricted access datasets.

All the reported experiments were devised by the present author in consultation with Dr. Eileen Johnson (Museum of Texas Tech University).

A bibliography of the previous dissemination of components of the broader project includes the following:

Peer-Reviewed Publications:

Backhouse, Paul N., Eileen Johnson, Alexander Brackenreed-Johnston and Briggs Buchanan 2005. Experimental Hearths and the Thermal Alteration of Caliche on the Southern High Plains. *Geoarchaeology: An International Journal*, 20(7):695-716.

Backhouse, Paul N., and Eileen Johnson 2007. Hearth Life: An Actualistic Examination of Site-Formation Processes Acting on Upland Hunter-Gatherer Camp Site Assemblages on the Southern High Plains. *Plains Anthropologist*, 52(202): 153-172.

Backhouse, Paul N., and Eileen Johnson 2007. Where Were the Hearths: An Experimental Investigation of the Archaeological Signature of Prehistoric Fire Technology in the Alluvial Gravels of the Southern Plains. *Journal of Archaeological Science*, 34:1367-1378.

Conference Papers:

Backhouse, Paul N.,
2003. Social Use of Space and the Role of Hearths. Paper presented at the 39th annual Southwestern Federation of Archaeological Societies, Midland, Texas.

2004. Conceptualizing the Social Use of Space Around a Prehistoric Hearth Pit. Paper presented at the 69th Society for American Archaeological conference. Montreal, Canada.

2004. A New Theoretical Approach in Interpreting Prehistoric Hearth Pits. Paper presented at the March meeting of the South Plains Archaeological Society, Lubbock, Texas.

2006. Lost on Ignition: Identifying Hunter-Gatherer Domestic Fire Technology in the Southern British Mesolithic. Paper Presented at the 12th Annual Meeting of the European Association of Archaeologists, Cracow, Poland.

2007. The Quest for Fire: State Level Databases and Secondary Data as the Basis for Meaningful Research. Presented in Small Pieces: Big Picture: Making results from contract archaeology illuminate the past, organized by Roger Thomas and Timothy Darvill. Paper presented at the 72nd Society for American Archaeological conference. Austin, Texas.

Backhouse, Paul N., Briggs Buchanan, and Alexander Brackenreed Johnston.
2002. Experimental Burned Caliche – Analytical Issues. Paper presented at the 60th Plains Anthropological Society conference. Oklahoma City, Oklahoma.

Backhouse, Paul N., Eileen Johnson, Alejandra Matarrese, Dan Rafuse, and Blake Morris
2005. Hearth Life: An Actualistic Examination of Post-Depositional Processes on Upland Hunter-Gatherer Camp Site Assemblages. Poster presented at the 70th Society for American Archaeological conference. Salt Lake City, Utah.

CHAPTER 1 - INTRODUCTION

Abstract

The hunter-gatherer use of fire has been identified as a necessary technology integral to post-glacial human communities. An apparent disjuncture, however, exists between the high frequencies of thermal features, such as hearths or ovens, observed during archaeological fieldwork and the disproportionately low level of research directed towards their interpretation. This situation is exacerbated as greater time-depth is considered. In response, a broad research goal for the investigation of these structures is outlined. A project that utilises four analytical stands of investigation is devised as a useful method for achieving this research goal. The project operationalises a holistic methodology whereby disparate research strands are explored and brought to bear at a regional scale. The Southern High Plains region of North America is identified as an ideal location in which to critically test and evaluate the research potential of this project. Lastly, the format of the report is set out in a brief chapter synopsis.

Fire, Hearths, Archaeology, and Hunter-Gatherers

To date, the earliest definitive evidence for the anthropogenic use of fire has been identified in Middle Pleistocene contexts from southwest Asia (Neumann 2004; Rincon 2004; Randerson 2004) and is arguably the technologically most distinct and iconic of human behavioural adaptations (Canti and Linford 2000:385). Evidence from disparate locations around the globe suggests that prehistoric hunter-gatherer groups routinely utilised fire in complex plant and game management strategies as well as in more domestic contexts (Pyne 1991; Smith 1993; Rolland 2004). Hunter-gatherer use of fire is therefore clearly a technological, diachronic, multi-scalar phenomenon, which can be approached by archaeologists from a number of perspectives. While some interest in the development of a "pyroarchaeology" has recently been explored (Gheorghiu 2002; Otte 2002) there has been little extension towards the systematic investigation of anthropogenic fire at regional scales.

The Acheulian site Gesher Benot Ya'aqov, located in modern Israel and dated to ca.790,000 B.P., is currently accepted as providing the earliest evidence for the controlled use of fire by humans (Neumann 2004; Rincon 2004; Randerson 2004). However, emerging evidence from sites in both eastern and southern Africa suggests the possibility of the use of fire by *Homo erectus* populations by 1.6 and 1.4 million years ago (Leonard 2002). The limited African evidence has led Richard Wrangham and colleagues to suggest that *Homo erectus* first developed the controlled use of fire around 1.8 million years ago as a means to process plant foods and therefore obtain higher calorific returns (Wrangham *et al.* 1999). Preservation issues, rather than human behaviour, appear to be the main barrier to establishing increasingly earlier dates for the development of pyrotechnic skill by humans (Leonard 2002).

It is very likely, regardless of the earliest date at which humans learned to control and use fire, that this technological innovation was developed prior to any migration of human populations into northern latitudes (e.g. Perles 1981). An existing continuum of knowledge and skill is therefore inherent in any post-glacial technological study of the human manipulation of fire. Data compiled for multiple Palaeolithic sites by Rolland (2004) suggests a significant range of fire technologies were represented at these sites (Rolland 2004, table I; see also Bellomo 1994, fig. 2). At the other end of the temporal scale, fire-based technologies are essential for all historic and modern societies (Pyne 2001). The continued need for fire technology as a necessary component of modern society renders it a somewhat familiar emic status, which stands in stark contrast to the unfamiliarity associated with discontinued technologies such as lithic reduction (Whittaker 1994).

The identification of localised thermal features (LTFs), such as hearths and ovens, during the archaeological investigation of a hunter-gatherer site or within a wider landscape survey is a relatively commonplace event. The terminology LTF is adopted here as the basic descriptor for the archaeologically visible component of hunter-gatherer domestic fire technology. The classification was selected because it is interpretively neutral and has not been used previously by other researchers. Fieldwork methodologies that facilitate their identification include, but are not limited to: pedestrian survey, geophysical prospecting, geochemical prospecting, and various forms of subsurface excavation.

Although commonly observed and recorded, there appears to be little serious dialogue between archaeologists as to what these features were used for, how to record them, what samples to take, and ultimately what they may have meant to the people who once constructed them. Concomitant research values appear to shift dependant on the recording context, but overall the investigation of thermal feature-related assemblages appears to have held only limited appeal for archaeological researchers. In part, this situation appears to be directly related to the nature of the evidence (i.e. predominantly hot-rock scatters) which historically appear, due to their great abundance, to have been assumed to hold "no real research value for reconstructing human behaviour or decision making" (Brink and Dawe 2003:86).

In contrast, a wealth of ethnoarchaeological studies have examined the relationships of human behaviours to thermal features and the resultant discard patterns that could be expected in the archaeological record (cf. Binford 1977; 1978; 1983; 1987; Yellen 1977; O'Connell *et al.* 1991; Kelly 1995). The strength of these ethnographic studies has historically been their ability to provide best-fit answers to site-level interpretation. This situation coupled with the general lack of interest expressed from an archaeological perspective has arguably resulted in the uncritical adoption of interpretative statements built largely on ethnographic generalisation.

A growing body of data suggests that this view is simplistic and furthermore, that increased emphasis can be placed on the material evidence by pursuing new and innovative directions in hunter-gatherer thermal feature research (e.g. Stevenson 1991; Dering 1999; Wandsnider 1997; Petraglia 2002; Thoms 2003, 2009). Despite these efforts, we are still a long way from the development of an explicit "pyroarchaeology" (Gheorghiu 2002). The presented research is therefore offered as a preliminary step in this direction.

Purpose of Research

The broad aim of this project is formalised as:

The design and implementation of a research framework for the investigation of the production and use of localised thermal features as built by Southern High Plains Hunter-Gatherer groups.

Features as constructed and used by historic and prehistoric hunter-gatherer groups for the controlled manipulation of heat, through the medium of fire, can be grouped together within the terminological class Localised Thermal Feature (LTF). This constructed feature class subsumes a massive range of internal variability in terms of the geographic distribution, type, technology, function, and attached cultural meaning(s).

The terminology LTF has been selected because semantically it encompasses a broad spectrum of features that have previously been classified in largely arbitrary typological frameworks. Examples of such features include: hearth pits, ovens, basin hearths, outdoor/indoor cooking facilities, and roasting pits. Localised thermal features are herein defined as non-portable, discrete, human-built structures (usually less than 3m in diameter), which have been utilised for a variety of social, economic, political, and ritual activities, all of which require fire as their catalyst. This definition is applicable across vast tracts of time and space (e.g. Thoms 2009).

Project Description

The archaeological investigation of prehistoric hunter-gatherer groups engages with limited material culture (and therefore archaeological evidence), badly preserved locations of activity, and vast spans of time (Kroll and Price 1991). Historically, the examination of these groups has been an examination of portable material culture and spatial distributions (Davies 2000). Resultant synthetic works trend toward an emphasis on materiality, focussing the analytical lens on change over long time scales.

A contrasting but ultimately complimentary research framework is presented here. The fundamental difference in approach is that the analytical lens is focussed on examining feature life-cycle data rather than material discard data. It is posited that the act of constructing an LTF necessitates at least a temporary cessation in the ongoing ambulatory movement of hunter-gatherers (Bender 2001). The physical imposition of these features on the local landscape can, therefore, always be construed as a dynamic locus for human activity (cf. Binford 1978). Activity may involve one or many people and may be as short as to be measured in moments or as long as to be measured in seasons. Thermal features contrast then with the majority of hunter-gatherer technology, in that the technological mechanism is not transportable and therefore has be studied *in situ*. Identification of LTF(s) at an archaeological site can be situated within a qualitative framework as a constituent component in the archaeological formulation of the term campsite. This realisation underscores a high potential for their investigation.

The research model examines variability in these feature types by exploring multiple lines of evidence derived primarily from four complimentary analytic strands of research (taxonomic, ethnographic, experimental, and archaeological). The relationship among these components is open-ended so that new lines of inquiry can emerge throughout the project (e.g. Wylie 2002). The archaeological record of the Southern High Plains of North America is selected as a suitable region with which to examine in detail the potential of this approach. The outcome is the foregrounding of LTFs as active structures, on analytical terms equal to the traditionally emphasised stones and bones artefact classes, in their potential to inform our knowledge of prehistoric human behaviour.

Aims are pursued through eight formal objectives, which together comprise a cohesive and achievable structure, within which to work toward the previously identified broad research goal:

1. Document the main types and distribution of localised thermal features.

2. Qualify the recorded archaeological signatures of prehistoric localised thermal features found on the Southern High Plains of North America through the application of a taxonomic key.

3. Assemble data on the processing of specific subsistence resources of the kinds represented at sites with localised thermal features.

4. Examination of the role of hot-rock technology in thermal features and site structures.

5. Carry out fieldwork to investigate research driven questions developed from a scrutiny of published material and experimental studies.

6. Analyse the archaeological dataset, qualified in terms of the formal, material, spatial, and ideological relationships involving localised thermal features and the material evidence for hunter-gatherer activities within campsites.

7. Critically assess the relevance of the overall interpretative project to the understanding of localised thermal features as identified in the archaeological record of the North American Southern High Plains.

8. Assess the relevance of the interpretative project and its findings to the broader understanding of prehistoric hunter-gatherer societies.

In order to practically achieve the stated research objectives, a panoptic research methodology was employed. Methodological considerations were contextually dependant on the stage of research and the task at hand. Six tasks are identified as being methodologically distinct, and together comprise the major components of the research project.

A background literature review consisted of the thorough examination of existing documents with descriptions or analyses relating to localised thermal features as identified in the archaeological record. Literature consulted ranged from brief archaeological site reports through to voluminous edited monographs. The breadth of research, although largely anglophone in scope, scaled down from a global perspective to that of the study region(s) and the individual site within these regions.

The development of a Geographic Information System (GIS) project facilitated the incorporation of a range of diverse spatial datasets (e.g. geographic, geologic, hydrologic, political) with which to examine, interrogate, and set geographic limits on the archaeological dataset.

A preliminary small-scale pilot study was undertaken in order to assess the range and type of data commonly recorded at a site level. Subsequently, a relational database was designed in order to store site data as sets of variables considered useful to characterizing the variability observed in the regional dataset. The completed database was designed to be queried independently and in concert with the GIS project.

Data collection of recorded sites required access to two cultural resource databases operated independently by federally funded agencies in New Mexico (New Mexico Office of Cultural Affairs, Santa Fe) and Texas (Texas Historical Commission, Austin). Searches were performed on both datasets using common criteria with which to extract comparable sites record data. The GIS project allowed for sites to be included in the sites database if LTFs were identified at a site and if the geographic location intersected with the study area boundaries.

Variation in the regional dataset was classified through the application of a hierarchical key. This practical framework establishes the vocabulary necessary in order to classify LTFs and allow broad patterns to be identified in the regional dataset.

Heuristic research methods were developed specifically to explore the research potential of the trends in the data identified through the application of the hierarchical key. *Experimental research* was initiated to examine physical processes from a materials perspective, e*thnographic research* examined the range in technological variation documented by contemporary hunter-gatherer groups utilizing thermal features, and *fieldwork* was initiated to target sites with high research potential, identified in the regional database.

Project Context - Southern High Plains, USA

The Southern High Plains are located in mid-continent North America and comprise a distinctive geographic sub-region of a vast north-south corridor of land collectively termed the Great Plains (Hunt 1974; Finch 2004; Wishart 2004). This primarily grassland corridor, east of the Rocky Mountains, stretches from the central Canadian provinces of Alberta and Saskatchewan in the north through to northern portions of Mexico where the grassland finally gives transitions to a desert environment (Figure 1.1).

The High Plains physiographic section of the Great Plains comprises a vast elevated plateau covering portions of the modern states of Wyoming, Nebraska, Colorado, Kansas, Oklahoma, Texas, and New Mexico (Figure 1.1). The Southern High Plains (also called the Llano Estacado) region of the High Plains consists of a relatively flat upland, which has been geographically separated from the rest of the High Plains by the west-east incision of the Canadian River (Johnson 2008).

The Southern High Plains covers an area estimated to be ca. 31,000 square miles (Stafford 1981; Johnson 1987; Wood *et al.* 2002). The physical geography of the region is the result of the deposition of aeolian sediments on top of a uniform layer, known locally as the 'caprock' (Holliday 1988a; 1988b; 1990; Johnson and Holliday 1995; Nickels 2002). The only topographic variation is provided by playas, dune fields, and dry valleys (Figure 1.1).

Previous research is characterised by interdisciplinary programmes which have sought to investigate the the development of the present landscape in terms of geomorphology, palaeoenvironment, and cultural chronology (e.g. Dillehay 1974; Wendorf *et al.* 1975; Johnson 1987, 2007; Johnson and Holliday 1989, 1995, 2004; Holliday 1987, 1995, 1997; Neck 1995; Meltzer 1999; Wood *et al.* 2002). The most recent synthesis of the available data suggests the region has been a prairie grassland throughout the late Quaternary with key indicator species fauna comprising bison, pronghorn antelope, and prairie dog (Johnson 2007:15).

Hunter-gatherer occupation of the Southern High Plains encompasses a period currently recognised as stretching from 11,500 RCYBP through to Anglo-American settlement in the late nineteenth century (Johnson 1987, 2006; Johnson and Holliday 1995, 2004). Late Pleistocene-early Holocene sites are however comparatively rare in the regional record (Holliday 1997). At Lubbock Lake, a regionally significant type site, the first well-dated appearance of LTFs do not occur until Middle Archaic contexts (ca. 5500 - 5000 RCYBP) some 6,000 years after the earliest evidence for human activities. Evidence for human activity and the commensurate numbers of recorded LTF appear to exponentially increase in Late Archaic deposits. Later contexts confirm this trend extends through to the Protohistoric period (ca. 500-300 RCYBP) when mobile bands were directly observed by early explorers to the region (Newcomb 1963; Miller Morris 1997).

Chapter Synopsis

Hunter-Gatherer localised thermal feature technology is investigated with particular reference to the archaeological record of the Southern High Plains Region of North

America. A broad-scale survey of localised thermal features (Chapter 2) indicates that they are often encountered in the archaeological record and furthermore appear to be a fundamental component of hunter-gatherer sites. Previous studies of these features are typically characterised by atomistic and parochial perspectives focussing on site scale recording and analysis. A clear need is identified, for a regional-scale LTF 'toolkit' that could provide a common structure from which to actively investigate these important feature types.

In response, a research design is devised (Chapter 3) with which to foreground the investigation of this feature class and integrate data across a range of scales. The first operative stage in this framework is the integration of previously recorded site data at a regional scale in order to form a baseline for the development of qualitative research structures. A multi-part research methodology formalises the research methods and maps out the steps necessary to complete the project.

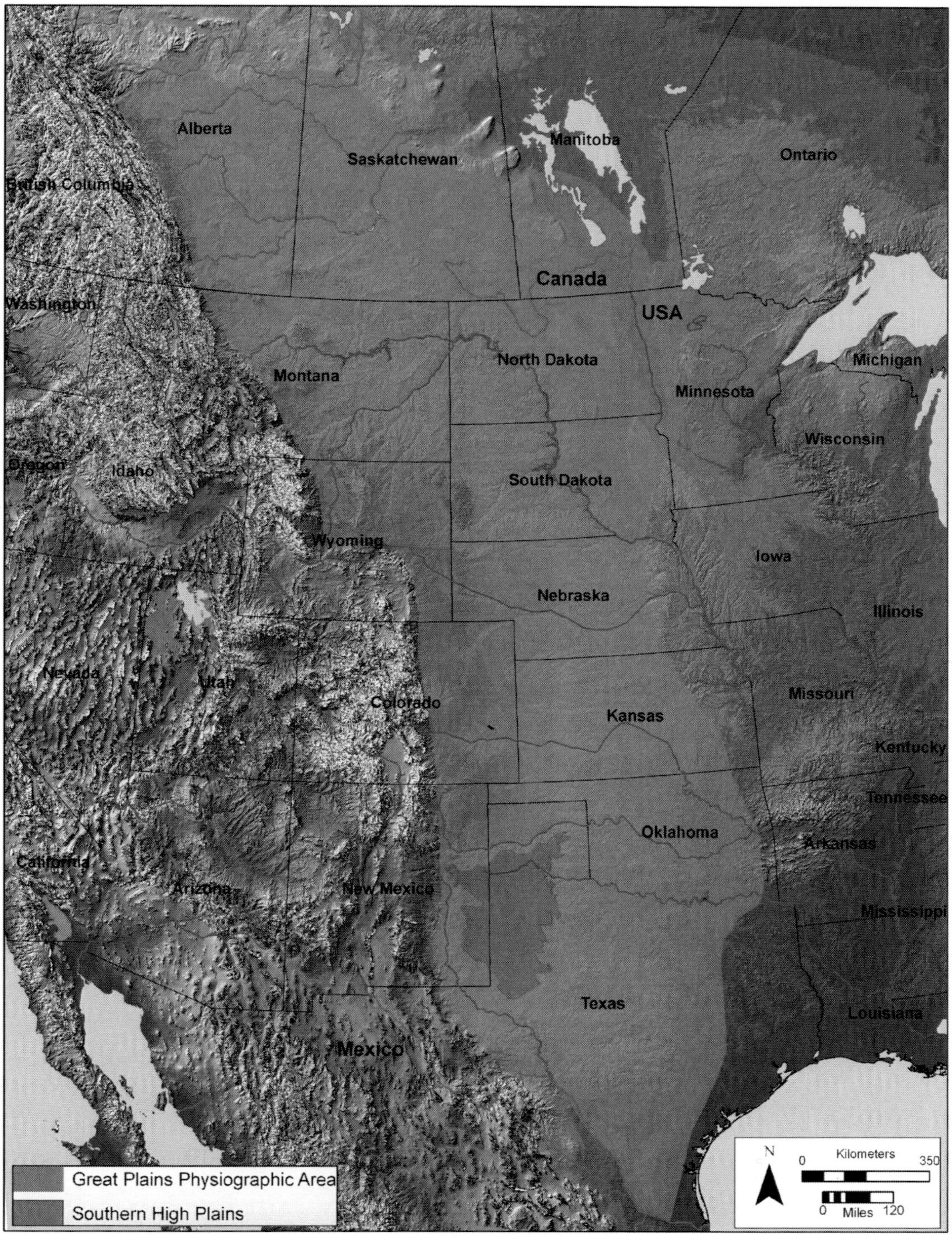

Figure 1.1. Location of the Southern High Plains: a) in relation to the Great Plains physiographic province.

The Southern High Plains of North America (Chapter 4) is an ideal location in which to model the usefulness of this approach. Detailed presentation of relevant cultural, geographic, geologic, and palaeoenvironmental data are placed in a regional context. The current state of knowledge regarding anthropogenic thermal feature technologies is explored and forms the baseline for the present research project.

The construction of a simple taxonomic key (Chapter 5) is designed as a practical theoretical tool for assessing the compiled regional sites dataset, identifying archaeological signatures with the potential to contribute to interpretative statements. In addition to its utility in the current project, the key establishes a common vocabulary that will allow future fieldworkers to identify and interpret LTF within a broader perspective of hunter-gatherer domestic fire technology. A three-part heuristic research programme is developed to practically test hypotheses drawn from the application of the hierarchical key to the regional database.

Results of this programme are presented in ethnographic (Chapter 6), experimental (Chapter 7), and archaeological (Chapter 8) contexts. Ethnographic results (Chapter 6) explore the range of physical structures, processes, and concomitant cultural activities and meanings associated with LTF from a broad anthropologic perspective. Subsequently, the overall model is supplemented by empirically testing research hypotheses through experimental (Chapter 7) and fieldwork (Chapter 8) based research. The results illustrate the potential of exploring multiple strands of evidence and provide an extremely useful practical response to examining physical variables no longer available through engagement with the archaeological record. The spatial distribution of sites with identified LTF features is assessed at a regional scale (Chapter 8) and quantitative analysis of this site distribution data is examined in relation to external geographic and political datasets. The resultant patterns are assessed in terms of their usefulness in developing predictive models for geographically prioritising areas for future research. The characterisation of the regional dataset through the application of the hierarchical key forms the baseline of this assessment.

The relative successes of the project (Chapter 9) are examined and discussed, and the potential for extension of the project to other regions tested. Drawing together disparate research strands and building on existing datasets, the project represents a robust method for the regional investigation of LTFs. The variability identified in the ethnographic and archaeological datasets indicates that the investigation of LTF is far more complex than is commonly accepted and further that this variability underscores a technological complexity that can yield important behavioural information. The Southern High Plains dataset is contextualised by examining the Southern British Mesolithic dataset and determining how applicable the research model is in understanding hunter-gatherer thermal feature technologies in a radically different geographic region. Lastly, significant conclusions (Chapter 10) are outlined and suggestions for further research posited.

CHAPTER 2 - BACKGROUND

Abstract

Hunter-gatherer manipulation of fire and specifically its constrained utilisation in localised thermal features represents a major technological strategy that can oftentimes be identified in the archaeological record. Examination of the distribution of these structures (both spatial and temporal) reveals they are a ubiquitous component of hunter-gatherer archaeological sites. Variability in feature morphology and archaeological examination is explored through three separate research strands: micro-scale examination of individual features, macro-scale investigation of features within a site context, and wider synthetic studies dealing with cultural, temporal, and regional frameworks. Background research indicates that localised thermal features are well represented on archaeological sites with evidence for prehistoric hunter-gatherer activities. The archaeological signature of these features is highly variable and their investigation and interpretation have historically been largely on a piecemeal basis. The high rate of encounter in the archaeological record and potential for technologic analysis highlight the need to critically develop contextual approaches for their investigation.

Introduction

The controlled use of fire is a singularly human characteristic (Canti and Linford 2000:385) and an integral technology of both sedentary and hunter-gatherer human communities (Otte 2002; Wrangham and Conklin-Brittain 2003). The range of behaviour associated with fire technology (Figure 2.1) identifies a complex suite of activities that vary from site-scale production of heat and light necessary for survival through to the wide-scale management of landscapes at a continental scale (e.g. Bellomo 1994; Stewart 2002; Rolland 2004). While the production and manipulation of fire appears to be a human universal, its examination is herein limited to non-sedentary human populations subsisting primarily through hunter-gatherer lifeways. The culturally embedded nature of fire technology underlines the range and intensity of processes available to any one hunter-gatherer group and it should be assumed that this complexity extends, in greater and lesser forms, to any attempt to investigate hunter-gatherer communities regardless of temporal or geographic positioning.

Localised thermal feature (LTF) are a small-scale site component within the wider framework of hunter-gatherer fire technology (Figure 2.1). Localised thermal features are defined here as non-portable, discrete, human-built structures (usually less than 3m in diameter), which have been utilised for a variety of social, economic, political, and ritual activities, all of which require fire as their catalyst. This definition deliberately subsumes the huge range of physical variation that has to date been recorded in archaeological surveys and excavation. Examples might be square, round, deep, shallow, deflated, stone-lined, or choked, but are all simply characterized by in-situ evidence for burning.

Research directed toward investigating LTFs is relevant to archaeology as hunter-gatherer subsistence strategies account for more than 99% of past human behaviour (Kroll and Price 1991:3). An examination of the available anthropological literature indicates a large corpus of information has been generated on these feature types at a global scale (e.g. Binford 1977; Frison 1983; Kelly 1995; Black and Ellis 1997).

The global distribution of LTFs identified through archaeological investigation is largely determined by two interrelated variables: 1) the palaeogeographic range of humans engaged in hunter-gatherer subsistence strategies and therefore potential for the identification of hunter-gatherer material culture within a particular site or region by archaeologists, and 2) the geographic distribution and research range of individuals and institutions with an agenda for investigating hunter-gatherer material culture. The intersection of these two variables has resulted in a situation in which westernised nations in Europe and North America have dominated the production, maintenance, and housing of the existing global dataset (Trigger 1986). This overarching political framework is crucial to situating research agendas, as it overwhelmingly fragments regions into smaller project areas, which require the production of a site report as proof of compliance with local and national governance (e.g. Roskams 2001).

Geographically, LTFs have been identified on all habitable continents and are common components of hunter-gatherer sites in both cave and open-air site settings. The temporal distribution of identified sites indicates that the *Homo* genus comprises a number of species capable of fire technology. *Homo erectus* (ca. 2.5Ma – 0.5Ma) is presently the earliest of these, although archaeologically visible fire technology has not yet been clearly identified (Wrangham *et al.* 1999). Later *Homo* species *H. heidelbergensis* (ca. 650 – 250Ka), *H. neanderthalensis* (ca. 250 – 29Ka), *and H. sapiens* (ca. 150Ka -) sites all exhibit strong evidence for fire technologies (Stringer and Andrews 2005). It is clear from this evidence that fire technology is observable through the physical presence of LTFs on archaeological sites and should be expected when investigating the archaeological evidence from any late or post-glacial human community.

The background for the investigation of LTFs identifies the signature and distribution of these features in the archaeological record. Variability in the examination of these structures is explored through three separate research strands: micro-scale examination of individual features, macro-scale investigation of features within a site context, and wider synthetic studies dealing with cultural, temporal, and regional frameworks. These diverse strands of hunter-gatherer investigation highlight the potential for the development of a holistic contextual approach to thermal feature research based on the development of a heuristic research programme.

Archaeological Evidence

It is obvious that the visibility of LTFs in the first place is dependant on the ability of hunter-gatherer groups to actively ignite and control fire. Prior to the invention of the modern match in 1805, the technologies used in this process are themselves complex with long concomitant histories of research (e.g. Hough 1890). For the purposes of the present study, ethnographic observations indicate that all known and historically recorded hunter-gatherer groups are proficient in the production of fire.

Ethnographies indicate that four methods of the human production of fire have been documented: 1) fire drills, whereby a stick is rapidly rotated in a depression in a second piece of wood; 2) fire saws, whereby an objective stick is rapidly drawn back and forth against tinder placed in a hollow in a second piece of wood; 3) fire plow, whereby a stick is vigorously rubbed in a notch in another stick; and 4) percussion, whereby two siliceous stones are struck together (Hough 1890; Davidson 1947; Gott 2002:651). It is useful to note that the production of fire may not have been considered a time-consuming task as observations of North American Apache groups show they could ignite fire in less than 10 seconds, using only traditional methods (Hough 1890).

Regardless of the variation in the technology by which fire was generated (which itself is rarely archaeologically visible), the identification of LTFs (i.e. the resultant physical remains of the fire lighting and maintenance) during the archaeological investigation of a hunter-gatherer site or within a wider landscape survey is a relatively commonplace event. Fieldwork methodologies that facilitate their identification include, but are not limited to: pedestrian survey, geophysical prospecting, geochemical prospecting, and various forms of subsurface excavation. It has been established that the hunter-gatherer use of fire has a long history and therefore it is no surprise when archaeologists encounter LTFs during research.

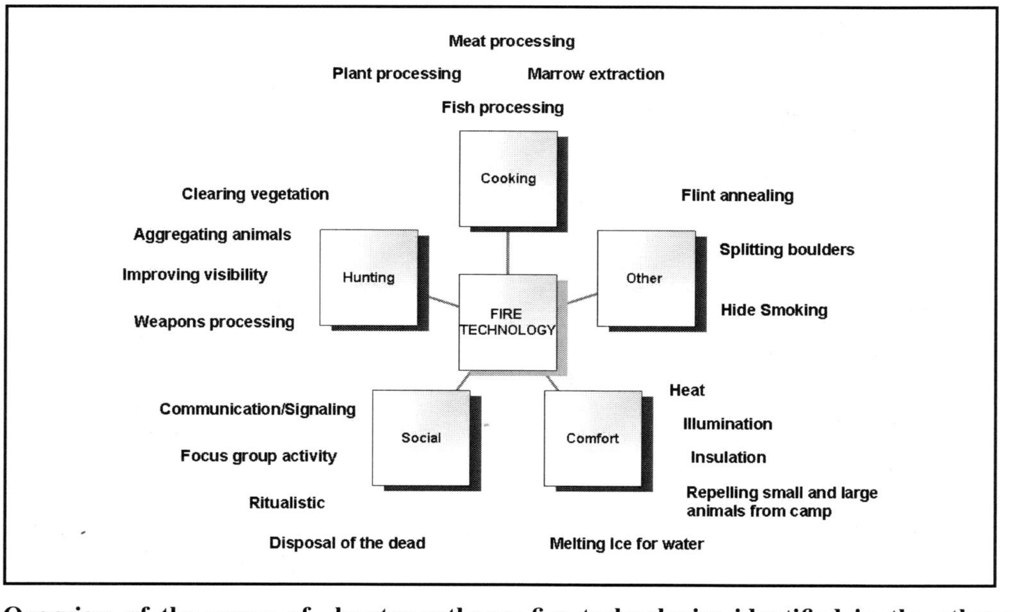

Figure 2.1. Overview of the range of hunter-gatherer fire technologies identified in the ethnographic and archaeological record (diagram after Rolland 2004:table I; Bellomo 1994, fig.2).

Very generally, archaeological research projects have historically approached the examination of LTFs at and across three different analytical scales or foci. The smallest unit of measurement is the micro-scale, which deals with individual features or components thereof. The macro or site scale may be composed of multiple LTFs and is also the most commonly discussed in the archaeological literature. The largest scale is also the most rarely dealt with and considers LTFs in terms of wider regional, cultural, or temporal synthesis and discussion. The following sections examine the archaeological evidence for LTFs from these three perspectives. Examples of individual features and groups of features within sites are called upon. The geographic focus of these examples is related to the current research project and therefore is heavily influenced by sites and features located on the Southern Plains of North America. It is hoped, however, that sufficient supporting evidence is presented to indicate that LTFs are a global phenomenon and a range of different sites and features should be expected within any geographic region (Figure 2.2).

Micro-scale Research

Site reports are the critical documents for providing detailed primary data on all aspects of the archaeological record. The identification and description of a feature is an excavation-oriented exercise. Descriptions generated in the field are, in general, unconditionally incorporated into site reports wherein some level of interpretation is afforded to and situated within a site discussion. The investigation of LTFs is primarily then a field-based activity.

Identification

The methodology of research conducted inevitably has an epistemological basis that conditions the way LTFs are engaged. Perhaps the most important distinction is between the vertical view encountered by excavation and the horizontal view observed by survey (Bradley 2003). Field methodologies therefore encounter and identify feature classes in different ways and at different intensities.

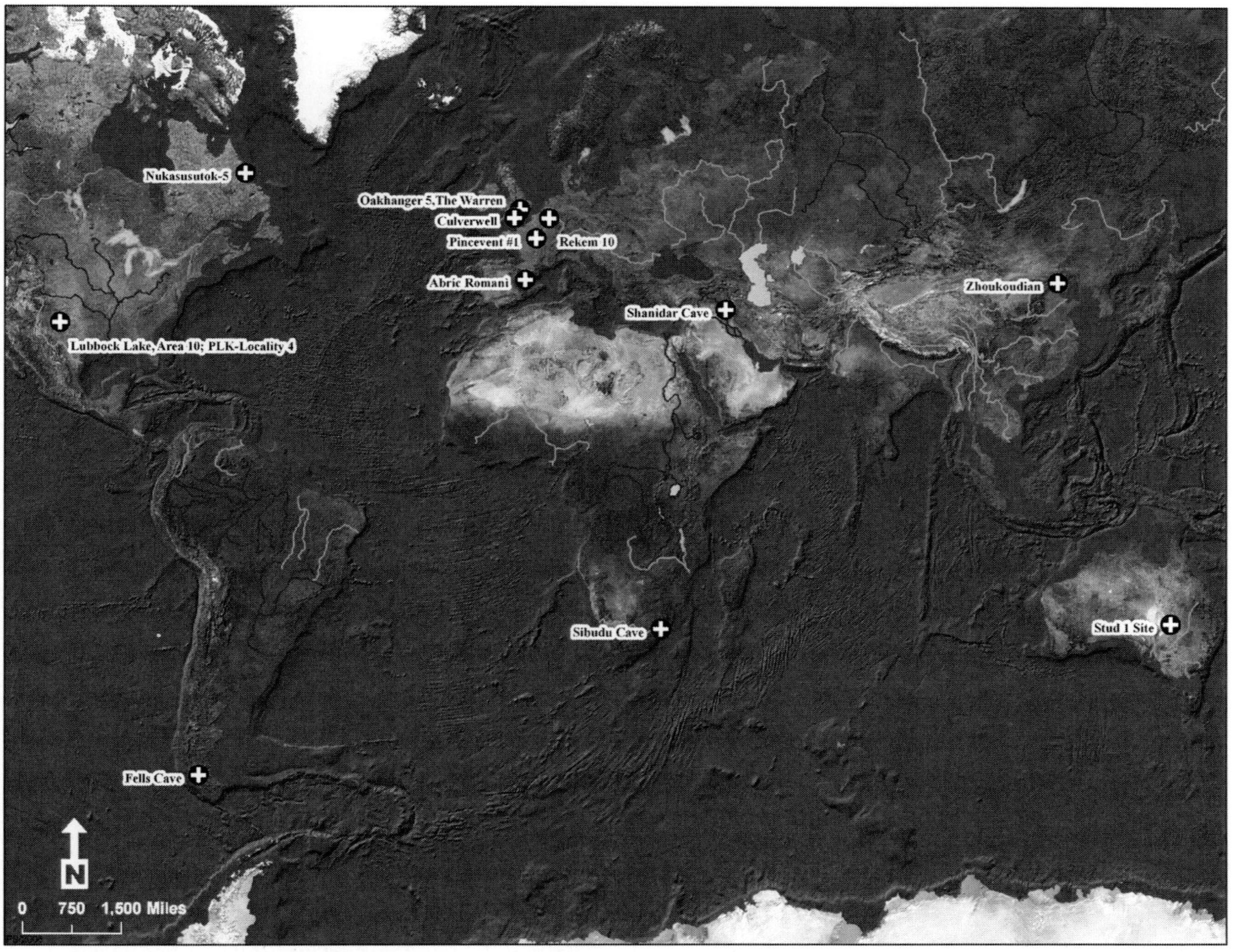

Figure 2.2. Location of hunter-gatherer sites discussed in Chapter 2.

The least intensive are the non-invasive approaches of feature identification. These include various forms of geophysical, geochemical, and remote observation-based prospecting. All three identify and therefore engage the feature in an off-site context. Geophysical methods have been successfully mobilised to identify LTFs and include soil resistance, magnetic, and radar methods (Bevan 1998). A useful example of geophysical survey methods is Jones and Munson's (2005) investigation of temporary campsites on the Northern Plains of North America. This research involved the magnetic susceptibility survey of 21 prehistoric campsites and the results indicated a strong association between the resultant signal and the size and type of buried feature (Jones and Munson 2005:35).

The potential of geochemical methods to detect the presence of LTFs has long been recognised by the archaeological community (e.g. Cook and Heizer 1965). Despite this realisation, the application of geochemical techniques to hunter-gatherer sites has largely been confined to the use of phosphate analysis in delineating site boundaries (Proudfoot 1976). The increasing availability of multi-elemental geochemical techniques such as Inductively Coupled Plasma Accelerated Emission Spectroscopy (ICPAES) and Inductively Coupled Plasma Mass Spectroscopy (ICPMS) has facilitated a suite of elemental characteristics associated with the archaeological signature of LTFs to be identified. Recent application of these techniques has been used with some success to test whether LTFs can be characterised by increased levels of K, Mg, and P (Knudson *et al.* 2004).

An emergent non-invasive technique with the potential for the identification of LTFs exposed at the modern ground surface is the manipulation of remote viewing data such as high-resolution aerial photography. The increasing availability of georeferenced orthoimagery (at resolutions up to 6in (15cm)) underscores this potential, especially considering archaeological expressions of LTFs typically range between 0.5 and 1.5m in diameter. The author is currently unaware of any published application of high-resolution aerial photography to the identification of LTFs; however, an examination of the location of two known experimentally created features, shows the potential of this method (Figure 2.3).

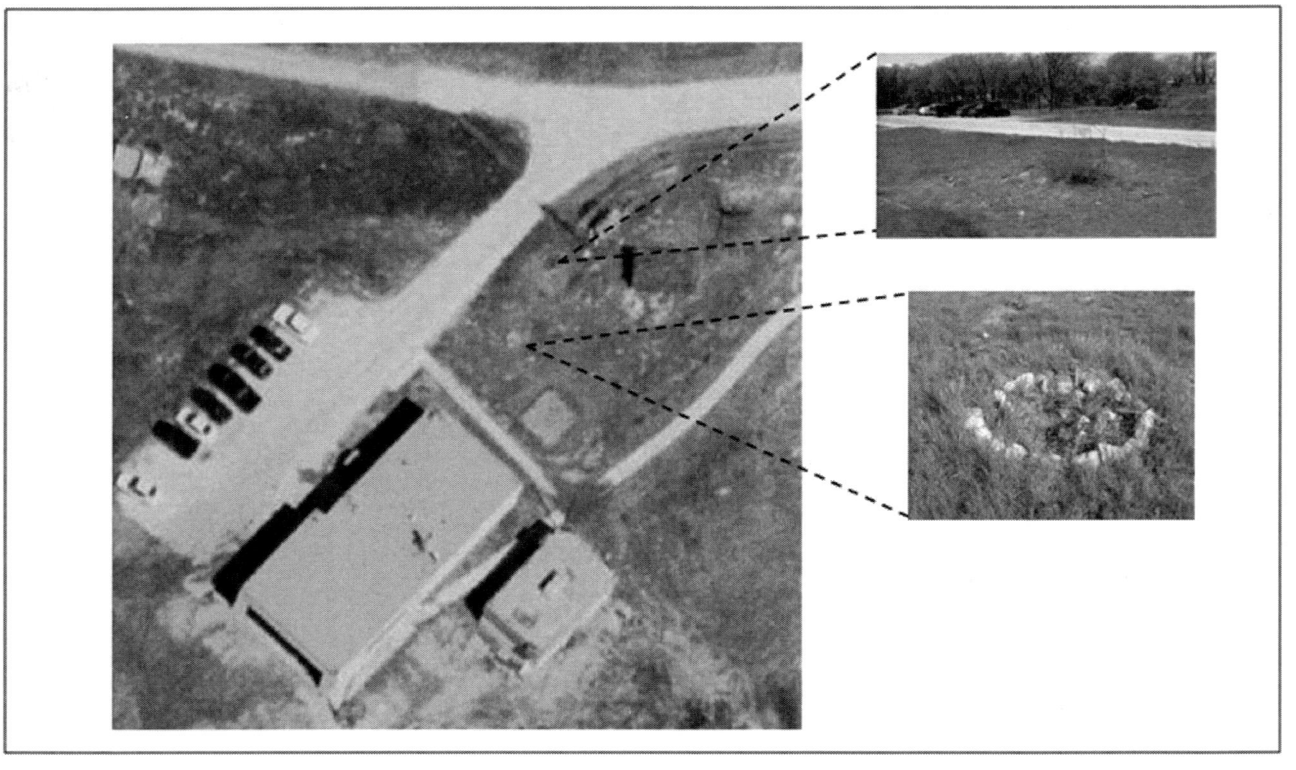

Figure 2.3. High resolution (6in/15.2cm pixel resolution) aerial photograph taken in March 2005 showing two thermal features within the experimental research area at the Lubbock Lake Landmark, Texas; top insert shows an oblique view of the corresponding feature which measures ca.0.5m in diameter and was also photographed in March 2005; bottom insert shows an oblique view of the corresponding feature, a large 1.5m diameter earth oven, photograph taken in June 2005 (aerial photograph available from: http://maps2005.ci.lubbock.tx.us).

Moderately intensive archaeological approaches to the identification and investigation of LTFs include (in order of destructiveness) pedestrian survey (or walk-over), shovel-testing, and trial trenching. Pedestrian survey can obviously only encounter exposed features either on or at the modern ground surface or exposed in a cut section such as river bank. Both situations are the result of either insufficient sedimentation, necessary in order to bury the feature following abandonment, or active erosional processes working to expose an already buried feature. In either circumstance, the encounter of LTFs is indicative of a relatively unstable environment and the feature has most likely already been adversely affected by several taphonomic processes. Pedestrian survey will therefore most successfully identify LTFs in highly dynamic, eroding, or slowly aggrading settings. Anecdotal evidence from long-time avocational archaeologists and 'arrowhead' collectors living in west Texas indicates that they often actively seek out LTFs, during their own pedestrian surveys, as LTFs are known to be useful visual reference points in the landscape for the recovery of smaller more 'exotic' items of aboriginal material culture.

The recent successful identification and recording of LTFs in submerged contexts underscore the ability of underwater survey to locate features and sites in drowned landscapes (Momber 2000; Hans 2002). Preservation conditions recorded during these surveys often result in the recovery of organic materials that can provide critical information not normally recovered by terrestrial approaches. For example, Dal Hans (2002) reports that the partially burned wood recovered within a submerged fireplace on a Ertebølle settlement exhibited evidence for gnawing marks from insects and fungi leading to the interpretation that dead wood was collected as the primary fuel for this feature (Hans 2002:32).

Shovel-testing and trial trenching both have the potential for encountering buried LTFs during their excavation. The success of these methods is obviously increased if coupled with the non-invasive methods already described. Although LTFs are perhaps the largest structural component of hunter-gatherer campsites, they are themselves relatively small in terms of the likelihood of encounter in a shovel-test or trial trench. Furthermore, the generally small size and ephemeral nature of hunter-gatherer material culture necessitates that matrix screening of the sediments excavated from either of these two methods be undertaken.

While shovel-tests investigate a lower volume of sediments and cannot be excavated to the depth of trenches, they have the advantage of a greater likelihood that an LTF would be recognised during the excavation processes. In contrast, the author's experience suggests that the reorganization of LTFs during mechanical trenching is a notoriously difficult process.

High-intensity archaeological approaches to the identification and investigation of LTFs involve hand excavation methodologies. Excavation facilitates: high-resolution mapping of the structural components of the feature; collection of samples for archaeometric analysis; recovery of material artefacts contained within the feature fill, such as hearthstones, bone, and ceramics; examination of geometry; firing history; fuel load and type; taphonomy; and function. The level of recording associated with hand excavation obviously facilitates a higher level of data recovery.

A range of modern archaeological methods, therefore, has the potential to encounter LTFs. Research questions developed for commercial archaeological projects operating within the remit of Cultural Resource Management (CRM) must plan to engage with the full range of cultural resources within a given area. The choice of field methodology will therefore be dependant on a complex set of factors not tailored specifically to the identification of LTFs. This realisation is important to the development of a strategy for investigating LTFs because any single feature type must be secondary to the overall aims of any research excavation project.

As an example, the vast majority of archaeological fieldwork (>90%) carried out in the southwestern portion of the United States are Phase I surveys. This is due to the large geographic size of project areas, typically erosional depositional settings, and type of archaeology expected to be encountered (limited evidence for highly mobile populations). Phase I investigations seek to characterise the extent of any cultural material within a project area. This is usually achieved through systematic pedestrian survey and limited shovel-testing with the objective being to identify and establish the boundaries of archaeological sites. Excavation occurs extremely rarely (less than 10% of investigations) in this region because the ability for developers to avoid sites identified during Phase I is very high. When excavations do occur, they are at first exploratory (phase II) and designed to test the significance of a site. Phase II investigations might consist of the excavation of several test-units across a known site. Phase III investigations happen only very rarely (estimated at a rate of less than five per decade) and comprise the excavation of large areas of a site. As such, Phase III investigations usually occur in advance of a site being destroyed or severely impacted. Field methodologies are dependant on the investigator and are scaled toward the phase of investigation undertaken.

Regardless of these methodological differences in identification and recording, LTFs are perhaps the most common structural feature type recorded on hunter-gatherer sites (e.g. O'Malley and Jacobi 1978; Biesaart 1985; Boyd *et al.* 1987; Johnson 1987a; Healey *et al.* 1992; Solecki 1995; Palmer 1999; Thoms 2009). Rather than adopting feature-based approaches, the excavation process has historically foregrounded the collection and analysis of transportable material culture, such as lithic and bone material, which can be analysed and displayed in off-site laboratory and museum contexts (Lucas 2001).

Morphology

On a global scale, significant variability is evident in the nature of the archaeological evidence for particular features at a specimen level (Table 2.1). Frequencies of encounter and preservation condition vary greatly by cultural group, age, type of feature, construction method, local geomorphology, modern and palaeoenvironmental conditions, and taphonomic issues. Ephemeral patches of stained sediment or clusters of burned artefacts are, however, perhaps the most often encountered LTF type.

Discernable features are also frequently encountered; these range in geometry from discrete areas of burned sediment staining, large excavated pits choked with hearthstones and exhausted charcoal nodules, shallowly excavated basins with fine charcoal lenses preserved at the base, and tightly clustered groups of hearthstones preserved where the surrounding sediments have long since departed (Figure 2.4). Regardless of the justification for originally building these features, there is clearly a huge range in the physical manifestation of LTFs that we observe during archaeological research. All these features and combinations thereof can be grouped in the terminological class here labelled as LTFs.

A typical example of an LTF encountered during fieldwork is FA14-1. Eroding out of the wall of an arroyo (deep gully cut by an intermittent stream), the feature was spotted by a rancher on his property in west Texas (Figure 2.5). Excavations were initiated by a crew of Museum of Texas archaeologists in order to record the feature prior to its complete loss to the ongoing erosional processes. The feature and surrounding occupation surface were entirely excavated, yielding some 225 artefacts (Backhouse 2003:70). Feature morphology consisted of an elliptical basin-like pit measuring 91cm horizontally with a maximum depth of 21cm. A ring of hearthstones were recorded rimming the edge of the pit in the excavated portion of the feature (Backhouse 2003:74).

Table 2.1. Example of the global variability in features interpreted as evidence for hunter-gatherer localised thermal feature technology.

Archaeological Evidence	Site	Archaeological Period	Geographic Area
Spatial clustering of burned artefact classes	Oakhanger 5, The Warren[1,2]	Mesolithic	Europe
	Stud 1 Site[9]	Late Holocene	Australasia
Burned sediment staining within discrete area	Shanidar[3]	Middle Palaeolithic	Middle East
	Abric Romaní[4]	Middle Palaeolithic	Europe
	Rekem 10[5]	Upper Palaeolithic	Europe
Excavated feature with indirect evidence of burning (i.e. spatial proximity to burned materials)	Culverwell[6]	Mesolithic	Europe
	Zhoukoudian[12]	Lower Palaeolithic	Asia
Excavated feature with direct evidence of burning	Lubbock Lake[7]	Archaic (US)	North America
	Fell's Cave[8]	Paleoindian (US)	South America
	Sibudu Cave[10]	Middle Stone Age (Africa)	Africa
	Nukasusutok-5 (HcCh-7)[11]	Martime Archaic	Artic

References: (1). Rankine et al. 1960; (2). Reyneir 1993; (3). Solecki 1995; (4). Vaquero & Pastó 2001; (5). Caspar & De Bie 1996; (6). Palmer 1999; (7). Johnson 1987a; (8). Bird 1988; (9). Holdaway et al. 2002; (10). Cain 2005; (11). Hood 1981; (12). Boaza et al. 2004.

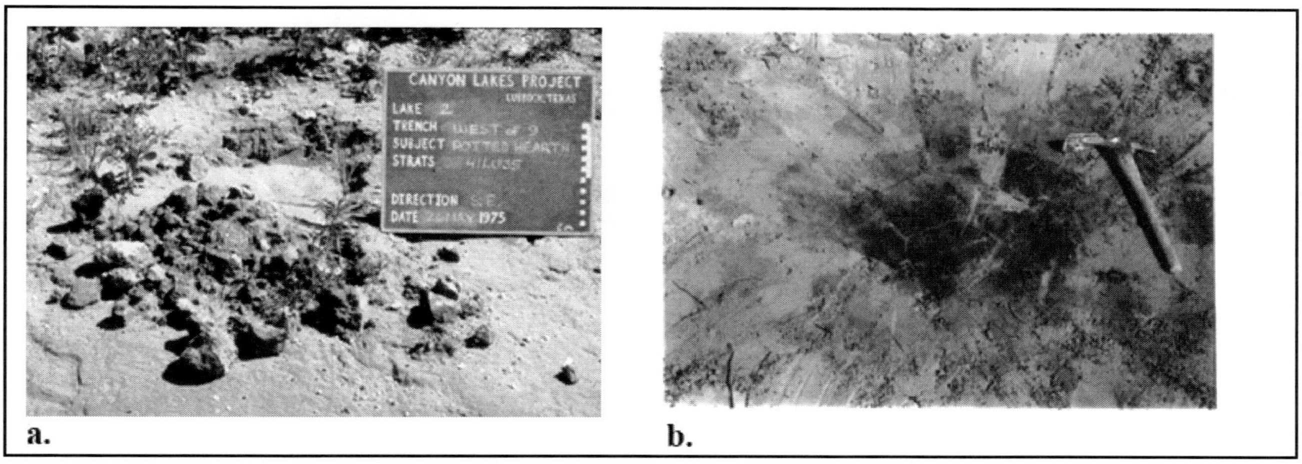

Figure 2.4. Example of some typical LTFs recorded during archaeological fieldwork: a) cluster of burned rocks exposed on the modern ground surface at 41LU35 in Texas, USA; b) discrete area of burning revealed on the modern ground surface after a blowout episode in a dune field, unnamed site, Southern High Plains, Texas. (figures Courtesy of the Anthropology Division, Museum of Texas Tech University).

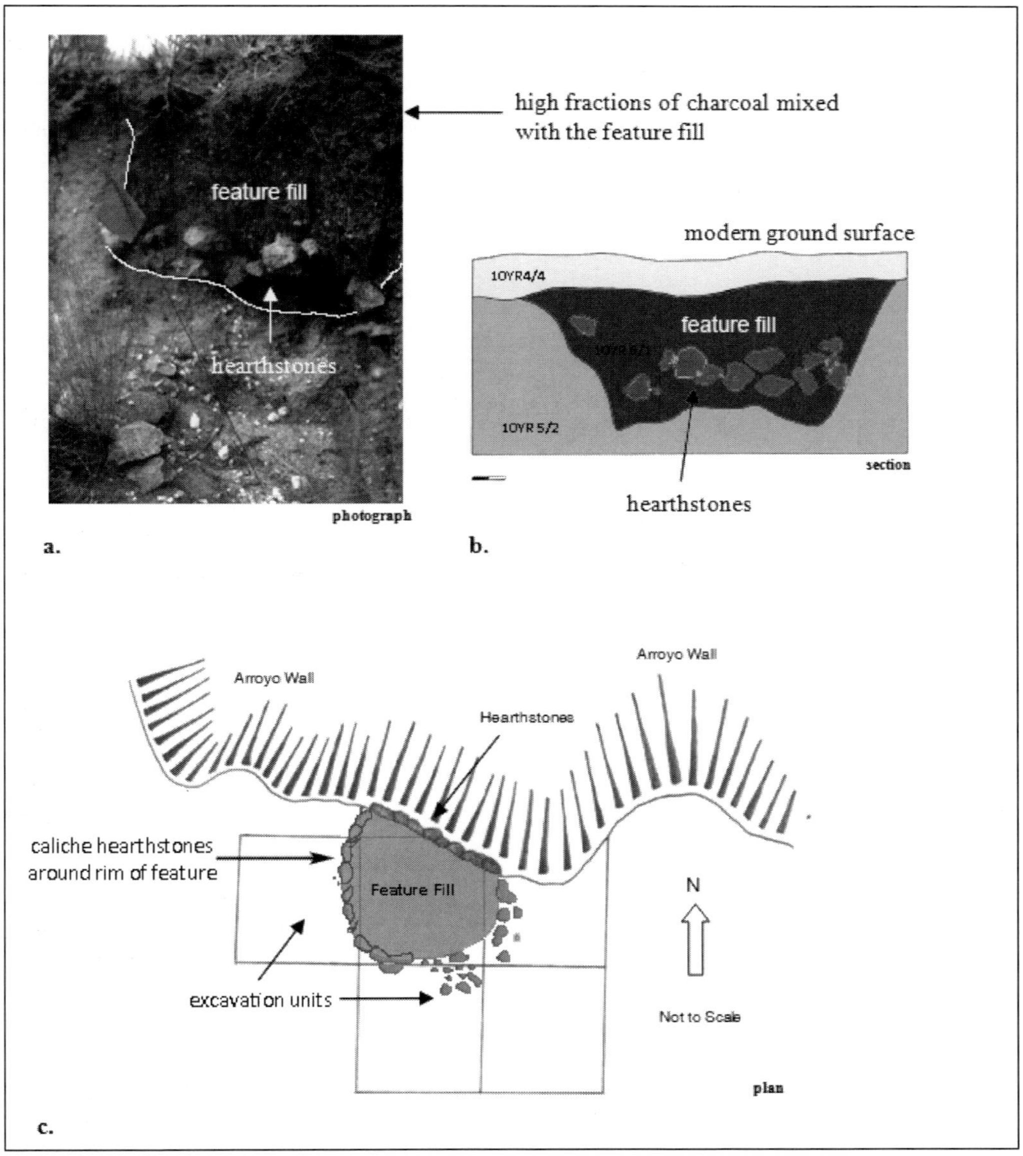

Figure 2.5. Excavation of thermal pit feature during the 2001 field season at Tahoka Lake (41LY52) Area 14: a) field photograph of the feature exposed in the arroyo wall; b) profile of feature with the locations of the hearthstones indicated within the feature fill; c) Plan view sketch of the excavation area. Note the hearthstones exposed in the wall of unit 2496N529W (figures modified from Backhouse 2003, fig. 6.2, 6.4, 6.3).

Hearthstones dominated the excavation assemblage (91.5%), all of which were macroscopically identified as caliche, a commonly available local rock type (Bretz and Horberg 1949). Preliminary colour and fracture analysis of the hearthstone assemblage suggested little breakdown of the individual nodules, indicating that the feature may have been used only once (Backhouse 2003:71). Carbonized wood recovered from the feature fill was subjected to tree identification analysis and indicated that the primary fuel source was Mesquite (*Prosopis glandulosa*) a native tree common in the area today. No macro-botanical remains were recovered within the feature fill and the structure was interpreted as a pit oven for processing vegetal matter. This interpretation was based on the recovery of large quantities of charcoal which are indicative of a low oxidising environment i.e. rapid burial following firing (Backhouse 2003).

Another less pronounced example of an LTF illustrates the range in morphology represented in the archaeological signature of these features (Figure 2.6). The feature (FAPLK4-1) was encountered during a large-scale pedestrian survey of 84,000 acres of ranch land carried out by Museum of Texas Tech University personnel on the Rolling Plains of west Texas and consisted of ca. 50 locally available rocks clustered in an area around 1m in diameter. All the rocks appeared burned although there was no evidence for foodstuff in the form of bones or carbonised roots and seeds. No charcoal was observed nor was there any evidence for staining (tiny carbonised particles) mixed with the sediment matrix in and around the feature. Significant frequencies of flaked lithic material in the vicinity of the feature indicated the presence of hunter-gatherer activity and added some contextual support to its interpretation as an LTF.

The surveyed landscape represents a highly dynamic geomorphic environment in which continuing aeolian transport of sediment and rapid flooding events contribute to expose dense concentrations of hunter-gatherer material culture on the modern ground surface. Due to these active processes, it is highly unlikely that the feature fill of any LTF will be preserved for investigation by archaeologists. Identified features such as the one pictured in Figure 2.6. comprise clusters of burned rock, which appear to be the result of deflated (or eroded) features that have themselves long since been eradicated. Identification of these features and differentiation from natural processes, such as range fires, is obviously challenging in these environments, and often requires careful examination of multiple lines of evidence before hearthstone clusters can be identified as LTFs.

Figure 2.6. Hearthstone cluster on the modern ground surface at PLK Locality 4 identified during pedestrian survey on a ranch in west Texas. Note the dark discoloration of the rocks. (P.N. Backhouse).

The two examples illustrate that the range of physical constructions available as LTFs are practically limited by local environmental factors, which serve to structure the technological options available to hunter-gatherer groups living in different geographic regions (e.g. Kelly 1995; Smith and Winterhalder 1992; Torrence 2001). Other practical constraints include: the physical setting, prevailing weather conditions, time of day, social context, human skill level, knowledge, available energy sources, raw materials, expected outcomes, and availability of tools (Torrence 2001:74). The active interplay among these variables facilitates that a broad range of LTFs may be constructed and potentially observed in the resultant archaeological record. It is therefore auspicious that LTFs are physical features that must conform to basic pyrodynamic principles in order to be useful in fire technologies. The range in feature variability and our ability to measure it is therefore always empirically relevant to archaeological research.

Classification

The identified variability in archaeologically derived field techniques for observation and the physical evidence presented by the LTFs themselves (Table 2.1) is commuted to the classificatory terminology used in reporting these features. A limited examination of a sample of 10 site reports identified 32 different descriptive terms commonly used in classifying LTFs at a site scale. All the identified terms can be grouped into one of three classes as either generic, descriptive, or functional (Table 2.2).

Table 2.2. Examples of typical terminology used in archaeological literature to classify LTF types.

Class	Classification
Generic	Hearth[1,6,7,8, 10], Fire Features[2], Fire Hearths[6]
Descriptive	Slab-Lined Hearths[2], Circular Clay Lined[2], Rectanglar Slab Lined[2], Circular Unlined[2], Roasting Pit[2], Oven[2], Fire Pits[3, 9], Stone Filled Fire Pits[3], Cylindrical[3], Cylindrical With Round Bottoms[3], Globular[3], Bell-Shaped[3], Basin-Shaped[3], Truncated Cone Shaped[3], Inverted Truncated[3], Basin[5], Pit[5], Wood & Fire[5], Oxidized Basin[5], Stone Lined Basin[5], Piles Of Heated Stones[5], Basin Hearth[6, 9], Pit Hearth[6], Scatters Of Ashes And Charcoal[6], Broad-scale Bands Of Ash[6], Smaller/Personal Hearths[6], Hearth Complex[7], Interior Hearth[7], Exterior Hearth[7], Disturbed hearths[10], caliche cobble lined hearths[10]
Functional	Ceremonial hearth[2,] Roasting pit[2], Earth Oven[4], Roasting Pit Hearth[6], Oven[10]

References: (1). Drass 1995; (2). Lowell 1999; (3). Frison 1983; (4). Campling 1999; (5) Ellis 1997; (6) Solecki 1995; (7) Timmins 1997; (8) O'Malley and Jacobi 1978; (9) Brown 1999; (10) Johnson and Holliday 1989.

Generic terms such as 'hearth' are the most frequently reported class and are also the most interpretatively ambiguous. This class identifies the presence of discrete thermal features and appears to be characteristically applied to old world Palaeolithic sites in both open-air and cave-site settings (e.g. Carr 1991; Solecki 1995; Vaquero & Pastó 2001; Schiegal *et al.* 2003). Generic terminology is the most passive of the three forms of classification in that the definition of the feature is largely taken for granted, as in-site furniture, or considered peripheral to attempts to understand human behaviour at a site scale.

The descriptive class is the second most common and also has the highest range in variability. Terminology at this level appears highly insular, often being regionally specific (e.g. Johnson & Holliday 1989; Radovanović 1996), site specific (e.g. Solecki 1995), or researcher specific (e.g. Frison 1983). The functional class is less often directly employed and the least interpretively ambiguous. It also has the disadvantage of equifinality, conflating the role of the feature to a specific technologic outcome (Thoms 2008). The preceding discussion indicates that the three identified classes associated with LTFs have clear implications when extension to different scales of analysis is attempted.

It should be equally obvious that the terminology LTF is also a generic classificatory term and in the light of the evidence presented thus far does little to enhance our understanding of hunter-gatherer domestic fire technology. This is an issue that will be returned to and expanded upon in the theoretical perspective; at this stage it is sufficient to indicate that the term LTF is itself original and therefore, for better or for worse, contains none of the theoretical baggage associated with the other identified classes. In addition, it should not be confusing as to which terminological scheme is being presented. Put simply, when I refer to features as hearths, pit ovens, or open griddle features, then I am using the investigator's terminology, whereas when I refer to different types of LTF, I am using my own schema.

Archaeometric Techniques

An increased emphasis on the application of archaeometric techniques to specific and largely internal elements of LTF analysis has dominated the recent literature (Figure 2.7). These studies range from the environmental examination of feature fill samples (Perry 1997; 1999; Albert *et al.* 2002; Elbaum *et al.* 2003), the application of archaeomagnetic methods to dating and understanding site-formation processes (Gose 2000; Gose *et al.* 2004; Lian and Brooks 2004), identification of fuel constituents (Théry-Parisot 2002), residue analysis recovered from hot-rock assemblages (Quigg *et al.* 2001; Buonasera 2005), geophysical detection methods in the field (Abbott and Frederick 1990; Frederick and Abbott 1992), taphonomic analysis of burned assemblages (Stiner and Kuhn 1995), and materials analysis, particularly of hot-rock assemblages (McDowell-Loudan 1983; Akins 1988; Frison 1983; Lintz 1989; Rapp *et al.* 1999; Brink and Dawe 2003; Thoms 2003, 2009).

The application of archaeometric techniques is clearly important in understanding different aspects of the technological function of LTFs at the scale of the individual feature. The resultant inferences generated from these observations have had important implications in understanding hunter-gatherer subsistence strategies. A useful recent example is David Perry's macro-botanical analysis of hearth fill samples (Perry 1997; 1999). The samples, independently dated to the Dutch Mesolithic (ca.8750 – 5850 RCYBP), were examined macroscopically and microscopically for any remaining biological material. The results indicated a strong presence of vegetable matter within the feature fill, allowing Perry to hypothesize that the role of plants as a subsistence resource in the Dutch Mesolithic has been historically downplayed, a situation he attributed to the high archaeological visibility of faunal relative to macro-botanical remains (Perry 1999).

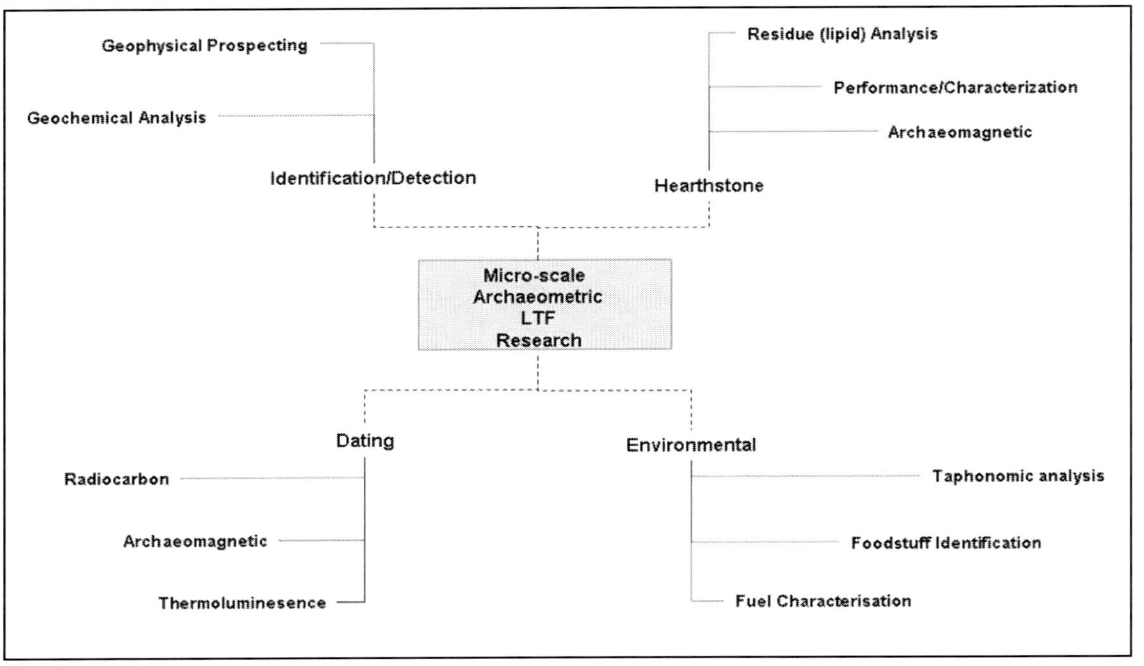

Figure 2.7. Overview of the range of archaeometric techniques applied to the investigation of LTFs encountered during archaeological fieldwork.

Another useful example is in the application of archaeomagnetic methods to understanding and modelling active site-formation processes operating on LTFs in dynamic landscape settings (Gose 2000; Gose *et al.* 2004). This approach uses palaeomagnetic techniques to examine the thermoremnant magnetization of hearthstones following an episode of heating. Results indicate that the magnetic orientation of individual hearthstones become aligned when heated (provided heating does not exceed the Curie temperature for the geologic rock type being utilised (Gose 2000:416)). Additionally, when analysed statistically the archaeomagnetic data can reveal discrete components of heating, which can be informative in reconstructing the original temperatures at which hearthstones were heated. Archaeomagnetic methods can therefore be utilised in the examination of site-level behaviours, temperatures achieved during heating, and in issues of feature integrity i.e. determining the presence of a fireplace versus a hearthstone dump. This method is also useful in an assessment of site-formation processes acting on the structure of fireplaces after a camp has been abandoned. The technique is destructive, requiring samples to be drilled from the original 'target' artefact. An advanced understanding of palaeomagnetic principles is also necessary in order to be successful.

Research directions identified at the micro-scale have largely focussed on increasingly atomistic approaches to the analysis of LTFs. While these approaches have undoubtedly advanced our ability to locate and understand the function of individual features, they all require significant investment in terms of developing an expert knowledge base, capital outlay for equipment and analysis costs, and time necessary to develop a particular technique. Many of the archaeometric methods identified here are available to external researchers on a consultant basis. This situation has arguably resulted in exclusivity in the application of these methods to large-scale well-funded projects, which consequently leave large gaps in our knowledge as modern construction projects are rarely distributed evenly over archaeologically significant regions.

Equally important is the realisation that all the identified micro-scale techniques require the acquisition of fresh field-data. For example, in order to analyse hearthstones by palaeomagnetic methods, the orientation of all field specimens needs to be recorded in the field by means of a special device or by casting a Plaster of Paris mould of the top of the targeted hearthstone (Gose 2000:411). It would be impossible to 'reorient' any hearthstones in museum collections and this technique therefore requires the creation of new projects to be a viable research option in the production of knowledge. Of obvious concern is the documentation and material evidence for LTFs that have been building up in public and private institutions over the previous century of archaeological research. These collections are largely not analytically suitable for modern micro-scale archaeometric research methods.

There is currently no attempt, among archaeologists, at a cohesive approach to the identification, excavation, analysis, or interpretation of specific thermal feature classes. The majority of micro-scale research is directed toward the scientific investigation of components of thermal feature technology. Interpretation of specific LTF features appears to be largely dependant on a number of factors more often than not situated within the dominant theoretical paradigm. North American LTFs are invariably examined in economic terms of optimisation i.e. in rates of calorific output, resource choice, and technologic efficiency (e.g. Smith *et al.* 2001). In contrast, the social cohesive aspect of group bonding, which has been frequently observed, ethnographically dominates European discussions of LTFs (e.g. Galanidou 1997; Vaquero and Pastó 2001).

Macro-scale Research

At the macro or site-scale, archaeologists have traditionally utilised the data from excavated LTFs in three distinct research directions: as direct evidence for the function of a particular site e.g. a hazelnut roasting camp; as a means for providing a temporal framework for a site through absolute dating techniques; and as a conceptual tool in site-structure research emphasising the relationship of features and discarded cultural material to infer spatial patterning of prehistoric behaviours. As with the micro-scale, the level of inference that these research directions provide is largely proportional to the scale and type of data recovery undertaken.

Interpretative Potential

Features encountered and recorded during archaeological investigation become candidates, to greater or lesser extents, in the production of interpretative statements aimed at identifying the dominant refuse-producing human behaviours occurring at the site scale. Due to their relative rarity in hunter-gatherer contexts, features take on an especial importance in providing critical information to the production of interpretative statements. LTFs are arguably the largest and most often encountered feature on hunter-gatherer sites. Other less often encountered features include various forms of post and stake holes, which are the result of the anchoring necessary in the erection of above-ground wooden structures such as the frames of habitations, brush fences, or attendant facilities like drying racks.

Feature encounter rates are not typically published by archaeologists working in a particular region, but it is useful to observe that a survey of the prehistoric sites located in the North American state of Texas recorded at least one hearth on 23.25% of the 20,220 identified sites, comprising by far the most frequently encountered feature class (Biesaart *et al.* 1985:28). Additionally, the second most frequent class (12.15%), 'burned-rock feature' would also be subsumed within the larger feature class LTF, further boosting the statistic to around 35% of recorded hunter-gatherer sites in Texas. The successful identification of features on an archaeological site commutes a level of integrity to the site and arguably adds an extra weight to the associated research value when considering preservation options. The identification, recording, and subsequent interpretation of LTFs are therefore crucial to the production of archaeological knowledge when considered within the site context. Single sites often contain numerous LTFs and a range of different technological processes and associated behaviours can often be identified. The aforementioned applicability of LTFs to radiometric dating methods in this instance allows a sequence of human behaviours to be developed at a site scale. In rare cases, the contexts of LTFs are so spatially confined that they intersect and cross-cut one another, allowing an extremely detailed sequence of site-formation processes to be constructed.

An excellent example of LTFs at a site scale is 41LU1, area 10 located on the Lubbock Lake Landmark in northwest Texas (Figure 2.8). Meticulous ongoing excavations at this area have recorded a sequence of 11 intersecting LTFs (Backhouse 2002). The features consist of subsurface pits and basins excavated into the sandy upland sediments, presumably in order to house various fire-based technologies. The features are choked with the remnants of this activity and include hearthstones, small burned-bone scrap, charcoal bits, and some lithic debitage. To date, around 30,000 individual artefacts have been recovered from excavations, the majority of which comprise caliche hearthstones (Backhouse 2002). Sediments in and around the features are highly carbonised, consistent with the interpretation of their active use in fire technologies. Radiocarbon assays on the features have yielded 21 discrete dates (Figure 2.9) and indicate that the area was repeatedly visited by hunter-gatherer groups during the late Ceramic period through Protohistoric periods (ca. 1000 – 500 RCYBP). Despite the excellent level of preservation and density of recorded LTFs, no solid evidence for habitation structures has yet been identified.

Figure 2.8. Intersecting LTFs excavated into the aeolian upland sediments at 41LU1, Area 10 in Lubbock County, Texas, USA. Note: field photograph taken after the complete excavation of several of the features (photograph courtesy of the Anthropology Division, Museum of Texas Tech University).

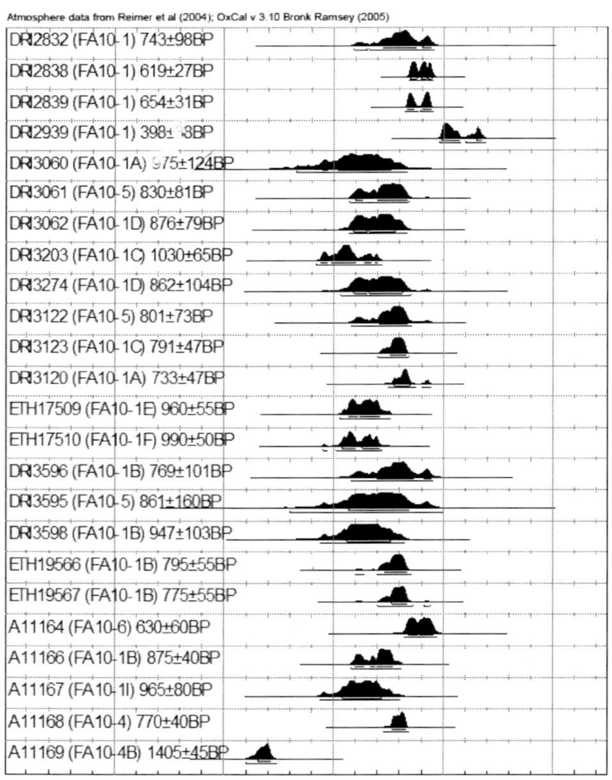

Figure 2.9 Results of samples sent for radiocarbon dating from 41LU1 Area 10 plotted using OxCal Viewer Version 3.9 (data from Buchanan 2002, table 7.1; Backhouse 2002, table 10.5).

Morphological analysis of the recovered hearthstones suggests significant fracturing, a product of intense heating or utilization in water-boiling activities. The massive frequency of broken burned and calcined bone that was recovered has been interpreted as evidence for specialist activities such as marrow extraction and grease rendering (Hamilton 1998). Regardless of the human behaviours and concomitant technologies represented at this area, it should be clear that the examination of sites containing one or more LTF is typical within hunter-gatherer research.

A second contrasting example of LTFs in a site context is taken from the well-published LTF assemblage recorded at the Middle Palaeolithic rock shelter site of Abric Romani in north-eastern Spain (Vaquero and Pastó 2001). This site exhibits significant evidence for Neanderthal fire-based technologies preserved in the well-stratified travertine layers. The horizontal location and size of the LTFs at this site are well established due to the underlying travertine construction surface, which has preserved thermal combustion events in the form of clearly delimited burn marks (Vaquero and Pastó 2001:1212). The features themselves are characterized as homogenous layers of ash and charcoal. By relating the accumulations of ash and charcoal with the burn marks on the travertine surface, the excavators were able to determine which features represented in situ LTFs and which had been displaced by behavioural or taphonomic processes from their original contexts.

The common factor in virtually all the available site-specific literature is the association of LTF data with the perceived economic function of the site. For example, Mithen (2000) seeks to understand the importance of hazelnuts to Mesolithic diet. Similarly, on the Southern High Plains of northwest Texas the recovery of a vegetable-roasting feature is interpreted as signifying greater reliance on plant resources during a period of perceived harsh climatic conditions (Johnson and Holliday 1989; 1995). At Pincevent No. 1. and many other Palaeolithic open sites, in Europe LTFs are interpreted in terms of tethering production and maintenance to specific spatial locations (e.g. Carr 1991). There are clear variances between the type, methodology, and theoretical framework of research conducted; moreover the focus of research at the site level tends to be artefact rather than context orientated.

Temporal Control

The applicability of LTFs to absolute dating techniques has arguably had a negative effect on their value to understanding site-level behaviours. This appears to be directly related to the general lack of opportunities to obtain absolute dates in hunter-gatherer camp site settings. In contrast, LTFs offer an increased potential for recovering carbonised organic material necessary for radiocarbon dating the mainstay of modern archaeometric methods. Furthermore, the widespread adoption and economic viability of Accelerator Mass Spectroscopy (AMS) has resulted in the realisation that very small amounts of organic material can yield a useful date (Gowlett 1987). This situation, although not widely discussed, has likely resulted in a reductive perception of LTFs as reservoirs for providing temporal control to human occupation within the wider site setting (e.g. Kruger 1957).

Systematic radiocarbon sampling of LTFs to provide a temporal framework of activity is a perspective positively embraced by Simon Holdaway and colleagues (2002) in their research conducted on the arid margin of south-eastern Australia. This perspective resulted from the realisation that the eroded thermal features and artefacts they were encountering during pedestrian survey probably represented different times and behaviours that had been "lumped together" by the dynamic nature of the geomorphic setting (low deposition rates and frequent erosional episodes (Holdaway et al. 2002:353). Instead of focussing analysis on the artefacts, Holdaway et al. build a model of human presence and absence within the investigated valley system as evidenced by periods of hearth-building activities, over long times scales. This perspective facilitates the identification of broad shifts in population dynamics and provides a useful temporal framework for more traditional technological analyses of the stone tools recovered within the study area. Where features are buried in well-stratified contexts, then geoarchaeological approaches can be usefully employed to determine the relative sequence of cultural events represented. LTFs should be expected to be particularly useful in this respect because if they are visible in deeply buried contexts, then it is unlikely that they have been displaced from their original location, as is often the case with individual artefacts (Metcalfe and Heath 1990).

Site Structure Research

The recognition that humans perform actions in different spatial domains allows the possibility for the identification of these actions in the form of discarded material culture and residues encountered during the process of archaeological excavation (Vaquero and Pastó 2001). The corollary of the increasing availability of precise artefact distributions is the theoretical engagement with concepts of the human use of space in both physical and metaphysical (or conceptual) sense (Robin and Rothschild 2002:162). Within the context of a hunter-gatherer campsite, the physical and conceptual space directly surrounding an LTF is an extension of the LTF, itself requiring detailed spatial analysis.

The spatial behaviour of individuals at a site scale is an expression of cultural identity (e.g. Binford 1978; 1983; Galanidou 2000). The epistemological basis for site-level spatial approaches has developed directly from ethnographic observation, which primarily sought to identify and isolate site-level behavioural expressions that could be directly related to the static archaeological record (e.g. Gould 1978; Binford 1978; 1983; Yellen 1977). Observations suggested that universal patterns of human behaviour may be solicited from the archaeological record and directly applied to groups in ecologically similar settings (Binford 1978). The normative statements this research generated had direct repercussions for the analysis of LTFs, an example of which is Binford's (1977) suggestion that:-

> "When people are working at a job which requires the use of a hearth, they tend to carry out the task according to a spatial pattern which appears to be universal" (Binford 1977:149).

Fom an archaeological perspective, two predictive models were generated with which to assess site structure and behaviour in relation to LTFs at a site scale. Leroi-Gourhan and Brézillon (1962) approached the problem inductively from the archaeological record, suggesting that the spatial variation in the recovery of discarded material was directly relational to the spatial location of LTFs. In contrast, Binford's deductive model of drop and toss zones around the periphery of an LTF, used direct analogy to interpret patterns encountered in the archaeological record (Binford 1978).

The bulk of the ethnographic observations used in generating Binford's model focussed directly on the physical dynamics of activity within a hunter-gatherer campsite. Specific 'universals' (or frames of reference) include: individuals sitting at right angles to an LTF when working directly with it, larger groups requiring more space with which to work around an LTF, small waste items are most likely to be dropped near the LTF whereas larger waste items are most likely to be tossed to the periphery of the area, large LTFs are most likely to occur in areas outside structures, people sitting outside around an LTF with no shelter alter their positions frequently in accordance with wind direction, it is easier to build a new hearth than relocate in relation to the old LTF when not confined by a building, areas around LTFs are predominantly used as communal activity areas, and the size and variability of accumulations (associated with LTFs) are functions of group size and therefore occupation duration (Binford 1977; 1978; 1983; 1987; Yellen 1977; O'Connell et al. 1991; Bartram et al. 1991; Stevenson 1991). These generalised statements are a useful baseline for assessing the physical constraints of working with an LTF, considering that the basic unit of analysis (the human body) remains largely unchanged (Binford 1977:163).

The early optimism that these types of study could provide explicit patterns of behavioural information from the archaeological record has been dampened and replaced by an approach that emphasises the role of formation processes (e.g. Schiffer 1983) in the interpretation of the spatial record (e.g. Stevenson 1985; 1991; Gamble and Boismier 1991; Wandsnider 1996; Galanidou 1997, Grøn and Kuznestov 2003).

Increasingly complex understandings of site-formation processes derived largely from heuristic research have incorporated concepts of the temporal structuring of human activity over the course of a site's occupation (e.g. Stevenson 1985). These approaches identify the social nature of LTFs as multipurpose locations of activity where deposits often comprise a palimpsest of behavioural information (Binford 1983; Carr 1984; Galanidou 2000, Vallverdú et al. 2005). Recent ethnographic research reveals that domestic tasks are only very rarely spatially segregated (Galanidou 2000:257). An example of this is the relationship of sleeping preferences to the location of hearths. Galanidou's study of ethnographic groups suggested that "sleeping areas are always adjacent to hearths, but not all of a site's hearths are used for sleeping by" (Galanidou 2000:250). Additionally, a strong correlation is identified between cultural identity and observed discard patterns around the hearth (Galanidou 2000:255). She further argues that extension of site-scale observations to inter-site scales may yield pertinent cultural signatures. An example of this kind of interpretive theory building from the archaeological record is Caspar and De Bie's (1996) spatial analysis of discarded lithic material in relation to hearth features at Rekem in Belgium. The results indicated human behaviours were spatially delineated by the technological stage of manufacture (Caspar and De Bie 1996:455). This spatially oriented thesis is currently being supported ethnographically (e.g. Dawson 2002; Grøn and Kuznetsov 2003) and archaeologically by increasingly detailed analyses of hunter-gatherer occupation floors, which need not be bounded by physical structures (e.g. Kent 1990; Whitelaw 1994; Caspar and De Bie 1996; Spikins 2000; Vaquero and Pastó 2001; Robin and Rothschild 2002; Yvorra 2003; Grøn 2003).

Recent research directions have concentrated on the empirical analysis of the material evidence rather than ethnographic analogy. An example is Marc Stevenson's (1991) size-sorting model, which predicts that size is an important variable in the displacement of discarded cultural material around an LTF. The larger the discarded object, the more likely it is to be kicked or scuffed during occupation of the site. This model results in a similar distribution, through differently identified cultural processes, to Binford's (1978) drop and toss model.

Two recent spatial models have adopted a more interpretative perspective on the material evidence. The statistical ring and sector model relates artefact density in relation to a central hearth and has been developed directly from Leroi-Gourhan and Brézillon's (1962) research at Pincevent No. 1. The model has been integrated in a suite of statistical tools (termed ANALITHIC) for the spatial examination of hearth-related assemblages at a site level (Stapert and Street 1997). Similarly, a contextual site-scale project, which seeks to integrate diverse research strands, has been proposed by Carr (1991:221) as a potential alternative to the narrowly interpretive models outlined above. Carr's re-examination of the Pincevent No. 1 site demonstrates the utility of a wider contextual approach in providing a critical synthesis of hearth-related behaviours at a site scale. It is unfortunate then that all the models, including those proposed by Carr (1991) and Stapert & Street (1997), involve an implicit relationship of the LTF to feature-oriented activities and behaviours. In all these projects, and as Michael Petraglia (2002:242) has pointed out, very little research has examined the construction, technological life cycle, and site-formation processes affecting the feature itself.

The contextual relationship between discarded material culture and abandoned LTFs is clearly problematic in environments with low rates of sediment deposition. This situation has led some researchers to dismiss the utility of site-centric scales of analysis. For instance, Holdaway and colleagues suggest, "People in the past dropped artefacts or abandoned hearths for many reasons, most of which archaeologists have little hope of understanding" (Holdaway *et al.* 2002:356).

The theoretical development of site structure analysis has been paralleled, and in many respects precipitated by, a significant improvement in practical field methodology, recording, and analytical techniques. The need for fine-grained three-dimensional data capture and integration with feature information at a site level has long been recognised (e.g. White 1980; O'Connell 1987). The increasing availability of precise mapping-equipment technology, such as total stations, has facilitated a high-resolution approach to material provenance (Spikins 2000; Roskams 2001) and recognition of the importance of a holistic approach to recording the spatial relationships of all artefact classes at a site level (Petraglia 2002). The analytical development of archaeobotanical (Balme and Beck 2002), material refit (Hoffman 1981; Caspar and De Bie 1996), geochemical (Frink *et al.* 2004), microrefuse (Metcalfe and Heath 1990; Simms and Heath 1990), statistical (Whallon 1973; 1974; Stapert and Street 1997), and taphonomic (Gifford-Gonzalez *et al.* 1985; Vallin *et al.* 2001) studies all have spatial properties that support the development of high-resolution site-structure analysis (Stapert and Street 1997).

Regional-Level Research

Attempts to synthesise LTFs as discrete technological entities at regional scales across long time spans have concentrated on the human adoption of fire as an evolutionary milestone in the development of the homo genus (Otte 2002; Wrangham and Conklin-Brittain 2003; Rolland 2004). Conversely, post-glacial approaches are largely absent in the archaeological literature. This situation appears related to the realisation that all known hunter-gatherer societies are capable of creating and utilising fire (Gott 2002) and due to its inherently expendable nature, cultural variation in the cultural uses for fire technology is difficult to observe in the archaeological record (e.g. Holdaway et al. 2002).

One attempt at a wider-scale evolutionary approach is Thoms' (2003, 2009) explicitly economic discussion of fire features as cooking facilities in North America during the Holocene. He proposes a working model in which perceived increases in the archaeological evidence for rocks as heating elements in LTFs is directly related to widening diet breadth as a function of time (Thoms 2003:87). The resultant technologic sequence is somewhat unilinear in progression, overpowering the culturally specific variability recognised from the ethnographic record (e.g. Gifford-Gonzalez 1989; Kent 1993; Jones 1993), assuming a priori that all LTFs were constructed as cooking facilities and therefore diminishing its potential for the interpretation of the archaeological record at a regional level.

Another example is the regional-scale analysis of slab-lined cylindrical basins in southwest Wyoming (Smith and McNess 1998), which focussed the analytical lens on LTFs, as important markers in the repeated occupation of specific sites over long diachronic timescales. This important study illustrates the utility of features as markers for persistent places, where material remains for multiple hunter-gatherer occupations are often recorded superimposed on one-another. Smith and McNess's research suggest that the features themselves are the determining factor in the reuse of specific locales. This interpretation is based on the assumption that the high initial energetic cost of building the features is dissipated by a long-term reuse strategy, with some examples apparently being used for up to 2000 years (Smith and McNess 1998:119).

Regardless of the validity of the energetic optimizing hypothesis presented by Thoms or the long-term planning hypothesis presented by Smith and McNess, all three researchers illustrate that LTFs need not be seen as passive site furniture. Instead, LTFs can be the central focus of a research model, which builds up from site-level observations to facilitate a useful interpretation of regional settlement systems over long time scales.

Thinking from Features, a Theoretical Background

The archaeological investigation of LTFs as constructed

by postglacial hunter-gatherer groups engages with David Clarke's prophetic 'bad samples' in a very real sense (Clarke 1973:16). A brief example, based on the southern British Mesolithic dataset illustrates the potential scale of this problem. If we let the population density be 0.25 persons per square mile (following Smith's 1992 estimates for ethnographically recorded populations) then 31,000 square miles (the size of southern Britain) could support a population of ca. 8,029 hunter-gatherers. Let us say that this population is aggregated into 321 bands, which average 25 members each (following Kelly's 1995:211 observations for ethnographically recorded groups) and let us also say that each of these bands constructs and uses one LTF per day. Given these, admittedly crude assumptions, then we can predict 321 LTFs were being constructed per day, resulting in ca. 117,165 LTFs a year, and ca. 585 million LTFs should be expected to have been built between 10,000 and 5,000 RCYBP (roughly the British Mesolithic). In contrast, there are currently around 100 known Mesolithic-age archaeological sites, within Southern Britain, with evidence for LTFs, yielding a sample of less than 0.00001% of the expected population. The foregoing gross calculation illustrates the relative importance attached to any archaeological discovery of post-glacial hunter-gatherer sites with secure contextual associations of LTFs. This situation also underscores the importance of developing a holistic approach, which can utilize the data already collected, whilst simultaneously integrating the results of modern investigations.

Portable Material Culture and the Traditional Perspective

An emphasis on research that prioritises the investigation of features contrasts with traditional modes of investigation, which tend to emphasise individual elements of portable material culture (such as stones and bones (See Straus 2006)). The reliance of archaeological approaches on these constituents is unsurprising, given their intrinsic value for onward transport to laboratories, museum displays, and classrooms. It is important to remember also, that hunter-gatherers rarely leave a large archaeological 'footprint' and the archaeological remains for hunter-gatherer campsites often consist of scattered ephemeral distributions of cultural material, where the identification of features is a relatively rare occurrence. Archaeologists must therefore, by necessity, derive the majority of the information that they collect from the most frequently encountered artefact classes.

Preservation factors favour stones and to a lesser extent bones as being more likely to survive over time and therefore be recovered by archaeologists. In contrast, the often fragmentary traces of activity that represent most hunter-gatherer features are highly susceptible to numerous processes that will destroy their physical structure or render them effectively invisible to archaeological investigation (Sergant et al. 2006). It is interesting to note that programmematic statements identifying the need for examining site-formation processes have also been artefact rather than feature oriented (e.g. Schiffer 1983). Of course, the behavioural linkage between human-made facilities (such as hearths) and surrounding artefact scatters are difficult to establish. This is because many hunter-gatherer sites are composed of palimpsests of discarded cultural material (Straus 2006:501). The nature of the record has, however, not greatly dissuaded lithic or faunal specialists from routinely making behavioural statements based on their analysis of recovered assemblages. In contrast, localised thermal features represent finite and identifiable expressions of human technology with concomitant use lives suitable for archaeological investigation.

Traditional modes of research developing from antiquarian traditions of collecting, only encountered specific features on an infrequent basis and therefore, by necessity, were forced to examine hunter-gatherer technology largely from the materials that were available, invariably flaked lithic tools and debris. Over 100 years of archaeological data collection, throughout the globe, has arguably resulted in a situation in which the identification and examination of hunter-gatherer features are no longer anomalies that should be considered in isolation. However, there is a perceived absence of systematic attempts to integrate the data already recovered and it would seem that archaeologists are condemned to repeatedly analyse these features in isolation, at the periphery of the core technological classes of lithics, ceramics, and faunal material. A brief consideration of non-portable structural features such as LTFs clearly illustrates the interpretative potential of emphasising the latter when examining the hunter-gatherer groups expressed in the archaeological record.

Interpretative Potential

Theory is a necessary a priori component of archaeological research, which structures the types of questions archaeologists ask and the ways archaeological data are interpreted (Johnson 1999). The process of archaeology is, however, circular and new theories and interpretations are developed a posteriori. Theoretical propositions are therefore politically influenced and exert significant sway on the range of possible interpretations applicable to the archaeological record (e.g. Hodder 1992; Boado 2001).

Research directions pertinent to the investigation of thermal features are clearly a component of this overarching epistemology. Furthermore, the relationship is rarely formalised, resulting in a reliance on implicit assumptions, which influence the way we understand the archaeological record. Critical examination of these relationships must be undertaken in order to construct an informed basis for thermal feature research. An example of this type of assumed relationship is Marcel Otte's exploration of fire as a catalyst for human development:

> "Fire constitutes a gathering agent of the social group, to exchange ideas, transmit knowledge and maintain individual cohesion. In the common meaning, the domestic "hearth" has more symbolic meaning than technological. This hearth is the place where one is reassured, protected and among friends. This "radiant" location around the fire constitutes from now on the basis of human society, where it is formed, consolidated and transformed" (Otte 2002:9).

The preceding statement makes a useful contribution to thermal-feature research in a heuristic sense engendering debate and refocusing attention on the social context of the structure. It is impossible, however, to isolate the prevalent cultural milieu within which Otte made this statement. Human perception of fire and its associated meanings must be identified as a diachronic phenomenon, which cannot be reduced to an essentialist perspective.

In contrast, the prioritisation of a single feature type has been attempted only rarely. A notable exception is Timothy Pauketat and Susan Alt's (2005) investigation of the postmold (post hole), a feature often recorded during archaeological research on residential Mississippi River basin sites. The theoretical basis they use for this research, which they term 'historical processual methodology', draws on multiple lines of evidence to assess the variability identified in a specific feature class (postmolds). The research stems from dissatisfaction with previous archaeological investigations of these features and the inherent problems largely mirror those identified for LTFs and are worth quoting at length:

> "Archaeologists struggle to define them (post molds), try not to overlook them, confuse them with animal burrows and, all too often, relegate them to the background of interpretations. Typically, once identified, counted, and measured, they are lumped together as a static type of archaeological feature or as a benign characteristic of houses, settlements, or monuments. *Treated this way, they are almost meaningless background noise in the larger interpretive schemes of archaeologists*" (Pauketat and Alt 2005:215, emphasis added).

Following from this realisation, Pauketat and Alt turn the conventional approach to analysis on its head by prioritising the practical examination and interpretative theorisation of postmolds. They use these features to explain human behaviour across time and space in the Mississippi River basin. The result is a highly original interpretation, which foregrounds the selected features through the holistic use of multiple scales and lines of evidence (Pauketat and Alt 2005:232). This interpretative approach, arguably compatible with more traditional modes of inquiry, appears highly suitable for the investigation of large heterogeneous datasets, such as those expected for hunter-gatherer LTFs. While labelling the present research project as a component of 'historical processual methodology' is perhaps unnecessary, the underlying interpretative perspective is adopted here in the examination of LTFs.

Summary

The preceding discussion suggests that a clear relationship exists between the scale of investigation and the ways in which the interpretation of LTFs can inform archaeological research. As analytical focus is drawn back from the micro-scale of the site to the wide-scale of the region or continent, then variability is subsumed in larger political structures, which trend toward economic explanation. Concomitantly, LTF research has concentrated on aspects of the particular, resulting in the development of an atomistic perspective, which has proved most fecund to site-level discussions. Site-structure analysis is arguably the only exception to this rule, in generating bridging models, which have multi-scalar diachronic potential.

Based only on the presented archaeological evidence, it is clear that LTFs are ubiquitous components of hunter-gatherer material culture on a trans-global scale and furthermore, that they present the opportunity to provide significant insight toward social, economic, and political research directions. Over the last 20 years, researchers have identified, excavated, recorded, and analysed LTFs in very different ways. Unfortunately, the information generated by this process has been largely restricted to a burgeoning library of descriptive site reports and a lesser number of state-of-the-art archaeometric analyses. Consequently, there has been very little development toward a critical understanding of hunter-gatherer fire technology in its own right. Considering the quantity and quality of information that can be derived from one feature or one hearthstone, it should be clear that the examination of hunter-gatherer fire technology through the LTFs observed in the archaeological record warrants serious attention. Heuristic approaches generated from and grounded in regional datasets represent a practical means by which we can begin to fill the gaps identified in the archaeological literature and therefore present a broader understanding of hunter-gatherer fire technologies.

The analysis of LTFs, as constructed and utilised by prehistoric hunter-gatherer groups, has been largely under-theorised by the archaeological community. The direct results of this situation have led to a perceived disconnection between the technological attributes of LTFs and their human manufacturers through the homogenisation and subversion of interpretation largely by implicit ethnographic and political assumption. In North America, this has largely resulted in a restrictive focus on economic production, empirically quantifiable in terms of (largely calorific) output(s). In Europe, a sense of historical 'humanness' appears to have pervaded the recent literature, arguably resulting in a past where hunter-gatherer groups made camp each night sitting at specified distances from a homely flame while regaling one-another with stories and song. Unfortunately, these interpretative positions are probably more informative of the differences in the ontological status of archaeological research within these two distinct geographic areas than they are of any real hunter-gatherer behaviour.

Hunter-gatherer domestic fire technology, identified in the archaeological record as LTFs, will be explored through a case study of the Southern High Plains (for which a detailed dataset is needed [Chapter 3]). Research progresses through four research strands: 1) a taxonomic key, to identify variation in the current dataset; 2) ethnographic research; 3) experimental research; and 4) fieldwork. The results of these studies will be used to create models of hunter-gatherer fire technology on the Southern High Plains and ultimately to asses the usefulness of such an approach in this setting and the relevance for its extension to other geographic regions.

CHAPTER 3 - APPROACH AND METHODOLOGY

Abstract

An integrated research project for the investigation of LTFs is presented. The project utilises cultural resource site records on file at government agencies as the basis for the development of a regional database. Compilation of the information required the creation of a third and combined record format. The assembled records were ordered into archaeologically useful groups by applying a classificatory structure. The resultant dataset was queried in order to examine variation in the physical evidence for LTFs in the archaeological record of the Southern High Plains of North America. A Geographic Information System (GIS) project was designed in concert with the data collection, in order to integrate the sites database, and by which to present the available data geographically. Subsequently, focussed research questions identified in the regional database, were investigated through a complementary phase of heuristic research. This phase engaged the physical record through ethnographic, experimental, and field-based research, focussing on site and feature specific problems that have the potential to be meaningful at a regional scale. The results of this problem-oriented phase not only contribute empirically to the understanding of hunter-gatherer fire technology, but also additionally open up new avenues of inquiry for further rounds of research. The outlined methodology is ultimately, therefore, recursive and forms the basis of a project for the continued research of hunter-gatherer fire technology on the Southern High Plains.

Research Approach

A research model for the understanding of the production and use of localised thermal features by hunter-gatherer populations is introduced. Practically, the research programme examines the physical evidence for LTFs in the archaeological record and situates the results in an active framework of ethnographic, experimental, and ongoing archaeological research. The programme is focussed at the regional scale and the Southern High Plains of mid-continental North America is selected as an ideal region (Chapter 4) within which to assess the usefulness of this approach to understanding hunter-gatherer fire technology.

Within this project, two task stages are identified as being methodologically distinct: data capture and heuristic research (Figure 3.1). Data capture involves the categorization of the Southern High Plains LTF dataset through the application of a simple classificatory structure (Chapter 5). Heuristic research (Chapters 6 & 7) targets selected research priorities identified during the data capture stage. The latter research feeds back into the overall development of the interpretative project and its relevance to understanding prehistoric fire technology on the Southern High Plains (Chapter 5).

Heuristic research projects, as presented here, comprise ethnographic, experimental, and fieldwork-based approaches (Darvill 2002) which offer critical insights for investigating exhausted technologies such as the LTFs recognized in the archaeological record. These conventional approaches allow archaeologists to engage with and conceptualise the dynamic properties of working with and controlling fire for various tasks and how these may be physically represented. Implementation of this model is practically achieved by following a standard sequence of research activities, which can be summarised as:-

1. Identification of variability in the regional LTF archaeological dataset;

2. Critical examination of the sources and context of this variability;

3. Development of research questions to qualitatively examine the identified variability;

4. Undertaking of practical investigations to address these research questions;

5. Discussion of the results in terms of the regional dataset; and

6. The development of suggestions for future research.

Research Method

A panoptic research methodology was developed, within the wider theoretical framework (Chapter 5) to practically achieve the stated research objectives (Chapter 1). Throughout the project, methodological considerations were contextually dependant on the stage of research and the task at hand. At the core of the research method is the necessity to successfully isolate and capture feature-specific information from state-level archaeological site records. These records, while not designed solely for recording LTFs, do represent the most comprehensive accounts for nearly 100 years of archaeological research on the Southern High Plains. Feature data extracted from these records therefore is critical, forming the base-data for the overall research project and informing the design of the subsequent ethnographic, experimental, and fieldwork research. Extraction of relevant data is achieved through the application of a taxonomic key (chapter 5) which seeks to classify the types of physical evidence identifiable in the state-level archives. The development of a computerised database and related geographic information system (GIS) project with which to practically store, search, and analyse site data is set out in the following section.

The methodologies utilised for specific elements of the heuristic research components were problem orientated and therefore not applicable to all datasets considered in the overall project. Therefore, general methodological considerations only are presented for these activities. Details concerning the specific methodologies of individual experiments are presented in the resulting published manuscripts (Backhouse 2004; Backhouse *et al.* 2005b; Backhouse and Johnson 2007a, c).

Data Capture and Coding

The first stage in the examination of aboriginal fire technology on the Southern High Plains was the assessment of the evidence gathered thus far by investigators working within the region. These data were then sifted, examined for geographic relevance, coded, and checked for errors prior to being available for output (Figure 3.2). Much of this phase of research was desk based and completed electronically.

Over 75 years of archaeological research has been undertaken within the study area (Chapter 4) and documents regarding these investigations are available in the form of: site reports, site-nomination records submitted to the state and/or national governmental organisations, the living memory of the investigators conducting the work, and the physical evidence for archaeological work occurring at a particular location. The most consistent and standardised form of information among these is the site survey form which, by law, is a necessary requirement in order to recognise the site as a specific entity at both state and national levels.

Unfortunately, the information contained on site-survey forms varies from state to state and has also changed over time within states. In addition, the quality of information reported is dependant on the skill and experience of the individual responsible for filling in the form as well as the amount of time, resources, and methodology deployed at the site being reported. These limitations result in site-survey forms that are often clearly less than ideal for assessing the regional archaeological record. They are, however, in many cases the only record available for individual sites and with over 1000 sites identified in the current research area, it would be practically impossible to visit all these sites individually. In any case, many of the archaeological features and material recorded on these sites and recorded on these forms are no longer available for inspection as they have been negatively impacted by archaeological investigation and construction work, or they have been eroded or buried by later sedimentation (Potter 2006). The forms are therefore critical historic records of prehistoric sites that oftentimes represent the only remaining available information.

The geographic focus of investigation intersects two states with roughly 75% of the area in the modern state of Texas and 25% in the state of New Mexico (Figure 3.3). Two datasets were identified, which contain site-survey information for all recorded prehistoric archaeological sites within this region.

The advantages of utilising data recorded on site-survey forms for the current programme of research are that they consistently contain brief and concise records of the types of features observed at individual sites. The cultural resource records for both Texas and New Mexico have recently been entered into electronic format and are searchable from off-site locations through restricted access databases. The electronic format greatly expedites the number of records that can be interrogated within a limited time-frame of data collection and without the expense of travelling to the government entity responsible for their curation (housed at the cities of Austin, Texas and Santa Fe, New Mexico respectively). The relative newness of the technologies involved in data capture i.e. internet-ready secure databases, means that the rapid assimilation of large datasets across two state boundaries would have been very difficult to achieve less than ten years ago. Currently, no national database of prehistoric site data exists for the United States.

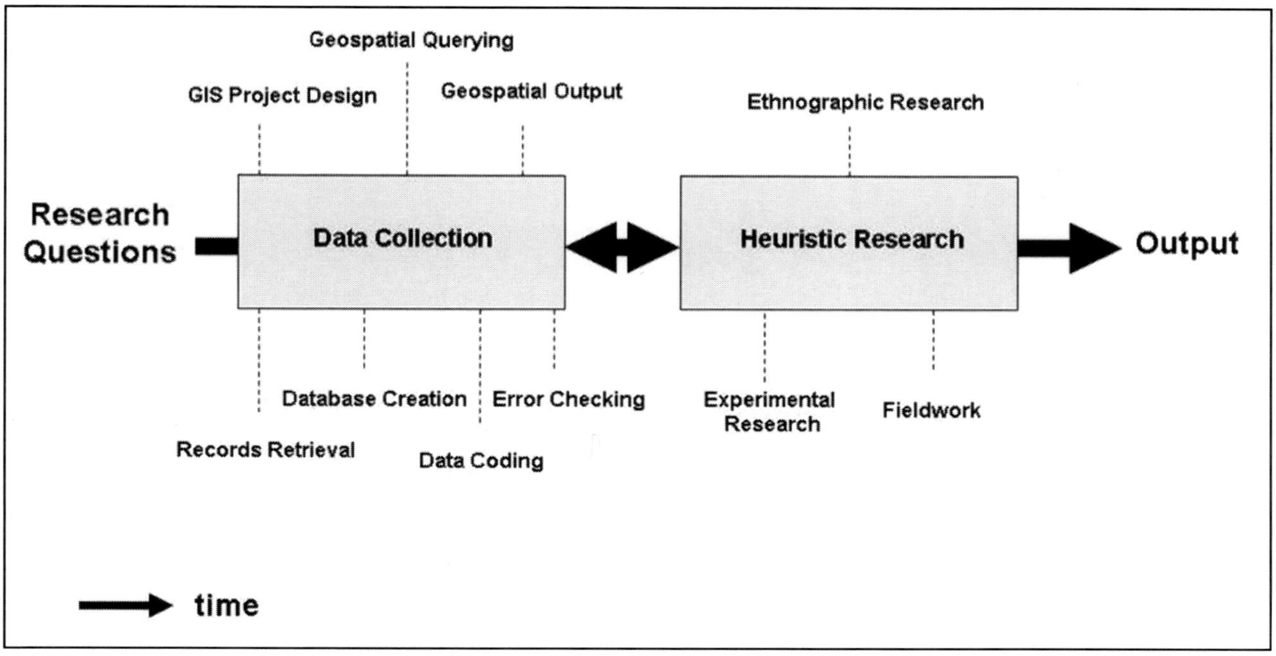

Figure 3.1. Conceptual timeline showing the main methodological components of the research project.

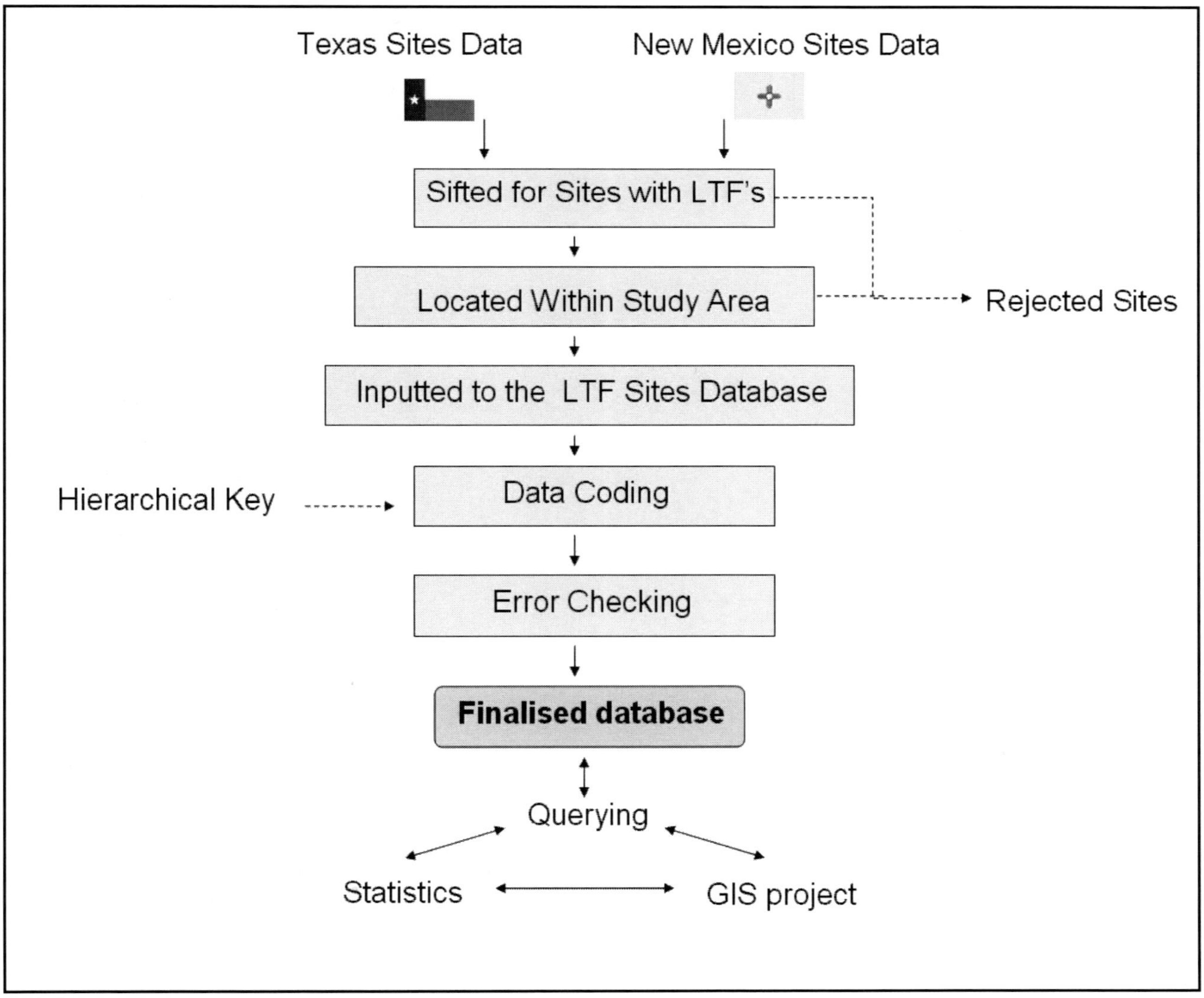

Figure 3.2. Simplified process chart showing the component steps of data capture.

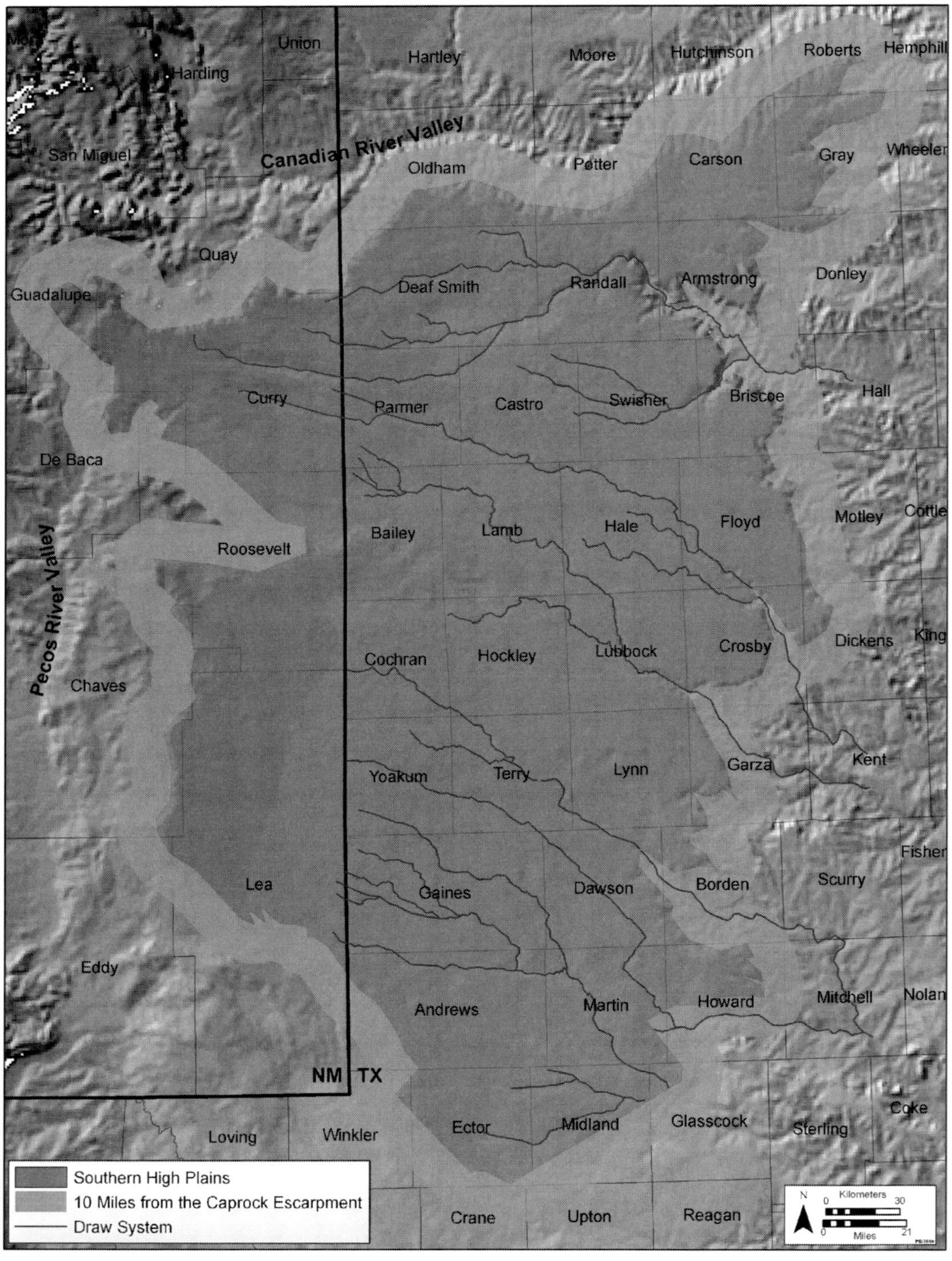

Figure 3.3. Southern High Plains region showing the County boundaries. Note: The ten-mile boundary around the physiographic region indicated in light gray.

Texas Section

Site-survey forms filed in Texas are held at the Texas Archaeological Research Laboratory (TARL) for the Texas Historical Commission, (THC) the state agency responsible for historic preservation. Filing of site-survey forms is a requirement of the National Historic Preservation Act (NHPA, 1966) and the Antiquities Code of Texas (1969) and is therefore a foundational aspect of compliance within modern Cultural Resource Management (CRM) research. It is important to note that both the identified laws governing archaeological research in Texas are geared toward sites located on federally owned lands. In contrast, 90% of Texas archaeological sites are privately owned and by law remain the personal property of the landowner (THC 2006:1). The archaeological process in Texas is therefore heavily weighted towards the identification and reporting of sites on public rather than private land.

Upon receipt of a site-survey form, the Texas Historical Commission assigns a unique trinomial designation to all recognised archaeological sites in the state of Texas. The trinomial designation consists of three separate elements. The first element is the number 41, which signifies that the site is located in the 41st state of the United States i.e. Texas. All sites in the database will therefore begin with the prefix 41. The second element is an abbreviation of the county within which the site is located, an example being LU for Lubbock County or PO for Potter County. The last element is a sequential designation based on the current number of sites identified in the particular county. For example if 121 sites had previously been recorded in Lubbock County, then the next designation would be 41LU122. The trinomial system is not used by all states in the continental United States but provides a unique reference to code archaeological sites, while retaining some information useful for quickly referencing sites to particular states and counties.

Data from completed site forms submitted to the Texas Historical Commission are regularly inputted into the Texas Historical Commission's Restricted Cultural Resource Information (RCRI) maintained at TARL. This compiled database is known as the 'Texas Archeological Sites Atlas' and is made available to off-site users via a secure internet connection (Note: The state of Texas uses the spelling variant 'Archeological' and this usage is retained when referring to Texas cultural resource infrastructure). Access to the database is dependant on professional qualifications and is not typically made available to members of the general public. Fears concerning looting and vandalism are appropriate in the often vast landscapes of North America and therefore security concerning the location of individual sites is a legitimate concern (McAllister 2000). Access was granted by Daniel G. Julien (Director of Texas Historical Atlas Program) on February 12th 2003.

The database has several options for querying site records and sites can be identified by map address, county search, United States Geological Survey (USGS) quadrangle (1:24,000 maps), trinomial search, or keyword search. For the current research, sites were queried by County. Texas counties are typically between 900 and 1500 square miles in extent. A total of forty-four counties containing nearly 4,000 archaeological sites were identified within the Texas study area (Table 3.1).

Table 3.1. Data for counties located within the Texas portion of the study area.

County	County Totally in Study Area	County Size in Study Area (Sq. Miles)	Total Number of Site Identified in County*
Andrews	N	1467	53
Armstrong	Y	914	14
Bailey	Y	827	9
Borden	N	726	14
Briscoe	N	856	548
Carson	Y	923	76
Castro	Y	898	27
Cochran	Y	775	1
Crane	N	160	30
Crosby	N	887	148
Dawson	Y	902	11
Deaf Smith	Y	1497	18
Dickens	N	354	26
Donley	N	401	26
Ector	N	893	19
Floyd	Y	992	85
Gaines	Y	1502	66
Garza	N	587	661
Glasscock	N	279	5
Gray	N	865	72
Hale	Y	1005	42
Hall	N	16	68
Hemphill	N	75	28
Hockley	Y	908	2
Howard	N	806	87
Hutchinson	N	132	220
Lamb	Y	1016	8
Lubbock	Y	899	125
Lynn	Y	892	9
Martin	Y	915	41
Midland	N	828	39
Mitchell	N	79	89
Motley	N	340	28
Oldham	N	878	260
Parmer	Y	882	7
Potter	N	577	348
Randall	Y	933	78
Roberts	N	536	130
Swisher	Y	900	38
Terry	Y	890	29
Upton	N	68	48
Wheeler	N	71	20
Winkler	N	241	77
Yokum	Y	800	2
Total		**31, 392**	**3,732**
Average		**713.5**	**85**

* all archaeological sites within the county (study area may be smaller), data accessed from TARL between Feb.03 and Dec. 05.

New Mexico Section

Site-survey forms filed in New Mexico are held by the New Mexico Office of Cultural Affairs, Historic Preservation Division located in Santa Fe, the state agency responsible for historic preservation. The Archaeological Records Management Section (ARMS) deals directly with site records. As was the case with Texas, the filing of site-survey forms is a requirement of the NHPA (1966). It is again important to note that the identified laws governing archaeological research in New Mexico are weighted toward the identification and reporting of sites on public rather than private land.

Upon receipt of a completed site-survey form, ARMS assigns a unique identifier to all recognised archaeological sites. The identifier consists of the prefix LA (Laboratory of Anthropology, Santa Fe) and a sequential numerator. For example, the site LA2455 would indicate the two thousand four hundred and fifty-fifth site identified in the State of New Mexico. The sequential system of site numbering contains no information as to where the site is located other than in the state of New Mexico.

Data from completed site forms submitted to ARMS are regularly inputted into the New Mexico Cultural Resource Information System (NMCRIS) database. This database is made available to off-site users via secure internet connection. Access to this service is again dependant on professional qualifications and is not typically made available to members of the general public. Access was granted by Karyn de Dufour (Archaeologist, ARMS) on July 25th 2003.

The database allows access to site records through a number of different search options. Sites can be identified by LA number, site name, other site number, United States Geological Survey (USGS) quadrangle (1:24,000 map), and Universal Transverse Mercator (UTM) coordinates. New Mexico counties are typically larger than their Texas counterparts and a total of 8 counties were identified within the New Mexico study area (Table 3.2). Because county data were unavailable for New Mexico, sites were instead queried by UTM coordinate ranges (Table 3.3). These searches necessarily covered rectilinear geographic areas that were not concurrent with the study area (Figure 3.4). A total of 7,896 archaeological sites were identified within this area (ARMS searches executed January 14th and 15th 2004). Because the UTM search area was significantly larger than the study area, sites identified were subject to a further cull following the development of the Geographic Information System (GIS) component of the project (Figure 3.5).

Search Criteria

Site forms for all prehistoric archaeological sites located within or adjacent to the study area were individually accessed and queried for references to LTF technology. Form inspection was on an individual basis and the type of search dependant on the recovery location (i.e. site located in Texas or New Mexico portion of the study area). In all searches, the physical evidence for hunter-gatherer fire technology included formally constructed features with evidence for burning, charcoal staining, hearthstones, burned bone, burned vegetable matter, or any feature identified by the original investigator as being culturally burned. The search criteria for the site data aims at identifying as representative as possible a sample of sites with evidence for hunter-gatherer fire technology and are therefore not an exhaustive list of archaeological sites with such evidence. In any case, as the regional record is clearly incomplete and the archaeological record a sample of the total population, such a list would obviously be impossible to construct.

Site forms submitted for prehistoric sites located in Texas usually contain information regarding cultural features observed, although this is often not the case on early site forms (pre 1970) in which information appears to be highly variable. On more recent forms, cultural feature information is assigned a separate entry field. Input in this field is generally non-standardised, allowing the free-text entry of an infinite variety of descriptive statements (Table 3.4).

Table 3.2. Data for counties located within the New Mexico portion of the study area.

County	County Totally in Study Area	County Size in Study Area (Sq. Miles)
Chaves	N	1025
Curry	Y	1406
De Baca	N	462
Eddy	N	153
Guadalupe	N	510
Lea	Y	3125
Quay	N	1540
Roosevelt	N	2296
Total		**10,517**
Average		**1,314.5**

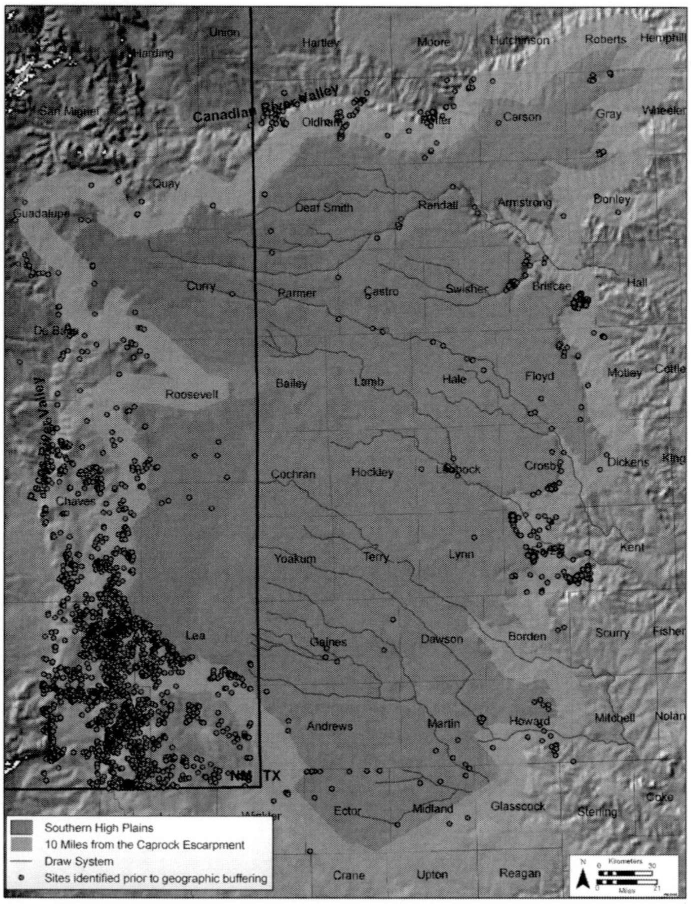

Figure 3.4. Distribution of identified archaeological sites prior to geoprocessing.

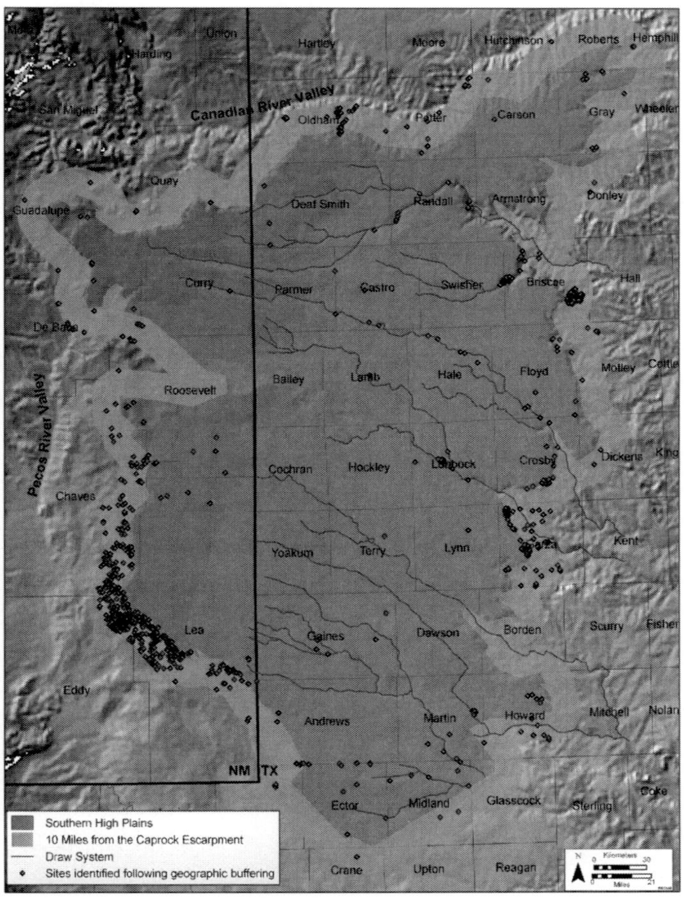

Figure 3.5. Distribution of archaeological sites within the investigation area following geoprocessing.

Table 3.3. UTM (North American Datum (NAD) 1927 datum) coordinates for the search area in the New Mexico portion of the study area.

Southwest Coordinate		Northeast Coordinate	
North	East	North	East
3540854	559008	3887427	673666

Table 3.4. Example of feature descriptions contained on site-survey forms in the Texas portion of the study area.

Site	Feature Description*
41AD22	hearth stone lined; hearths are possible, but badly eroded and scattered
41BI228	massive hearth area
41CB140	Two burned caliche hearths set on the eroded surface. They are both ca. 1 meter in diameter and parallel the caprock's edge. Only two flakes were found around the hearths.
41GR120	6 hearths see feature list for detail #1 partially exposed on surface 33" N/S by 29 " E/W sandstone slabs extends to 6 inches to 8 inch below surface #2 sandstone slabs, charcoal sample 14 inch diameter; #3 large sandstone slabs on top no dimensions. #4 some ashes (rock type not given) 18 inch L by 15" W deep. #5 covered with sandstone slabs 32 inches N/S by 30 inch E/W by 10 inches. #6 burned ash lens 16 inches across? Assoc. bone and charcoal. Charcoal sample from hearth #2 collected.
41LU27	Burned caliche forms two distinct concentrations that had similar elevations. The majority of caliche from this site is highly burned and two concentrations probably represent disturbed hearth features that were not identified in the field.

* all examples from the TARL database.

The free-text entries necessarily subsume all possible feature classes and are not just restricted to the description of LTFs. Additionally, evidence for LTFs is often present within other portions of the survey form and need not necessarily be confined to the feature description. For example, hearthstones, a useful indicator of thermal feature technology, may be recorded in the sites artefact manifest but not in the feature description. For these reasons, site-survey forms relating to sites in the Texas portion of the study area were subject to the following search criteria:

1. All forms were inspected with particular attention placed on the 'Cultural Features' field. If evidence for LTFs was identified during this inspection, then the site was marked for inclusion in the research dataset.

2. All forms were electronically searched using standard keywords:

a. 'Hearth'
b. 'Fire'
c. 'Hot-rock' or 'Hot rock'
d. 'Hearthstone'
e. 'Charcoal'
f. 'Fire Cracked Rock' or 'FCR'
g. 'Burned' and 'caliche'
h. 'Midden'

If no evidence for hunter-gatherer fire technology was identified during these searches, then the site was not included in the research dataset. Older sites with limited descriptive information or younger sites with poorly completed survey forms would probably not be included in the research dataset, regardless of whether evidence for hunter-gatherer fire technology had been observed at the site by the original investigator.

Site forms submitted for prehistoric sites located in New Mexico also usually contain information regarding cultural features observed. New Mexico site survey forms are more standardised than their Texas counterparts and a list of formal feature definitions is often utilised for site-survey form purposes. Although new features can be added to this list, over one hundred are formally defined (New Mexico Office of Cultural Affairs Historic Preservation Division 1993, appendix 7, 55-58). Of these, 11 are directly relevant to hunter-gatherer fire technology (Table 3.5).

Although not formally defined in the NMCRIS user manual, fire-cracked rock concentrations can be added to the above list. Searches of the New Mexico data were undertaken and sites included in the research dataset if one or more of the formal thermal feature definitions were identified. As with the Texas data, if no evidence for hunter-gatherer fire technology was identified during these searches, then the site was not included in the research dataset. Again, older sites with limited descriptive information or younger sites with poorly completed survey forms would most likely not be included in the research dataset, regardless of whether evidence for hunter-gatherer fire technology had been observed at the site by the original investigator.

The search criteria applied to both the Texas and New Mexico state databases attempted simply to identify the presence or absence of a record for LTFs at individual archaeological sites. While the result of the interrogation of the site forms has hopefully yielded a representative sample of the sites for which LTFs were observed and recorded by the investigators, it does not represent the sum total of sites for which such evidence exists. The atomistic nature of archaeological research suggests a disjuncture likely exists between the identified and 'true' number of sites with recorded thermal features. Biases include but are certainly not limited to: poor (or no) recording of the features in the field, the experience level of the investigator, the type of investigation, the geomorphic context of the site, the size of the site, the focus of the fieldwork, and the time and financial constraints of the project.

Nevertheless, only the sites with documentary physical evidence matching the search criteria are considered. Furthermore, it is anticipated that this focus may be informative as to any biases that may be skewing the visibility of hunter-gatherer domestic fire technology either geographically, technologically, temporally, or some combination thereof and that these biases may become evident during the analysis of the archaeological results (Chapter 8).

Geographic Information System (GIS)

In order to spatially present and interrogate the data, a GIS project was developed. The software package ArcGIS desktop (versions 8.3 and 9.1.) available from ESRI (Redlands, California) was used for all geospatial manipulations, analysis, and output. Base data were acquired from various freely available sources (Table 3.6.).

The geographic limits of the Southern High Plains physiographic region have been interpreted by various different authors as covering slightly different areas depending on where the southern and south-western boundaries are placed (Holliday, personal communication, August 2004). For the purposes of the present programme of research, the boundaries of the region were set out by tracing the northern, eastern, and western escarpments visible on freely available digital elevation models (DEM's) at a scale of 1:24000. A ten-mile extension was placed around the boundary of the region (discussed in detail in Chapter 5). The ten-mile extension boundary was created by using the buffer command located in the geoprocessing component of the ArcMap toolbox. The buffer was applied to the previously generated Southern High Plains region shapefile.

Table 3.5. Standard thermal feature definitions used for New Mexico site-survey forms.

Feature Type	Description	Related Features
Ash Stain	A distinct coloration of areas of soil on a site, usually bounded. Ash stains are distinctly grey in colour, and may include bits of charcoal	charcoal stain, dump, hearth midden
Burned Rock Midden	Large dense concentrations, often mounded, of fire cracked rock (FCR), usually associated with large-scale plant processing. Although other cultural material is usually present in the midden, FCR is predominant	hearth, mescal pit, ring midden, roasting pit, sweat lodge
Charcoal stain	A distinct dark coloration of areas of soil on a site, usually bounded. Use for all surficial stains which are believed to be cultural but which nature is indeterminate	ash stain, dump, hearth, midden
Dump	Formal or informal concentrations of historic trash, containing individual or multiple episodes of deposition. Use for all ash or coal-cinder dumps. Validation: historic components only	ash stain, charcoal stain, midden
Hearth	An extramural, localized area of controlled intentional burning, Encompasses all surficial fire-related phenomena including fire pits, formal hearths, fire rings, burned rock rings, fire deflectors, and slab-lined hearths. Hearth does not include standing fireplaces or chimney remnants. Such items should be entered as part of a House Foundation	ash stain, burned rock midden, charcoal stain, horno/oven, mescal pit, pottery kiln, ring midden, roasting pit, sweat lodge
Horno /Oven	An enclosed space used to heat objects placed within its bounds. Includes earth ovens, oven pits, mud ovens, and bread ovens	brick kiln, coke oven, hearth, kiln, lime kiln,mescal pit, pottery kiln, roasting pit
Kiln	Any of various ovens used for hardening, firing, burning or drying substances	brick kiln, coke oven, lime kiln, horno/oven, pottery kiln
Mescal Pit	A pit dug for the processing of succulent plants. Often leaves behind masses of fire-cracked rock and charcoal stained sediments	ash stain, burned rock midden, charcoal stain, hearth, horno/oven, ring midden, roasting pit
Midden	An archaeological deposit exposed on the surface of a site containing discarded artefacts and materials. Middens may have considerable depth, or may be entirely surficial (i.e. sheet midden). Midden deposits normally contain ashy or charcoal-stained sediments, and domestic trash such as sherds, lithic debitage, and bone. Use for protohistoric and prehistoric trash deposits	ash stain, burned rock midden, dump, charcoal stain, ring midden
Ring midden	A general donut-shaped or concentric burned rock midden	burned rock midden, hearth, mescal pit, roasting pit
Roasting pit	An excavated hole or pit for cooking without the direct application of fire, usually accompanied by concentrations of burned rock-related features	burned rock midden, dump, hearth, horno- oven mescal pit, midden, ring midden

* all definitions from NM Office of Cultural Affairs Historic Preservation Division 1993, appendix 7, 55-58.

Table 3.6. Summary of organizations providing GIS data used in this project.

Data Provider	Layer Types Used	Available From (URL)	Date Accessed
USGS Seamless Data Distribution System, Earth Resources Observation and Science (EROS)	boundaries, hydrography, orthoimagery, land cover, elevation	http://seamless.usgs.gov	Jan 04-Dec05
USGS Earth Surface Processes	flora	http://esp.cr.usgs.gov	Jan 04-Dec05
Texas Natural Resource Information System (TNRIS)	elevation, hydrography, land cover	http://www.tnris.state.tx.us	Jan 04-Dec05
Geocommunity	boundaries, elevation, orthoimagery	http://www.geocomm.com	Jan 04-Dec05

Both THC and NMCRIS use the Universal Transverse Mercator (UTM) grid system to plot the location of prehistoric archaeological sites and both entities also utilise the older North American Datum (NAD) 1927 rather than the newer NAD 1983 as the official reference ellipsoid for this geodetic network. The older system is retained largely because of its relevance to locating sites on USGS topographic maps, many of which have yet to be updated to the new system. The UTM system divides the earth into zones originating at the intersection of the equator and central meridian. Two longitudinal zones occur in the current research area. All New Mexico sites are located in UTM Zone 13, whereas Texas sites comprise sites in both UTM zones 13 and 14. In addition to zone information, all prehistoric sites have a six-figure northing and easting value, which represents the centre point of the site in meters within the overall UTM zone. This co-ordinate system and reference datum are retained for the current research project.

Developing a Relational Database

A database was designed using Microsoft Access (2003) with the purpose of drawing together the information contained in the two queried state databases. Sites identified as containing evidence for LTFs were inputted into the database. The formal structure consisted of 24 independent variables listed following this paragraph (Figure 3.6). The majority of data inputs are necessarily at nominal or ordinal scales, and consequently, the results of querying this dataset are mainly qualitative rather than quantitative nature.

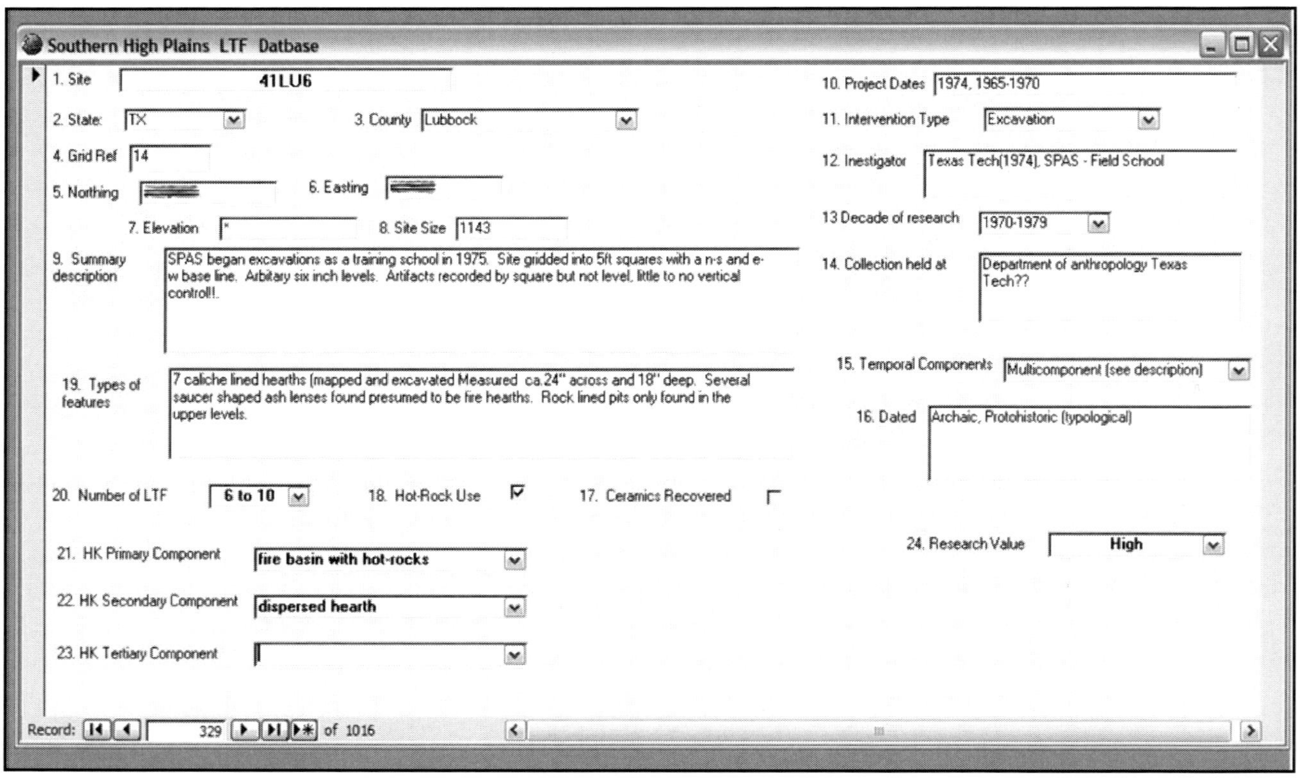

Figure 3.6. Example record output from the sites database. Note: the northing and easting are obscured in this example to protect the location of the archaeological site, site location data in Texas are confidential under state (§191.004 Texas Natural Resources Code (VTCA 1995)) and federal (Archeological Resources Protection Act of 1979, 16 USC §470aa et. seq.) laws.

1. Site. Key variable identifying each individual site. Sites in the Texas portion of the study use the trinomial system and sites in New Mexico are identified by the prefix LA (Laboratory of Anthropology). No overlap is possible between these two systems and therefore each entry in this field is unique.

2. State. Modern state in which the site is located, either Texas or New Mexico.

3. County. Location of the site within the wider politically defined county system. New Mexico counties (average 1314 square miles) are on average twice the size of their Texas counterparts (average 713.5 square miles).

4. Grid Reference. UTM zone in which the site is located, for the Southern High Plains all sites are either in zone 13 or zone 14.

5. Northing. Seven-digit grid reference locating the site centroid along the latitudinal axis using the UTM system (in meters) in reference to NAD1927.

6. Easting. Seven-digit grid reference locating the site centroid along the longtitudinal axis using the UTM system (in meters) in reference to NAD1927.

7. *Elevation*. Average elevation of the site in feet above Mean Sea Level (MSL). Where a range of values is expressed on a site form then the mean is recorded.

8. Site Size. Estimate of the overall extent of the site in square meters. For sites located in the modern state of Texas, this measurement is often infrequently recorded or recorded only in relative terms i.e. "site stretched ca. 25 yards N-S and 56 yards E-W". Where sufficient information is provided on the site form, then site size is calculated and recorded in the appropriate unit of measurement. No attempt is made to calculate the size of sites with ambiguous site boundaries e.g. "between the small creek and the large Arroyo" and these are left blank.

9. Summary Description. Free-text entry of a general description of the site. Typical observations include identified artefacts observed, location, setting, vegetation, reasons for recording.

10. Project Dates. Free-text entry indicating the time periods associated with research at the site. Where projects covered multiple years, then the range is expressed e.g. 1972-1976. Where more than one investigator is identified (11), then project dates are listed chronologically (oldest through youngest).

11. Intervention Type. Ordinal entry identifying the highest phase of investigation undertaken. Possible responses are: 'Survey' (phase 1); 'Testing' (phase 2), and 'Excavation' (phase 3); and 'unknown'. Where more than one investigation has taken place, then highest phase is recorded regardless of the project date.

12. Investigator. Free-text entry identifying the person(s), company, or companies responsible for recording the site. Where more than one entity has recorded the site, then the investigators are listed by age (oldest through youngest) e.g. University of New Mexico, SLP Consultants, Pierce and Associates.

13. Decade of Research. Interval measurement based on the response to variable 12. Where more than one decade is indicated, then the decade of the latest investigation is recorded.

14. CollectionHeldAt. Free-text entry identifying the entity responsible for curating any recovered documentation and prehistoric artefacts. Where more than one entity is identified, then the institutions are listed by chronological order of assemblage deposition (where known).

15. Temporal Components. Ordinal entry identifying the cultural time period assigned by the original investigator. All responses are either 'typological' or 'dated' dependent on the relative or absolute dating methods applied. Possible responses are:

1. Unknown	2. Paleoindian (typological)	3. Paleoindian (dated)
4. Archaic (typological)	5. Archaic (dated)	6. Ceramic/Mogollon (typological)
7. Ceramic/ Mogollon (dated)	8. Protohistoric (typological)	9. Protohistoric (dated)
10. Multicomponent		

16. Dated. Free-text entry listing the results of any absolute dating assays, or explanation of the multicomponent category identified in (Q17).

17. Ceramics Recovered. Boolean value based on the presence or absence of prehistoric pottery at the site.

18. Hot-Rock Use. Boolean value based on the presence or absence of hearthstones, Fire-Cracked Rock (FCR), burned caliche, burned sandstone, or other rock types that have been previously associated with LTFs.

19. Type of Features. Free-text entry summarising the thermal features identified on site based on the information provided by the original investigator, e.g. 'Eroded hearth with burnt deer mandible buried by 15 cm of over burden, fire-cracked rock and burnt caliche, manos and metates of sandstone' (Site Survey form for 41BI461 on file at the Texas Historical Commission).

20. Number of LTFs. Ordinal value based on the observations of the original investigator. Possible responses are:

1. unknown	5. 11 to 20
2. 1	6. 21 to 30
3. 2 to 5	7. 31 to 40
4. 6 to 10	8. 41+

21. Hierarchical Key – Primary Component. Ordinal value characterising the LTF(s) most prevalent at the site by the application of the hierarchical key (Figure 3.7) to the available descriptive data. Note that the categories '1. no feature' and '15 insufficient data' are two separate outcomes. Possible responses are:

1. No feature	9. Disturbed Pit
2. Hot-Rock Cluster	10. Disturbed Pit with Hot-Rocks
3. Dispersed Hearth	11. Fire Basin
4. Dispersed Hearth with Hot-Rocks	12. Fire Basin with Hot-Rocks
5. Hearth	13. Fire Pit
6. Hearth with Hot-Rocks	14. Fire Pit with Hot-Rocks
7. Ephemeral Hearth	15. Insufficient Data
8. Ephemeral Hearth with Hot-Rocks	

22. Hierarchical Key*–Secondary Component.* Ordinal value characterising the second most common LTF(s) at the site by the application of the hierarchical key (Figure 3.7) to the available descriptive data. Responses are identical to (21).

23. Hierarchical Key – Tertiary Component. Ordinal value characterising the third most common LTF(s) at the site by the application of the hierarchical key (Figure 3.7) to the available descriptive data. Responses are identical to (21).

24. Research Value. Ordinal value summarising the overall quality of the site in terms of number/type of LTF represented, investigation method, and collections generated. Possible responses are 'Low', 'Moderate', and 'High'.

Applying the Classificatory Structure

The database structure contains fields (21-23) with which to identify the three major types of LTF represented at any one site. Assignment of features into one of the possible categories is facilitated by application of the hierarchical key identification device (Figure 3.7 and explained in detail in Chapter 5). Statements routinely reported as part of the site-survey process are the main inputs in this system and are generally derived from responses to the descriptive variables (responses 10, 19-21). For example, the data for 41LU6 indicate that seven LTFs were encountered by the South Plains Archaeological Society (SPAS) during their 1970's excavations (Figure 3.7). All these features were "caliche lined" and the stated measurements of the features was around 24" across and 18" deep. In addition, several "saucer-shaped ash lenses" are recorded and "presumed to be fire hearths" (Site-Survey form for 41BI461 on file at the Texas Historical Commission). The relationship between the seven hearths and the ash lenses is somewhat unclear. At 41LU6, two separate LTF components therefore are assigned. The primary component consists of the seven caliche-lined hearths. The secondary component appears to be an unknown number of discrete ash lenses.

Application of the hierarchical key is undertaken for each of the two identified components. In this case, the decision making process for the primary component is fairly straightforward (Figure 3.7). The features have depth measurements indicating that they have subsurface components (decision process 1), the features are spatially bounded by the caliche lining and this lining also provides good evidence that they retain some structure (decision process 2 and 3). According to the field measurements, the structures are not deeper than their diameter (decision process 4), and caliche hot-rocks are already identified in the caliche lining (decision process 5). The primary LTF component at 41LU5 are therefore type-12 fire basins with hot-rocks (Figure 3.7). It is unknown whether the secondary LTF component of "ash lenses" has any sub-surface expression (decision process 1) and the negative path is therefore followed. They do appear to be spatially bounded discrete deposits (decision process 2) as they are reported to be "saucer-shaped". The structural integrity of these features is unknown (decision process 3) and therefore the negative path is followed and while hot-rocks are noted at this site, they are not noted within these features (decision process 4). Following the simple decision process, the second LTF component "ash lenses" are characterised as type-3 dispersed hearths (Figure 3.7). The archaeological evidence for LTF technology at 41LU6 then is identified as a primary component of fire basins with hot-rocks and a secondary component of dispersed hearths.

The decision to include three components (primary, secondary, and tertiary) in the characterisation of the archaeological evidence for LTFs at any one site was based on a pilot investigation that indicated that there was rarely sufficient information supplied with site-survey forms to identify more than one type of LTF and that sites with three identifiable LTF types were regionally very rare. Sites with a substantial diversity of well-recorded feature types will obviously not be well represented by the present methodology. This unfortunately does not appear to be the case for the Southern High Plains and most likely is a result of the limited range in the physical manifestation for hunter-gatherer fire technologies in general.

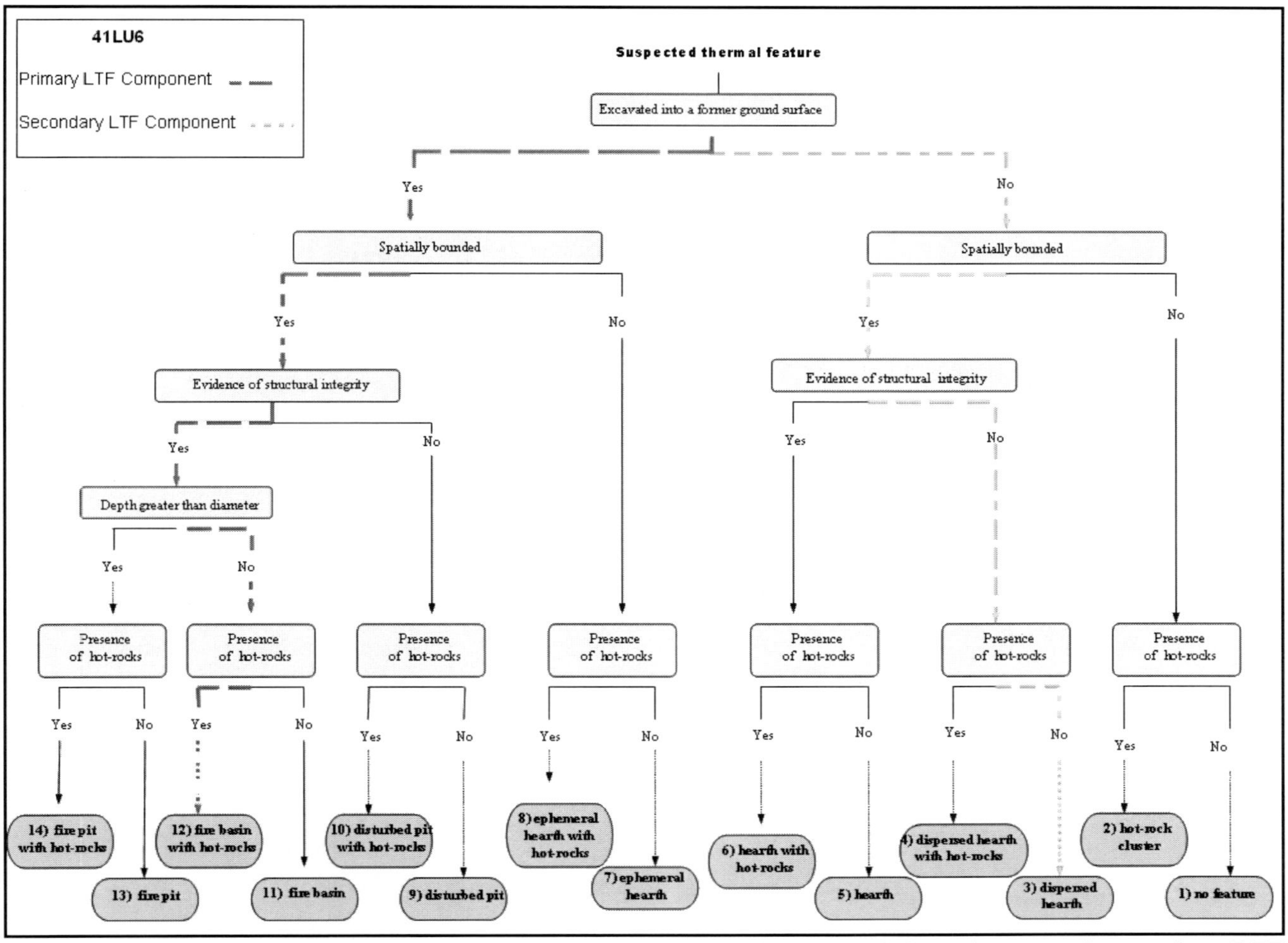

Figure 3.7. Example of the decision making process in the assignment of the primary and secondary LTF components at 41LU6 based on the descriptive data compiled from the site survey record.

Heuristic Research

The heuristic stage of investigation involved the examination of the regional record through three distinct heuristic methods: ethnographic literature, experimental research, and archaeological fieldwork. Each of the phase-two components was able to approach and therefore examine the characteristics of the regional record in differing but ultimately complementary ways.

Ethnographic Research

The objective of a scrutiny of the ethnographic record is to provide much of the central contextual detail necessary for exploring and interpreting the range in technological variation present when confronting hunter-gatherer fire technology from an archaeological perspective.

There is little doubt that ethnographic research has strongly influenced archaeological interpretations of prehistoric hunter-gatherer behaviour (Wobst 1978). In European Old World contexts, ethnographic analogy appears to implicitly underlie the interpretation of most LTF structures, apparently largely in an effort to historicise them (cf. Proctor 2003), whereas, much North American ethnographic research is explicitly applied to the investigation of LTFs from a perspective derived from behavioural ecology (e.g. Kelly 1995). The North American framework is in stark contrast to the European approach emphasising the economic importance of the feature rather than the social meaning. While both approaches appear to tolerate one another, very little dialogue is evident between these two positions. This situation appears primarily to be a reflection of the contrasts between the development of European and American archaeologies and their general position within the wider academy (Shott 2005:5).

From a European perspective, an excellent example of ethnographic influence, can be discerned in a recent interpretative illustration rendered by staff at the Museum of London Archaeological Service (MOLAS). The illustration depicts events that may have occurred at a Mesolithic campsite, excavated by MOLAS in the Thames river valley (Figure 3.8). The figures depicted in this reconstruction perform in a regulated arena within which a clear activity area can be discerned around the central thermal feature (*c.f.* Binford 1977; 1978; 1983; 1987; Yellen 1977; O'Connell *et al.* 1991; Fisher and Strickland 1991; Kelly 1995). A single person, the cook perhaps, reaches into this void to tend the LTF. The feature itself has little or no apparent purpose other than to spatially arrange the assembled hunter-gatherers in relation to it. The surrounding participants sit or interact in economic two-person task groups reminiscent of a modern production line. Children and animals are excluded at the periphery of the area and are the only actors not engaged in the production process.

The limitations of this interpretative project are largely self-evident, and are primarily the result of the powerful coercion implicit in an uncritical use of ethnographic and theoretical propositions. Ultimately, the behavioural actions of humans around a fire in the Thames Valley during the Early Holocene look much like we might expect them to. This situation is not limited to the interpretation of British datasets and has long been recognised in the North American literature (e.g. Wobst 1978).

For this study, ethnographic research was primarily document based and focussed on identifying evidence for North American aboriginal fire technologies in the centuries following sustained contact with European nations (i.e. sixteenth through twentieth centuries). Aboriginal fire technology was not the primary area of interest for any of the ethnographic accounts consulted, although it is probably over-represented in this literature in the descriptions of ceremonial events. Domestic fire technology, in contrast, is invariably underrepresented by anthropological investigation. Early accounts of European incursions into North America are rare and for the most part written in Spanish. It is therefore fortunate that the details of these expeditions have been the subject of intense scholarly research, which is more readily available than the original texts (e.g. Wade 2003). Due to constraints in access, time, and language, much of the historic ethnographic data examined in the current research project are therefore derived from secondary sources. This is obviously a less than ideal situation but necessary given that ethnographic research is not the central thrust of the current project. The author suspects, however, that more detail on aboriginal fire technology could be gained by returning to the original documents.

Figure 3.8. Reconstruction drawing of hunter-gatherer residential activities in relation to a centrally placed LTF (figure Sidell et al. 2002:20, fig. 15, copyright Museum of London Archaeological Service).

While ethnographic literature from North America is examined, geographic emphasis is placed on identifying the fire technologies of the mid-continental Plains region and in particular on the Southern Plains area. No direct historic link between historic and prehistoric populations is implied by this methodological emphasis on the Southern Plains. It is, however, reasonable to assume that the range of possible fire technologies and the physical expression thereof is constrained by the environmental setting of the study region. The well-studied and largely homogenous Southern Plains biome is therefore an ideal setting for such an investigation (See Chapter 4).

Because the outcome of such an investigation is a body of data detailing the range of historically documented uses of fire in the study region, a cultural baseline of possibilities was reflexively generated with which to assess the prehistoric archaeological record. Details concerning the physical construction, use-life, materials, methods of cultural utilisation, function during use-life, type of fuel, discard or abandonment practices, and associated activities occurring in the vicinity of the feature were all treated as of especial importance in the scrutiny of the documentary records.

In addition to the documentary evidence, an informal interview was conducted with Ray Olachia, a Native American participant (part Mescalero, Jicarilla Apache, and Yaki). The historically recorded range (post 1540's) for these groups is along the Pecos and Rio Grande river valleys, the latter valley forming the western boundary of the Southern High Plains study area (see Chapter 4). Mr. Olachia's knowledge of earth oven construction and cooking, taught to him by his father and used for subsistence during his childhood, provides a useful glimpse at the practices, gender roles, and construction techniques associated with earth oven technology.

The interview was conducted on the 17th of June 2005 during an educational demonstration of earth oven cooking methods in the experimental research area at Lubbock Lake Landmark (Figure 3.9). A full record of the interview was recorded using a digital voice-recording device (RCA digital voice recorded) and supplemented by photography (Sony Cyber-shot camera). Both forms of media and recording were given express consent by Mr Olachia prior to the commencement of the interview.

Figure 3.9. Ray Olachia instructs local high school teachers, students, and professional archaeologists in the construction of an earth oven at the Lubbock Lake research area. (P.N. Backhouse)

Experimental Research

The objective of the experimental research programme was to increase our understanding of hunter-gatherer domestic fire technology by practically exploring the physical processes associated with physically building and using different feature types and morphologies. Various LTFs were constructed using evidence identified in the archaeological/ethnographic record for the Southern High Plains. The four experiments quantitatively tested specific research hypotheses, relating to the physical processes and transformations of materials that occur when they are subjected to intense heating,

Experimental research has arguably contributed very little to the theoretical development of the wider discipline (Odell and Cowen 1988). This is clearly a result of the historical position of experimentation in the positivist tradition as a method to objectively evaluate physical processes and predict the concomitant probability of these being responsible for the objects and features we observe in the archaeological record (Sraydar and Shimada 1973; See also Chapter 1, Thermal Features and Experiment).

The dominant theoretical position is most simply stated by Odell and Cowan (1988) and the experimental projects undertaken during the current programme of research also follow this basic framework:

> "a systematic control of relevant variables in order to test the effects of each on the observed phenomena. Its purpose is to present a particular archaeological problem and investigate its solution through the experimental method" (Odell and Cowen 1988:196).

In addition to the testing of physical properties and processes, a learning experience is also a fundamental embedded component of experimental research (Telles 2000). Sequences of experiments conducted over time engage researchers with unfamiliar physical structures and technological processes not normally encountered in purely archaeological contexts (e.g. Townend 2002). Conducting experimental research has no psychological analogue with people that may or may not have made similar structures in the past (e.g. Thomas 2000). However, this type of research has a pedagogic element for the researcher and therefore influences the wider programme of research being conducted (Gheorghiu 2002). This implicit relationship is rarely explicitly acknowledged, despite obvious linkages between experimental archaeology and hunter-gatherer technologies. A good example of this process can be seen in the development of the sub-discipline of lithic technology where individual archaeologists have become experts at flint-knapping (e.g. Odell 2001). The development of this skill-set has clearly influenced the ability of individual archaeologists to be able to provide interpretative commentary on various aspects of prehistoric stone tool technology.

Experimental research then offers two distinct and complementary theoretical propositions. First, the facility to test the physical properties, interactions between properties, and physical phenomena associated with thermal feature technology. Second, the pedagogic development of experience and a skill-set from working directly with a dynamic fire-based technology. While the first more often acknowledged result is crucial to the development of middle-range research, the second provides a heuristic mechanism through which novel ideas and news lines of enquiry are often initiated. Furthermore, the theoretical linkages between these two propositions ensure that any time the materials-based testing occurs, then pedagogic learning is also taking place.

For this study, experimental methodologies were tailored to particular research questions and are presented as self-contained reports (Backhouse 2004; Backhouse *et al.* 2005b; Backhouse and Johnson 2007a, c). Tested processes included the thermal properties of hot-rocks (Backhouse *et al.* 2005b; Backhouse and Johnson 2007c), site-formation processes associated with LTF after abandonment (Backhouse and Johnson 2007a), and physical use of space around an LTF (Backhouse 2004).

Common to all the experimental research was the direct engagement with open flames and the active construction of structures in 'real-world' (or actualistic) settings. Quantification and replication were important to the experimental programme, but so too was the important pedagogic element that introduced students and archaeologists, used to looking at the static result of fire (charcoal and burnt stones) to the dynamic processes that control their transformation. All the experiments were undertaken in controlled conditions at the Lubbock Lake Landmark Experimental Research Area.

The setting of the area inside the main entrance to the research compound at the Lubbock Lake site was inadvertently beneficial in the methodological development of the experimental research programme. The location actively engaged research staff, visiting professionals, tour guides, and members of the public in an ongoing dialogue that often generated positive criticism.

The degree of control exerted on the experiments ranged depending on the specific research question(s) being examined. The experiments all sought to test clearly defined hypotheses and were therefore inherently empirical in nature. The methodological development of an experimental programme involving the Native American and local communities, archaeological staff, and volunteers in a learning experience directly benefited the appreciation of prehistoric fire technologies as an interesting topic worth researching and as a vehicle for learning, engaging, and communicating the past.

Field Research

The objective of the ongoing programme of fieldwork was to develop qualitative datasets for specific types of LTF that were identified in the regional dataset, but that are poorly understood. Fieldwork projects ran intermittently, but were a perennial feature of the overall research project. It should be noted, however, that no fieldwork was initiated with the specific research design of examining LTFs in isolation and neither is such a proposition proposed by the research project presented here. The opportunity to actively investigate the physical remains of former LTFs was a continuing theme, which integrated and complemented the wider research design of the Lubbock Lake Landmark Regional Research Program (LLLRRP; as outlined in Johnson 2005).

Standard Lubbock Lake methodology was followed for all investigations (e.g. Johnson 1987a, 1989, 1990, 1993, 1995, 2002, 2005). Three broad scales of examination were utilised within this framework. Survey involved field-walking, site survey, limited artefact recovery, and exploratory trenching. Testing involved focussed research at a previously identified location or site to examine its potential significance. Limited excavation, geophysical survey, and focussed recording typically take place at this stage of research. Finally, full-scale excavation involved the extensive hand excavation of a large portion of the identified site (Figure 3.10). Large-scale data recovery, intensive sampling assays, and detailed geoarchaeological research are typically completed.

Figure 3.10. Excavation block at 41LU118 during full-scale excavation. (P.N. Backhouse)

In total, six projects were undertaken as part of the LLLRRP identified sites with one or more LTFs between the period 2000 through 2006. As previously stated, the field methods were consistent among these projects; however, the scale of research varied significantly (Table 3.7).

Table 3.7. Summary of field work undertaken between 2002 and 2007 by the LLLRRP at hunter-gatherer sites with LTFs on the Southern High Plains.

Project/Site	Scale	Date	Reference
41LU1, Area 10	Full-Scale Excavation	2000, 2001, 2006	Johnson 2002
41LU118	Full-Scale Excavation	2000-2005	Johnson and Backhouse 2001
41LY52	Full-Scale Excavation	2001	Backhouse 2003
41LU129	Testing	2002-2003	Backhouse *et al.* 2005a
Canyon Lakes Resurvey	Survey	2002-2006	Johnson 2005
U-Lazy-S Ranch	Survey	2005-2006	Backhouse and Johnson 2007b

Summary

The development of a research project to examine the archaeological evidence for LTFs at a regional scale necessitated a structured research approach whereby a desk-based component of data collection, compilation, and analysis (Chapter 5 and 6) is complemented by a field-based component of data gathering and experimentation (Chapters 7 and 8). The four avenues of analysis represented by this research (taxonomic, ethnographic, experimental, and archaeological) were developed in parallel with one another. The results of each strand are, however, presented sequentially. Synthesis of the diverse evidence is subsequently drawn together and critically discussed as the basis for a broad diachronic model for understanding hunter-gatherer domestic fire technology within the study area (Chapter 9). The geographical context of the Southern High Plains is presented (Chapter 4) as a background for the main analysis (Chapters 5, 6, 7, and 8), which unfolds the detail of the methodology.

CHAPTER 4 - GEOGRAPHIC FOCUS

Abstract

The Southern High Plains, located in mid-continental North America, is an excellent location for the assessment of hunter-gatherer domestic fire technologies over deep time scales. A rich history of interdisciplinary research has resulted in detailed modelling of the evolution of the landscape and extant flora and fauna throughout the late Quaternary period. The presence of hunter-gatherer groups within the region spans the entire indigenous prehistory of the North American continent and local conditions within the region appear to have been less than conducive for the shift to agricultural production. The result is an almost uninterrupted regional record of hunter-gatherer occupation that stretches back into the late Pleistocene. No wide-scale examinations of thermal feature technology have been attempted on the Southern High Plains even though the potential for such research is extremely high. A detailed background for the region indicates that the periphery of the area contains critical resources that are necessary to the continuation of hunter-gatherer lifeways practised within the interior. Therefore, an expanded regional approach to the examination of hunter-gatherer domestic fire technology is proposed. The approach incorporates the regional borderlands within a formal boundary designed to yield the maximum possible Localised Thermal Feature dataset. This dataset provides the opportunity to examine changes in technological strategy through time and across the vast flat spaces afforded by the unique topography of the Southern High Plains.

The Southern High Plains or Llano Estacado as it is locally known is a vast flat plateau located in the central southern portion of the North American continent. Geologic investigations have indicated that the region is relatively young and the key geomorphic development largely the result of the deposition of massive quantities of gravels and sediments during the Tertiary period, a product of the rising Rocky Mountains to the west (Holliday 1997). Part of the North American Great Plains and surrounded on three sides by large escarpments, the region is characterized as dry and semi-arid (Osterkamp et al. 1987). Palaeoenvironmental records indicate that it appears to have been a grassland throughout the Quaternary (Johnson 2007). Climate oscillations since the late Pleistocene have resulted in a general shift from more moist conditions to a drier climate present today (Johnson and Holliday 2004). During the late Pleistocene, megafauna species such as mammoths and camels appear to have coexisted with human populations (Johnson 1987).

A century of archaeological research has yielded several sites of national and international significance, including the type site Blackwater Draw #1, which was first excavated in the 1930's and provided the earliest evidence for the peopling of the Americas (Hester 1972). Artefacts recovered from this site were placed together in a cultural group termed the Clovis culture after the nearby town in New Mexico. Subsequent regional investigations have revealed a rich archaeological record of populations subsisting largely through hunter-gatherer lifeways (e.g. Johnson and Holliday 2004). The temporal sequence of these occupations is perhaps best expressed in the late Quaternary valley fill at the Lubbock Lake site, where the well-stratified sediments contain evidence for human occupation that stretches from the earliest Paleoindian period through to the Anglo-American settlement over 12,000 years later (Johnson 1987a).

A regional approach to the archaeological record is necessitated by the current research programme and an additional ten-mile extension to the geographic limits of the Southern High Plains is applied to the current study. The buffer serves to contextualize the Southern High Plains within a wider landscape in which the movement of peoples onto and off of the Southern High Plains has been well established from the archaeological record. The buffer therefore includes dynamic areas of transition and important resources where hunter-gatherer sites are highly visible. The Southern High Plains of west Texas and eastern New Mexico offer sufficient archaeological features, a long temporal sequence of peoples living by hunter-gatherer means, and a history of investigations by archaeological and related disciplines. The region is therefore an ideal laboratory for investigating hunter-gatherer LTF technology and has the concomitant potential to contribute globally, as well as regionally, to understanding this important feature class.

Physiographic Setting

North American archaeology has historically dealt with

the enormity of the continent by dividing the examination of the development of human cultures into nine distinct geographic regions (Fagan 1991). The historic Plains region comprises a vast corridor of grassland that stretches through the central portion of the continent (Hunt 1974; Wishart 2004; Figure 4.1.). Lying east of the Rocky Mountains and west of the Appalachian Mountains, the Plains stretch from the central Canadian provinces of Alberta and Saskatchewan in the north through to dry arid landscape of northern Mexico in the south,

Great Plains

The Great Plains physiographic province is a component of the larger Plains region (Hunt 1974). Covering approximately 575,000 square miles, the region slopes gently from west to east, with elevations decreasing from ca. 5,500 feet above sea level in the west to ca. 2,000ft in the east (Hunt 1974:334). The ca. 375-mile corridor is bordered by the Rocky Mountains to the west and grades into the Central Lowlands to the east (Osterkamp *et al.* 1987). Apart from within valleys, the area is largely devoid of trees and has been a grassland throughout the Quaternary. The otherwise flat landscape of the plains are interrupted to the north by the development of mountains; notable amongst these are the Black Hills of South Dakota (Hunt 1974).

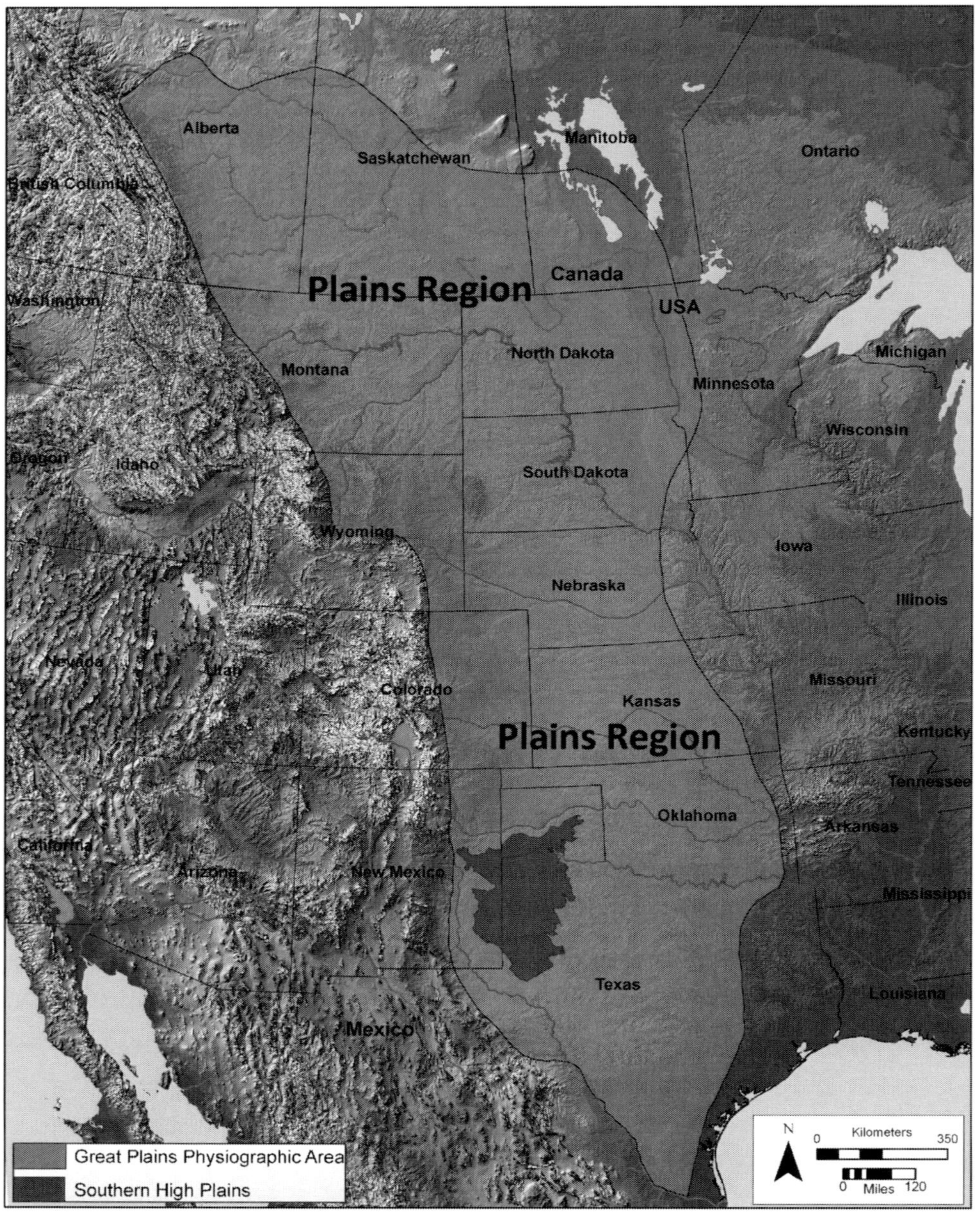

Figure 4.1. Physiographic location of the Southern High Plains on the North American Plains overlaid with the location of modern political boundaries.

The underlying geology of the region consists of a generally horizontal depositional sequence of Paleozoic, Mesozoic, and Cenozoic Formations (Hunt 1974). These formations were pushed up by the emerging Rocky Mountains to the west. The subsequent eastward displacement, from the Rockies, of sediments during the Tertiary period comprises the underlying landform for much of the modern landscape. The rate of deposition of sediments and associated age of the landform is time dependant and varies from the north to the south of the region. The three principal landforms in this sequence are the Alberta Plain, the Missouri Plateau Section, and the High Plains (Hunt 1974:338). Maximum Pleistocene glaciation extended only as far as South Dakota and sections of the Plains to the south remained free of ice throughout the Pleistocene (Osterkamp *et al.* 1987:166).

High Plains

The High Plains section of the Great Plains comprises a vast elevated plateau covering portions of Wyoming, Nebraska, Colorado, Kansas, Oklahoma, Texas, and New Mexico. The plateau is formed by a thick blanket of Pliocene deposits, of which the gravels and sands which comprise the Ogallala Formation, form the majority of the eolian and alluvial sediments (Hunt 1974). Overlying these deposits are various aeolian deposits, which make up the majority of the modern ground surface. The deposition of these aeolian sediments is largely controlled by the predominant westerly winds and the lack of significant topographic barriers. The High Plains are generally accepted to be "the most persistently windy inland area of North America" (Osterkamp et al. 1987:195).

The eastward incisions of various drainages cut across the High Plains. Because the upper Tertiary sediments are more permeable than the older deposits underlying them the surface is conducive to the development of shallow intermittent streams (Hunt 1974) and throughout the High Plains section, the smaller geomorphic features such as intermittent stream draws and playa lakes appear to be closely related to and controlled by the dynamic groundwater system (Osterkamp *et al.* 1987). The High Plains section is effectively divided in two by the east west-west incision of the Canadian River during Pliocene-Pleistocene times (Wood *et al.* 2002). This separation of the Northern and Southern High Plains has allowed increased landscape stability in the development of the southern portion of the area.

Southern High Plains (Llano Estacado)

The Southern High Plains, or Llano Estacado as it is locally known, is a vast flat plateau covering an area estimated to be 31,000 square miles (Hunt 1974; Stafford 1981; Johnson 1987a; Wood *et al.* 2002; Figure 4.2). The region has a virtually featureless surface formed by the deposition of thick, widespread aeolian sediments (Osterkamp *et al.* 1987; Holliday 1988a; 1988b; 1990; Johnson and Holliday 1995). The homogeneity of this geomorphic context is facilitated by the development of a thick protective calcretic layer which largely resists ongoing errosion processes and has formed at the top of the Tertiary Ogallala Fm. This layer is known locally as the caprock (Reeves 1976; Figure 4.3). This layer provides a stable surface onto which the younger aeolian sediments have been deposited. The English translation of Llano Estacado is 'Staked Plain' and one interpretation of this name suggests that it was necessary for early travellers to drive stakes into the ground in order to traverse across the otherwise featureless landscape (Hunt 1974:340).

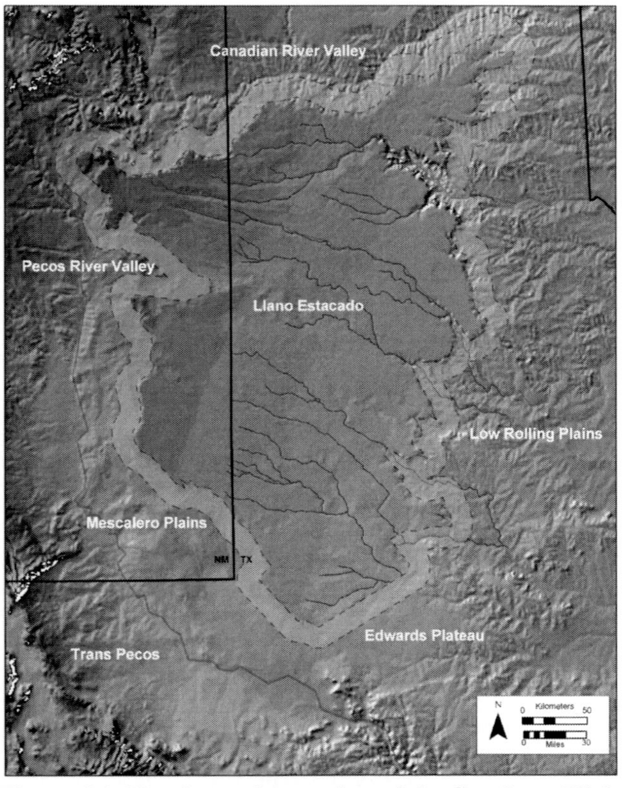

Figure 4.2. Physiographic setting of the Southern High Plains (Llano Estacado) with neighbouring geographic areas indicated.

The area is sharply delineated to the north, east, and west by steep erosional escarpments (Figure 4.3). These escarpments form the boundaries to the Rolling (Osage) plains to the east, the Canadian River valley to the north, the Pecos River valley to the west, and the Mescalero escarpment to the southwest (Figure 4.2). These escarpments represent a dynamic system in which the eastern frontier is steadily creeping westward (Gustavson 1986:6; Wood *et al.* 2002). The sharp difference in elevation is most visible in the northeast of the region where at Palo Duro Canyon it approaches 300 meters (Hughes 1989:6). A gentle reduction in elevation occurs from the northwest (c. 5,000ft) to the southeast (c. 2,300ft) of the region and the southern boundary seamlessly enters the Edwards Plateau province of Central Texas (Holliday 1997). The only topographic variation is provided by the playa basins, lunettes, dune fields, and the dry southeast-trending valleys known as draws, that dot the landscape. Approximately 25,000 small playa basins today provide the only available surface water sources (Sabin and Holliday 1995).

The development of the cultural and physical landscape during the late Quaternary is well understood by a series of major inter-disciplinary research programmes (e.g. Wendorf 1961; Dillehay 1974; Johnson 1987a, 2007; Johnson and Holliday 1989, 1995, 2004; Holliday 1987, 1995, 1997; Neck 1995; Meltzer 1999; Wood *et al.* 2002).

Figure 4.3. Geomorphic development of the Southern High Plains; photograph of the eastern escarpment showing the exposed upper caprock section above the low Rolling Plains. (P.N. Backhouse)

Environmental

Despite the presence of large trapped water sources below the Southern Plains, the region as a whole is water deficient (Osterkamp *et al.* 1987:182). The modern Southern High Plains climate is classified as dry semiarid (steppe) (Holliday 1997: 9). Explanatory geographic relationships between precipitation and temperature regime indicate that precipitation generally increases from west to east (Figure 4.4). In contrast temperatures gradate from highs in the northwest to lows in the southeast (Holliday 1997). Cyclic periods of drought are common events in this landscape and often linked with range fires in the control of the spread of trees onto the grasslands (Johnson 2007; Stewart, 2002). Rainfall occurs mainly in the spring and fall, although violent summer thunderstorms are also common (Johnson 2007). Strong winds are a perennial feature of the Southern High Plains and for much of the year the wind direction is from the southwest. Life-threateningly cold winds locally known as *northerners* can occur during the winter months (Hughes 1989, Flores 1990).

Climatic Record

The palaeoclimatic record of the Llano Estacado indicates a long-term warming and drying trend and significant fluctuations occur throughout the late Quaternary (Figure 4.5). Despite a large amount of research, proxy indicators for specific events have proved difficult to identify in the Southern High Plains record. Holliday's (2000) geoarchaeological examination of upland sediments and valley-fill sequences incorporating both stable-carbon isotope, and paleontological data is the most comprehensive synthesis of climate data currently available for the terminal Pleistocene.

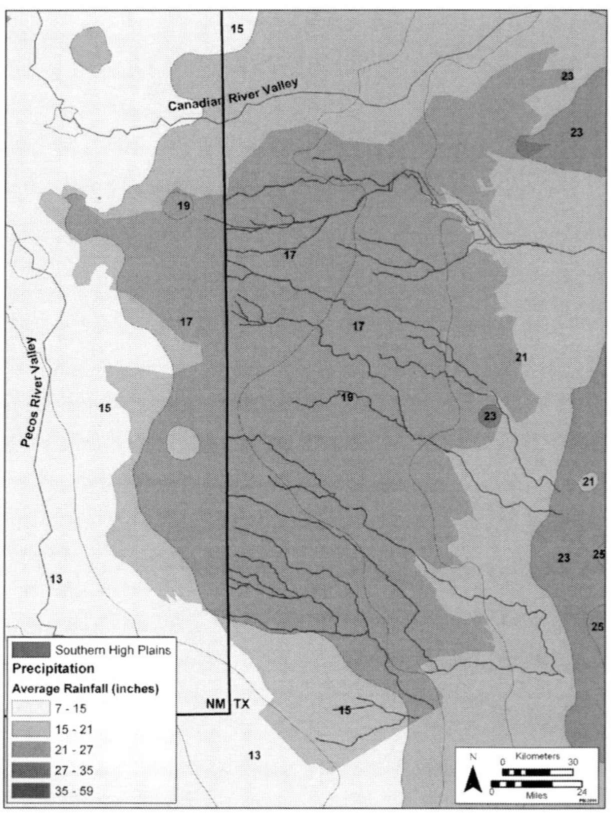

Figure 4.4. Modern precipitation pattern for the Southern High Plains Regions. Note that rainfall increases from southwest to northeast across the region (data National Oceanic Atmospheric Administration).

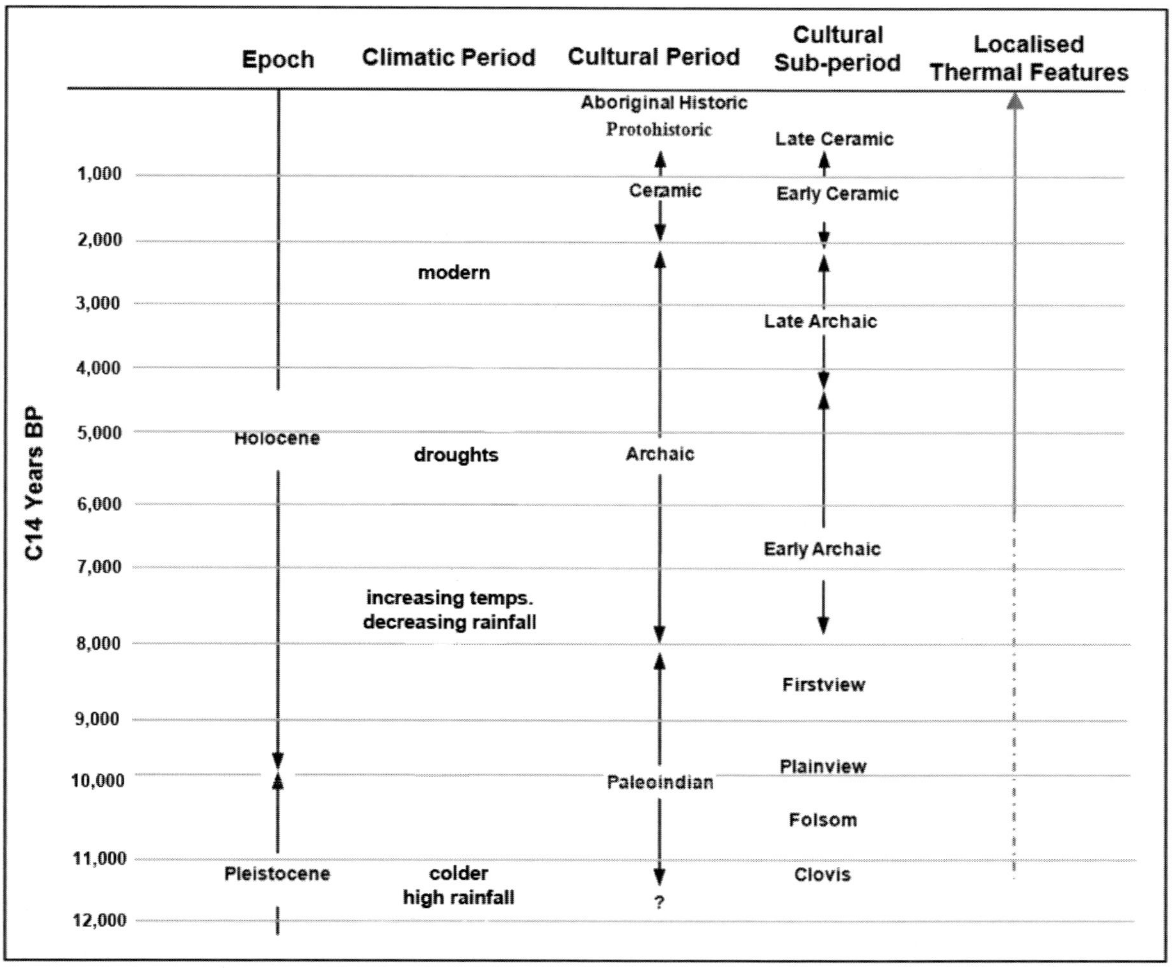

Figure 4.5. Generalised temporal framework for the Southern High Plains showing climatic and cultural periods, with the approximate appearance of localised thermal feature technology indicated in red (modified after Johnson 1987a, fig. 2.7; chronostratigraphic conventions follow Holliday 1997:16).

This research indicates that cooler and wetter conditions existed prior to c.11,000 RCYBP, contemporaneous with the first evidence for humans in the region (Figure 4.5). During this time, water still appears to have flowed in the draws and mean annual temperatures were between 2°C and 5°C cooler than today (modern mean annual temperature 15°C) (Johnson and Holliday 1995, 2004; Holliday 2000). Subsequently, the climate appears to become less stable and an underlying warming and drying trend is evident. By c.10,000 RCYBP, water in the draws had ceased to flow and ponds fed by springs appear (Johnson and Holliday 1995, 2004; Holliday 2000). Subsequently, several phases of drying have been identified between c.11,000 and c.8,000 RCYBP (Holliday 2000).

The middle Holocene record following c.7,500 RCYBP is marked by increased temperatures and lower rainfall, culminating between c.6,500 RCYBP and c.4,500 RCYBP with the most severe period, known as the "Altithermal" (Holliday 1989; Meltzer 1991, 1995, 1999). During this time, the evaporation of surface water and lowering of the ground water resulted in a substantial decline in ecological productivity throughout the region.

A shift in climate patterns emerges in the period following c.4,500 RCYBP and the region returned to cooler more mesic conditions (Johnson 1987a, 2007). The late Holocene record indicates that by c.2,000 RCYBP, droughts are common. This pattern continues today with drought cycles measured by slight fluctuations in intensity, duration, and frequency (Johnson 1987a, 2007; Wendland 1995). Strong winds that generate large spring-time dust storms, the direct resultant of modern agricultural practices, are common events (Stout 2001).

Flora and Fauna

The Southern High Plains is part of the larger Kansan biotic province (Hagmeier and Stults 1964) and considerable research has been undertaken in developing a model of this evolving grassland ecosystem (e.g. Wendorf, 1961; Dillehay 1974; Johnson 1987, 2007; Neck 1987, 1995; Johnson and Holliday 1989, 1990, 1995, 2004; Holliday

1987, 1995; Lewis *et al.* 2000; Finch 2004). Early examinations of the paleoenvironment (e.g. Wendorf 1961) were plagued by problems in interpreting the often poorly preserved and misleading pollen record, which appeared to indicate the presence of a Boreal forest during the late Pleistocene. This interpretation has been proven untenable by modern geoarchaeological investigations (e.g. Holliday 1987), which suggest that the area has remained a grassland throughout the late Quaternary. In this landscape, only isolated communities of non-coniferous trees occur along the escarpments, and within the system of draws (Holliday 1987). In areas where native vegetation has not been removed by agriculture, a short-grass prairie exists, comprising both blue grama (*Bouteloua gracilis*), and buffalograss (*Buchloe dactyloides*) (Johnson 2007:11).

The most recent regional synthesis of the various plant and animal communities is supplied by Johnson (2007) who draws on multiple lines of evidence from 20 localities across the Llano Estacado to build up a detailed picture of the shifting communities of plants and animals represented in the late Quaternary record.

Although the Southern High Plains are too far south to have been directly affected by glaciation, the indirect effects appear to have been dramatic. Termination of the Wisconsinan conditions around c. 11,000 RCYBP marks the end of cooler, wetter conditions. During this time, environmental evidence suggests cool-season grasses, aquatic plants, and isolated or small groves of trees existed (Johnson 2007:12). Typical species include some, now extinct, megafauna such as Colombian mammoth (*Mammuthus columbi*), horse (*Equus mexicanus* and *Equus francisci*), camel (*Camelops hesternus*), llama (*Hemiaucheina*), and bison (*Bison antiquus*) (Johnson 1987a). The extant faunal evidence suggests a lush grassland was well established by the late Pleistocene (Johnson 2007).

Following the climatic warming shift that marked the transition to the Holocene (c.10,000 RCYBP), megafauna species diversity was drastically reduced and only the larger now-extinct form of bison (*Bison antiquus*) appears to have survived. Zooarchaeological evidence suggests smaller grassland animals do not appear to have been as affected by this transition (Johnson 1987a).

By c. 8,000 RCYBP, environmental evidence suggests that a short-grass ecosystem has developed. Bison bones recovered from deposits of this age suggest that this species adapted to the changed environmental conditions by significantly reducing body mass. Skeletal measurements indicate that modern bison (*Bison bison*) is evident in the record by ca. 7000 RCYBP (Johnson 2007:19). Pronghorn antelope (*Antilocapra americana*) also appear well suited to this new ecosystem, as does their principal predator the Coyote (*Canis latrans*). Again, the smaller grassland animals appear to have adapted quickly to the changing environmental conditions.

It is most likely that the indigenous species of the Southern High Plains were under severe stress during the middle Holocene Altithermal (Meltzer 1999). Nevertheless, zooarchaeological evidence again indicates all three key species, bison, antelope, and prairie dog appear to have successfully adapted to the desert-like conditions (Johnson 2007). Honey mesquite (*Prosopis glandulosa*) also appears in the record at this time. By the late Holocene, the climate improves and this appears to have resulted in an expanded range of fauna (Johnson 2007:22). The three dominant species, however, remain unchanged.

The greatest diversity of flora and fauna is encountered at the physiographic boundaries of the region (e.g. Sebastian and Larralde 1989; Flores 1990, Boyd 1995). This is particularly true in the canyonlands that border much of the eastern escarpment (Hughes 1989, Boyd 1995). In these areas, numerous springs and creeks still flow out of the Ogallala Fm. aquifer providing precious habitat for numerous plants and animals (Figure 4.6). The boundaries of the region then provide a relatively stable resource base (in terms of food and water) as compared to the interior of the region, where both resources are relatively uncommon and not distributed evenly (Boyd 1995).

The Anglo settlement of the Southern High Plains in the late nineteenth century had both quick and radical effect on the native flora and fauna of the region (Flores 1990). Bison and antelope were quickly removed from the landscape to make way for fences and beef stock. Today, of the three key grassland species, only the prairie dog and occasional antelope remain. The modern landscape is dominated by agriculture (Figure 4.7) with cotton providing the largest regional crop.

Very little research has examined the distribution of edible native plants on and around the Southern High Plains. Because these resources are most likely to be critical to the interpretation of thermal feature technology, a very brief summary is undertaken here (Table 4.1).

Archaeological

The historic development of archaeological research on the Southern High Plains is crucial to structuring the state databases. The relationship among the diverse elements that comprise archaeological investigation include the history of research, location of research institutions, access to land, methodology of researchers, specialist interest of researchers, access to appropriate field and laboratory equipment, researcher knowledge and skill, time and money invested in a project, and even the propagation of local informants are all crucial in structuring the regional dataset. Only, therefore, by understanding the limitations of the present database, can we hope to infer meaningful conclusions from the present course of study. It is with these caveats that a brief history of archaeological research on the Southern High Plains is examined.

Figure 4.6. The modern environment on and around the Southern High Plains: a) short-grass prairie in a virtually featureless landscape; b) a prairie dog emerges from his subterranean burrow excavated into the prairie; c) the desert scrubland just off of the south eastern escarpment, note the honey mesquite to the right of the photograph and prickly pear cactus in the foreground; d) looking back at the western Mescalero escarpment across the dune fields of New Mexico; e) water flowing in a perennial stream in the eastern canyonlands; f) Stands of juniper growing along the eastern escarpment and rough canyonlands. (P.N. Backhouse)

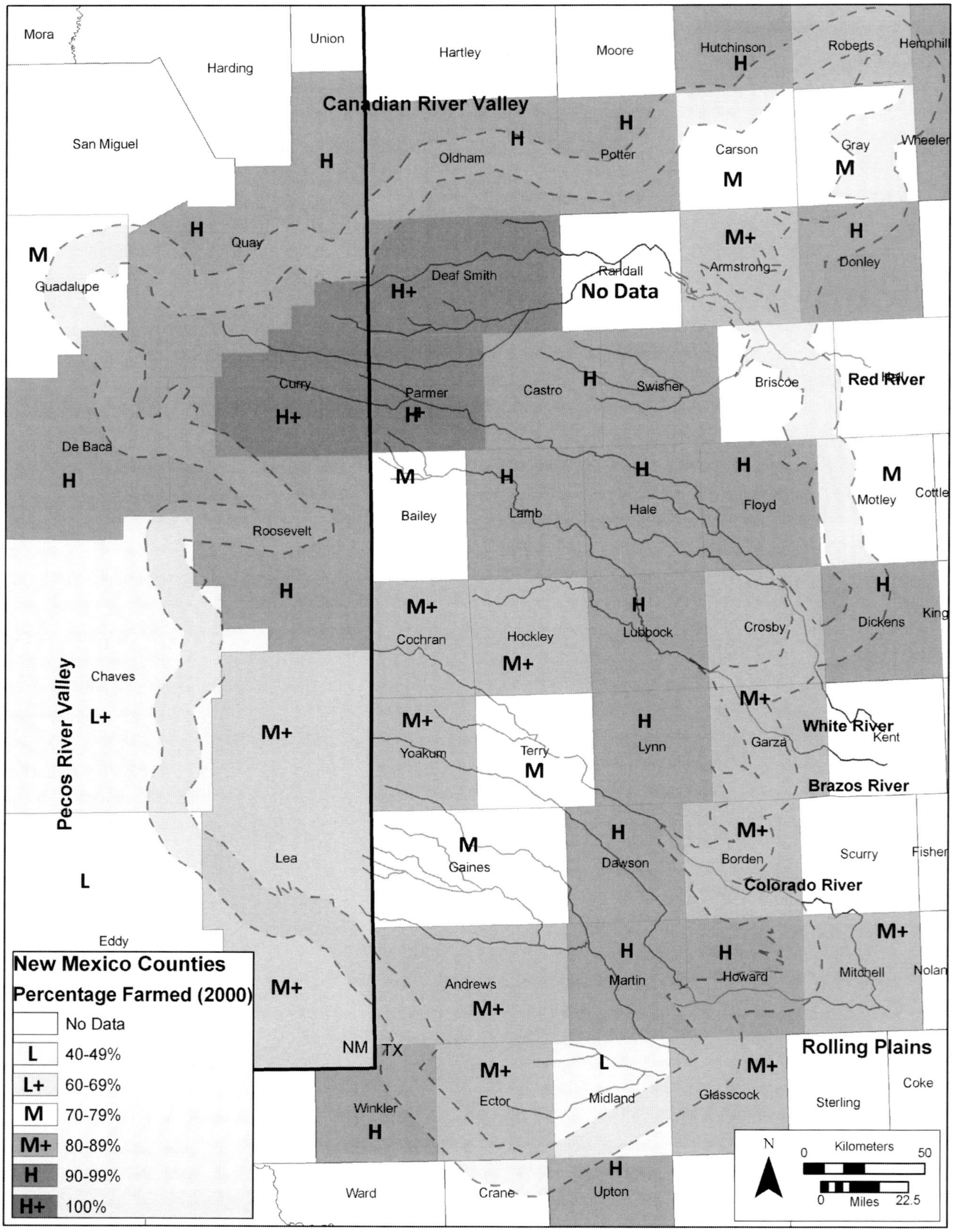

Figure 4.7. Modern farming intensity by county (data U.S. Census Bureau, Census 2000).

Table 4.1. Edible native plants identified in archaeological contexts on the Southern High Plains and in adjacent areas.

Common Name	Scientific Name	Context	General Location	Reference
Bottle gourd	Cucurbitacaea	Sam Wahl, Boren No. 2.	Eastern Canyonlands	Boyd 1995
Brome grass	*Bromus* sp.	Sam Wahl	Eastern Canyonlands	Boyd 1995
Buffalo gourd	*Cucurbita foetidissima*	Modern observation	Eastern Canyonlands	Boyd *et al.* 1994
Bullrush	*Scirpus* sp.	41LU1	Interior	Thompson 1987
Carpet weed	*Mollugo verticillata*	Sam Wahl	Eastern Canyonlands	Boyd 1995
Cholla	*Opuntia* sp.	Modern observation	Eastern Canyonlands	Boyd *et al.* 1994
Devil's claw	*Proboscidea* sp.	41LU1	Interior	Thompson 1987
Dropseed	*Sporobolus* sp.	Sam Wahl	Eastern Canyonlands	Boyd 1995
Goosefoot	*Chenopodium* sp.	Kent Creek, Sam Wahl, Boren No. 2.	Eastern Canyonlands	Boyd 1995
Hickory	*Carya* sp.	41GR484	Eastern Canyonlands	Boyd *et al.* 1994
Honey mesquite	Prosopis glandulosa	Sam Wahl, Boren No. 2.	Interior, Eastern Canyonlands	Thompson 1987; Boyd 1995
Lotebush	*Zipiphus obtusifolia*	Sam Wahl, Boren No. 2.	Eastern Canyonlands	Boyd 1995
Netleaf hackberry	*Celtis Reticulata*	Sam Wahl, Boren No. 2.	Interior, Eastern Canyonlands	Thompson 1987; Boyd 1995
Oak	*Quercus* sp.	Kent Creek	Eastern Canyonlands	Boyd 1995
Panic grass	*Panicum* sp.	Sam Wahl, Boren No. 2.	Eastern Canyonlands	Boyd 1995
Paspalum	*Paspalum* sp.	Sam Wahl, Boren No. 2.	Eastern Canyonlands	Boyd 1995
Pigweed	*Amaranthus* sp.	Sam Wahl	Eastern Canyonlands	Boyd 1995
Prairie turnip	*Psoralea* ssp.	Modern observation	Eastern Canyonlands	Boyd *et al.* 1994
Prickly pear	*Opuntia* sp.	Gobbler Creek Bridge Site	Eastern Canyonlands	Boyd 1995
Prickly poppy	*Argemone* sp.	Sam Wahl	Eastern Canyonlands	Boyd 1995
Purslane	*Portulaca* sp.	Kent Creek, Sam Wahl, Boren No. 2.	Eastern Canyonlands	Boyd 1995
Skullcap	*Scutellaria* sp.	Boren No. 2.	Eastern Canyonlands	Boyd 1995
Sotol	*Dasylirion texanum*	Modern observation	Eastern Canyonlands, Edwards Plateau	Boyd *et al.* 1994
Water lilies	*Nymohaea* sp.	41LU1	Interior	Thompson 1987
Wild onion	*Allium drummondii*	Modern observation	Eastern Canyonlands	Shroeder and Holliday 1994
Yucca	*Yucca* spp.	Modern observation	Eastern Canyonlands, Edwards Plateau	Boyd *et al.* 1994

History of Research

Jack Hughes (1989) has observed that the archaeology of the United States may have its origin on the Southern High Plains. This claim is based on a bison bone bed at Silver Lake, in Hockley County in the southern portion of the region, which may have been described by the Spanish Coronado expedition in 1541 (Hughes 1989:2). Despite this early example, it is not until the twentieth century that the type of investigations we today would recognise as archaeological begins to emerge.

For close to a century, archaeological research has been conducted on and around the Southern High Plains. Disparate and largely unrelated programmes of research characterise the earliest investigations (prior to 1930). Initial research focussed squarely on the investigation of the standing structures representative of populations that had once lived along the Canadian River Valley in the Texas Panhandle. Earliest amongst these, was perhaps T.L. Eyerly's excavations at the Old Buried City Ruins (Hughes 1989:3) located on the north side of the Canadian River Valley and therefore slightly outside of the current research area.

The second programme of research focussed on the Southern High Plains with the goal of recovering evidence for the earliest humans in the New World (Holliday 1997). The success of Edgar B. Howard's excavations near Clovis at Blackwater Draw #1 in the 1930's achieved this goal and has arguably resulted in a lingering legacy of research oriented towards spectacular Paleoindian sites. This second programme of research can largely be characterised by low numbers of high-profile excavations by researchers affiliated with academic institutions. Their goals were concerned with the recovery of lithic tools in association with extinct Pleistocene megafauna. Holliday (1997) provides an excellent background to this research, indicating that multidisciplinary research has a long history within the region.

Examination of a representative sample of archaeological sites located within the current study area and on file at the appropriate state-level cultural heritage repositories (the Texas Historical Commission (THC) in Texas and the New Mexico Office of Cultural Affairs in New Mexico) indicate the general intensity of archaeological research throughout the twentieth century. The results (Figure 4.8) show that the first sixty years of research, while prolific, account for less than 1% of the total investigations undertaken to date. In contrast and largely as a result of cultural resource legislation imposed on public lands during the second half of the twentieth century, an exponential growth rate in archaeological research continues up to the present day (Hughes 1989; Holliday 1995).

The dominant pattern of growth in the intensity of archaeological investigations undertaken between 1960 and today is inversely proportional to the number of investigations reaching the excavation stage of research. Put simply, earlier investigations were far more likely to be excavation based, whereas later investigations are more likely to be survey based. The relationship between intensity and methodology clearly has affected the structure of the regional archaeological record. Furthermore, the continuing development of modern scientific techniques can be added to this mix underscoring the realisation that no studies of hunter-gatherer thermal-feature technology have yet been undertaken on the Southern High Plains. This is important as it can be assumed that the primary research goal of all the previous investigators was not the examination of thermal-feature technology.

Key Sites

A handful of sites has emerged from the thousands identified, as key to the interpretation of the regional archaeological sequence (Figure 4.9). Among these, sites located in the well-stratified draw systems have provided the most complete sequences of cultural occupation (Johnson and Holliday 1995, 2004). Lubbock Lake in Yellowhouse Draw and Blackwater Draw No. 1 in provide the best evidence of the presence of hunter-gatherer groups over deep time-scales (Johnson and Holliday 1995, 2004).

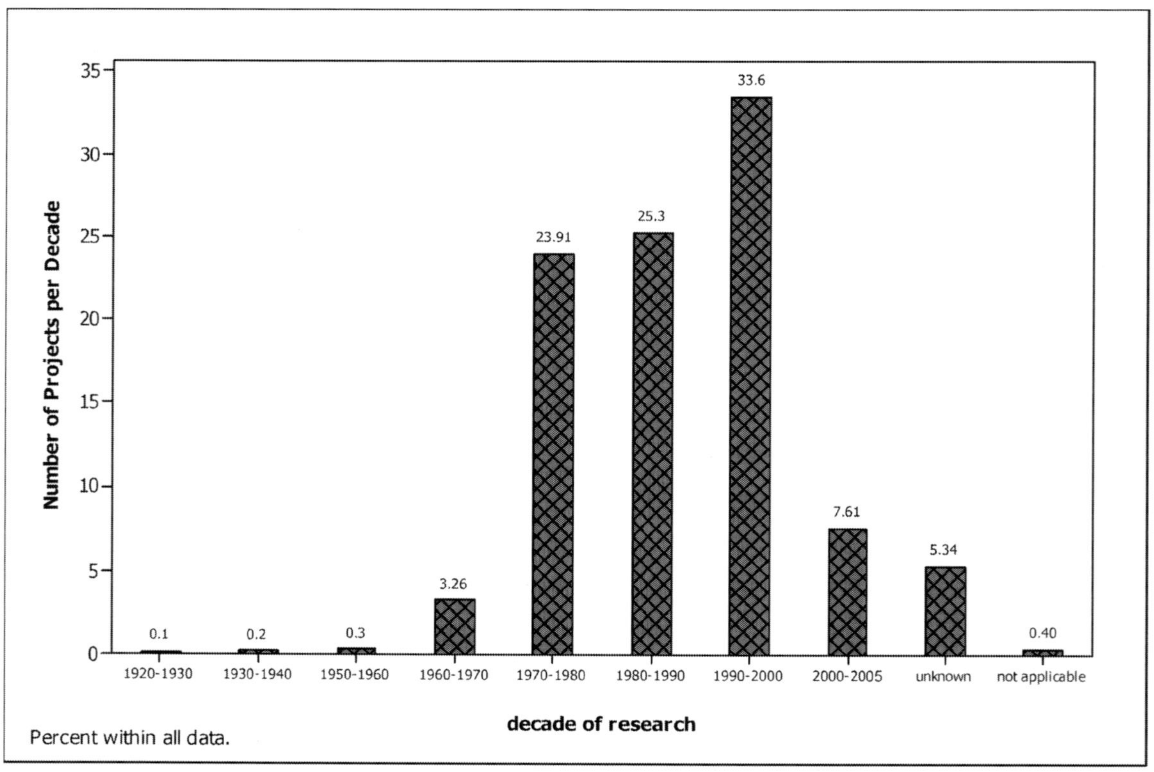

Figure 4.8. Intensity of archaeological research on the Southern High Plains presented by decade (data sampled from TARL and ARMS and accessed September 2005).

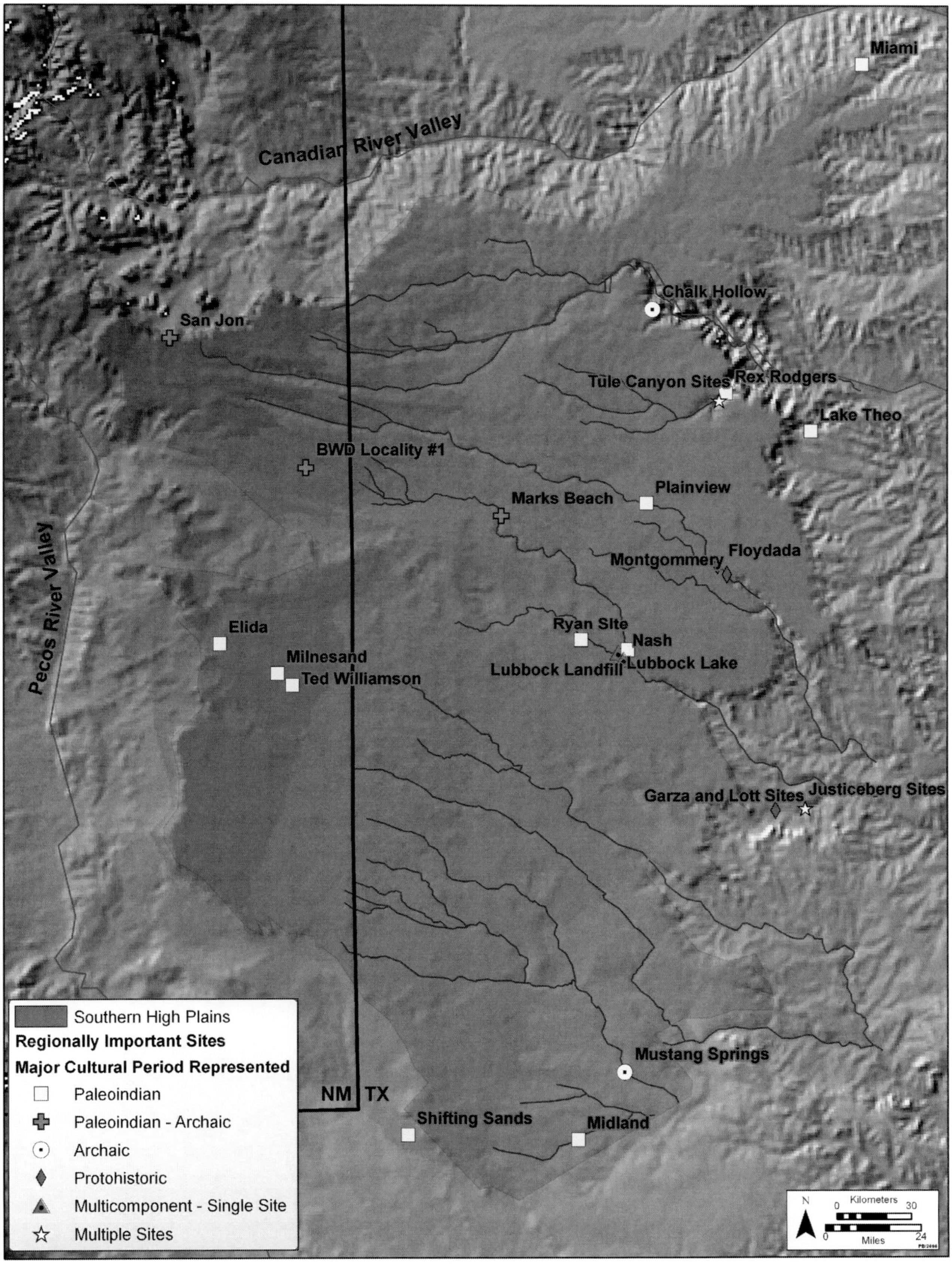

Figure 4.9. Key sites identified on and in proximity to the Southern High Plains.

Lubbock Lake (41LU1) is a National Historic Landmark archaeological preserve comprising some c.330 acres of Yellowhouse Draw in the southeast of the Southern High Plains. The site exhibits the most complete and intensively studied sequence of cultural activity available in the region (e.g. Stafford 1981; Holliday *et al.* 1985; Johnson 1987a, 1989, 1990, 1993, 1995, 2002; Johnson and Holliday 1981, 1986, 1989, 1990, 1995, 2004; Holliday 1988a, 1988b, 1995, 1997). The deeply buried and well defined strata at this site span the last 12,000 years (Figure 4.10) containing evidence for the repeated return of hunter-gatherer groups to the fresh water springs that once existed at this location (Johnson 1987a). Evidence recovered to date indicates that throughout much of the late Quaternary, this predictable source of water acted as an oasis attracting both larger grazing herd animals and their human predators.

Archaeological work was first undertaken at this site in 1936 with the recovery of Folsom-aged artefacts after an abortive attempt by the city of Lubbock to dredge the valley in order to rejuvenate the then dried-up springs (Holliday 1997). The current programme of research under the direction of Dr. Eileen Johnson of the Museum of Texas Tech University began in 1972 and to date, over 80 discrete areas and 17 sites have been identified within the boundaries on the property (Johnson 2002). The results of this programme of research indicate the wealth of information that is available in the region's draw systems where they remain relatively undisturbed by modern development. To date, less than 1% of the cultural material-bearing deposits contained at Lubbock Lake have been systematically examined.

Lubbock Lake is particularly important in the regional sequence, as the dominant late Quaternary models that characterise the region derive much of their analytical foundation from this site (e.g. Johnson and Holliday 1989, 1995, 2004). At this site, the strong commitment to interdisciplinary research has primarily concentrated on geoarchaeological investigations and vertebrate faunas in the construction of these models.

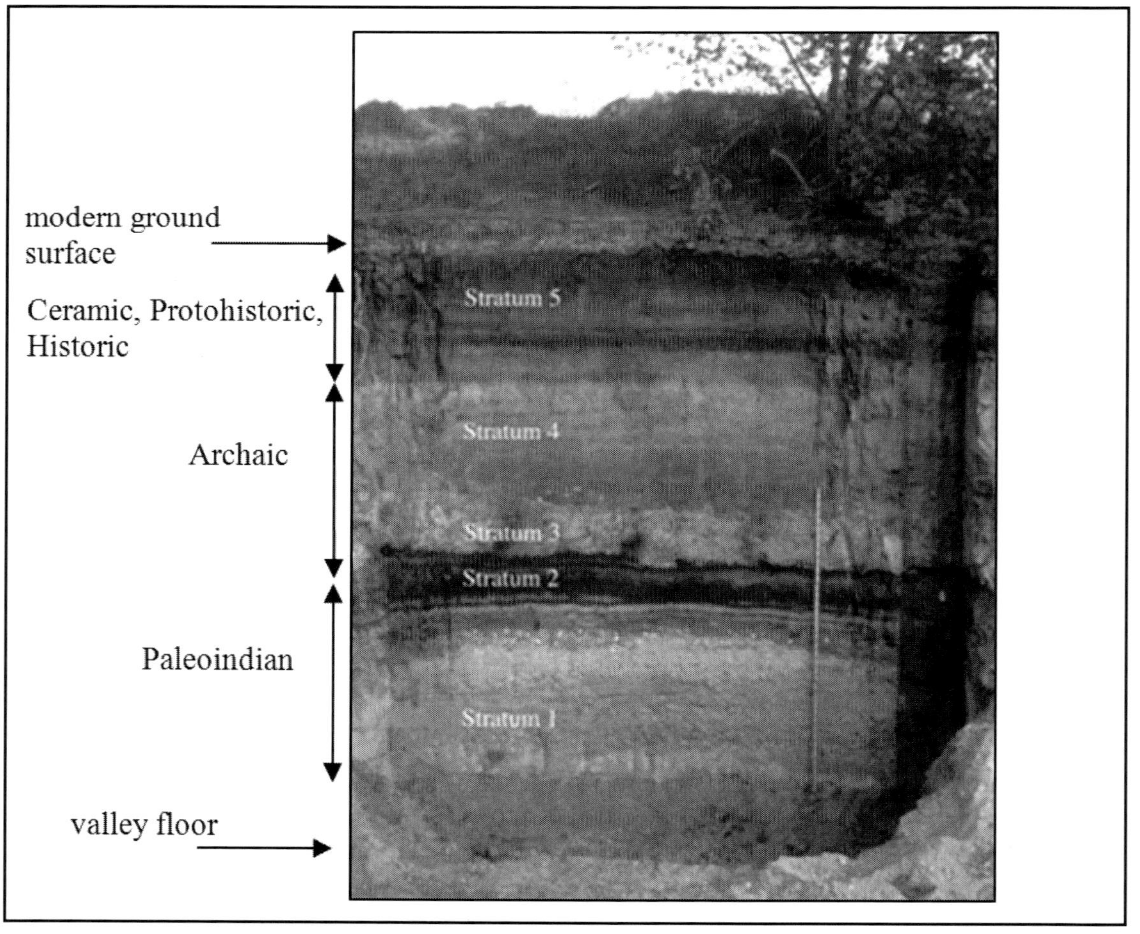

Figure 4.10. Late Quaternary valley fill exposed in the type section at Lubbock Lake (41LU1). Note the stratigraphic sequence spans the late Pleistocene through Holocene (photograph courtesy of the Anthropology Division, Museum of Texas Tech University).

The previously mentioned spectacular recovery of spear points with proboscidean vertebrae by Howard's team excavating on the Southern High Plains near Portales in New Mexico is deeply emblematic of the regional archaeology. The site Blackwater Draw #1 or the Clovis Site, as it is otherwise known, gives its name (Clovis, NM is the nearest major town to the site) to the wider culture of hunter-gatherers who may or may not have been the first colonisers of the new world around 12,000 years ago (Meltzer 1993). Like Lubbock Lake, work at this site also has been ongoing since the 1930's and today this research tradition is continued by Eastern New Mexico University (Carlson 2005). However, unlike Lubbock Lake, excavations at the Clovis site have consistently focussed on the Paleoindian levels, largely in response to salvaging material in advance of commercial gravel extraction operations (Holliday 1997). The results of research at the Clovis Site have perhaps always been most germane at continental rather than regional scales.

In addition to these two archaeological colossuses of the Southern High Plains, numerous smaller less intensively investigated sites have been identified (Figure 4.9). These sites have contributed significantly to our interpretation of the regional record, providing a valuable cultural context with which to assess human activity within the dominant ecological perspective (e.g. Johnson 1991, 2007). While any list of regionally significant sites is in itself a subjective exercise, it can quickly be ascertained that the majority of these sites are again Paleoindian, again underlining the historic focus of much research (Miami, Elida, Millnesand, Ted Williamson, Ryan Site, Lake Theo, Midland, Plainview, and Shifting Sands) all fall within this category and all yield distinctive styles of projectile point armatures. Important Archaic sites have been identified at San Jon on the northwestern escarpment and at Mustang Springs at the opposite end of the region. At the latter site, David Meltzer has interpreted scores of small circular pits or hollows as evidence for well digging during the harsh conditions of the Altithermal (Meltzer 1991). Later Holocene sites with well-excavated stratigraphic sequences are regionally rare. These sites appear to be confined to the eastern portion of the region within the draws and canyonlands where the Garza, Lott, Floydada Country Club, and Montgomery sites all have yielded exceptional evidence for the later occupation of the region (Johnson and Holliday 1995, 2004).

Cultural Chronology

Construction of a cultural chronology for the aboriginal peoples living on and around the Southern High Plains has been the focus of much previous research (e.g. Hoffman 1989; Hughes 1989; Shelley 1994; Johnson and Holliday 1989, 1995, 2004; Boyd 1995; Nickels 2004; Carlson 2005; Johnson 1991, 2006). It is clear from this work that non-sedentary hunter-gatherer subsistence was the dominant lifeway throughout much of prehistory. This reliance on hunter-gatherer subsistence can in part be explained by the environmental setting of the Southern High Plains.

The energy stored in the grassland ecosystem that dominated the region throughout prehistory (e.g. Johnson 2007) is largely unavailable to humans, as it is stored in the inedible portions of plants (Hill 2006). This energy can be retrieved through intensive processing (e.g. grinding and baking) or by targeting the animals that subsist on these plants (Hill 2006). On the Southern High Plains, bison were the key primary consumer of the available grasses. These animals then effectively packaged and stored energy that could be reliably targeted by highly mobile hunter-gatherer groups. Therefore, in environments such as the Southern High Plains where critical resources such as food and water are widely spaced, then behavioural ecology models suggest that highly mobile foraging lifeways should be expected (Kelly 1995, Bird and O'Connell 2006). The archaeological record of the Southern High Plains backs up this prediction and further suggests that during times of high environmental stress, hunter-gatherers were forced to increase residence time at critical resources such as water (Shelley 1994). This is demonstrated in the abundant Middle Archaic record at sites with subterranean freshwater springs, such as is the case at Lubbock Lake, or Mustang Springs (Johnson 1987; Meltzer 1991, 1995, 1999).

Because the archaeological record is dominated by hunter-gatherer groups, the construction of cultural groupings is largely the result of projectile point types in the late Pleistocene through the early and middle Holocene. Other lines of evidence such as pottery, trade items, and art have also been important in the late Holocene record. Because of the high predicted mobility, a limited archaeological 'footprint', and other difficulties associated with assigning cultural affiliation (see Wobst 1978), identification of cultural groups should be treated with caution. Interpretation of the archaeological record of the Southern

High Plains has been heavily influenced by Culture History models, which have sought to identify complexes, focuses, and phases (Figure 4.11). These models are based on extremely limited datasets and a great many assumptions (L. Johnson 1986) and are therefore largely not examined in detail in the current programme of research.

The cultural chronology set out by Johnson and Holliday (1995, 2004), is utilised for the present programme of research as it is based on detailed multi-disciplinary research at deeply buried well-stratified sites. The chronology classifies hunter-gatherer adaptation into five distinct cultural periods and eight sub-periods based on regional hunter-gatherer adaptation to the dynamic late Quaternary environment (Figure 4.5). Brief descriptions of the five main cultural periods serve as a background to this framework:

Paleoindian Period ca. 11,500 – 8,000 RCYBP

The Paleoindian period includes from oldest to youngest the Clovis, Folsom, Plainview, and Firstview sub-periods. The Paleoindian period is defined by a series of projectile points, the earliest of which Clovis and Folsom, exhibit distinctive basal flutes (Prewitt 1995). These points have generally been functionally interpreted as tips for spears (Shea 2006:328). Johnson and Holliday (1995, 2004) identify a narrowing shift of the subsistence base from the Clovis through to later Paleoindian sub-periods; this is due in part to the departure of megafauna from the archaeological record in contexts younger than Clovis. Paleoindian subsistence on the Southern High Plains appears to be largely meat-related (Johnson 1991; Johnson and Holliday 1995, 2004); however, the small dataset is most likely to be skewed by the relative abundance of kill versus domestic contexts thus far investigated.

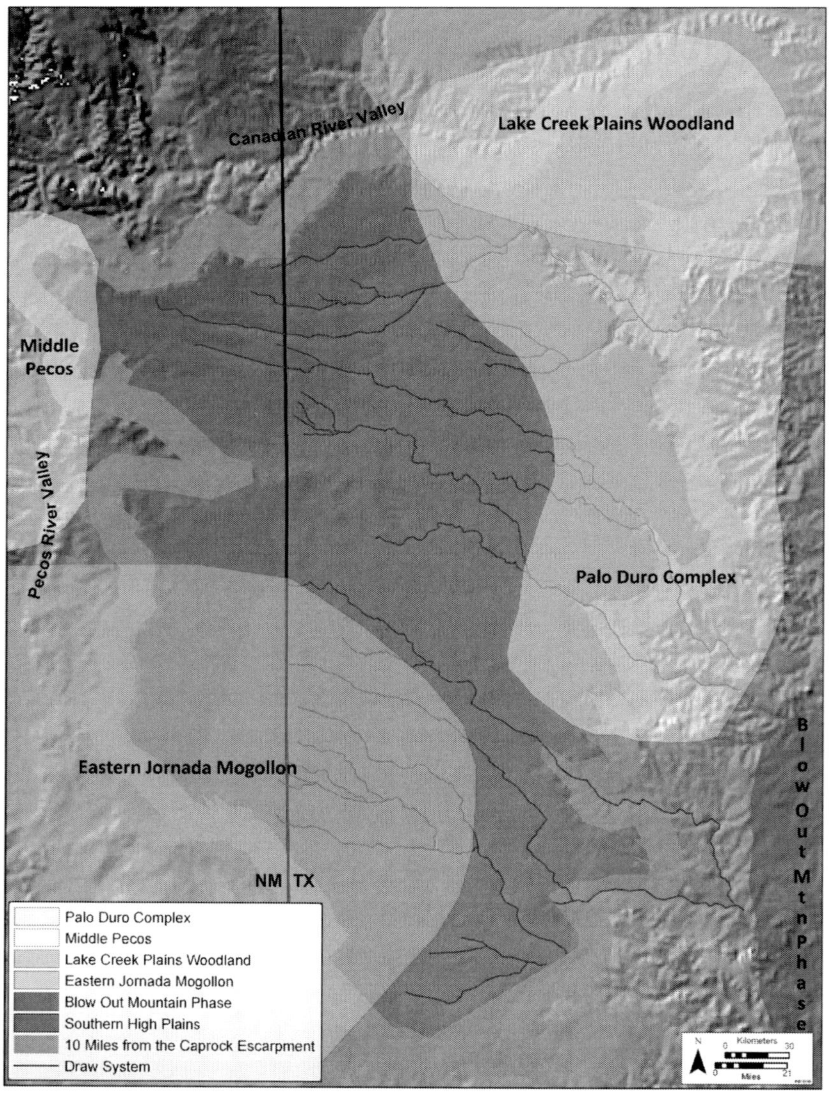

Figure 4.11. Boyd's Late Prehistoric (c. 750 – 1450 B.P.) culture complexes and phases for the Southern High Plains region (modified after Boyd 1995, fig. 3).

Archaic Period ca. 8,000- 2,000 RCYBP

The Archaic period is divided into three sub-periods. These are Early (8,000-6,500 RCYBP), Middle (6,500-4,000 RCYBP), and Late (4,000-2,000 RCYBP). Culturally very little is known about the Early and Middle Archaic sub-periods and they are represented at very few sites on the Southern High Plains or surrounding regions (Sebastian and Larralde 1989; Johnson and Holliday 1995, 2004). Several distinct types of projectile point are again common at Archaic sites; these include Ellis, Trinity, Marcos, and Bulverde types (Prewitt 1995). Archaic points are generally characterised as being smaller than Paleoindian counterparts and larger than the later Ceramic period arrow-points; they most often are interpreted as being attached to a dart and launched with a spear thrower (Shelley 1994:384). The Middle Archaic coincides with the previously mentioned Altithermal, during which environmental conditions may have tethered human populations to locations with critical resources (Shelley 1994). An increasing reliance on plant-based resources is identified at this time (Johnson and Holliday 1986) and may be a function of longer residence times at these oases or refuge locations (e.g. Meltzer 1999). A shifting subsistence economy of greater reliance on plant resources during times of environmental stress and on bison during other less severe times (Johnson and Holliday 1986, 1995, 2004) is a useful working model with which to examine the Southern High Plains Archaic (Shelley 1994).

Ceramic Period ca. 2,000- 500 RCYBP

The Ceramic period is divided into an early (ca. 2000 to 1000 RCYBP) and late (1000 to 500 RCYBP) sub-period based on the appearance of Mogollon puebloan trade wares in the later period (Johnson and Holliday 1995, 2004). The appearance of the bow and arrow and ceramic technologies during this period is significant. It is also during this time that peoples settled along the Canadian River Valley to the north (known as the Antelope Creek Phase) of the region and along the Pecos River Valley to the west of the region (Katz and Katz 1985; Sebastian and Larralde 1989). These groups practised agriculture and appear to have became at least semi-sedentary (Hughes 1995). This change in subsistence does not appear to have extended throughout the Southern High Plains and it was not to be a long-term enterprise with people again returning to hunting and gathering lifestyles in later periods (Johnson and Holliday 1995, 2004). These authors hypothesise that the late Ceramic marks the final appearance of peoples indigenous to the Southern High Plains region (Johnson and Holliday 1995, 2004).

Protohistoric Period ca. 500- 300 RCYBP

The Protohistoric period marks the displacement of local populations by the incursion of Athabaskan peoples, dominated by the Apache, moving down the Plains region from the north and the contact between European peoples and indigenous populations (Boyd 2001; Johnson 2008). Much of the cultural interpretation regarding this period has been extrapolated back from the descriptions made by Spanish conquistador Francisco Vásquez de Coronado who travelled through the region in 1541 (Carlson 1998). During this trip, Coronado described the peoples he encountered as Teyas and Querechos who often moved about the landscape hunting bison, living in tipi-type structures, and utilising dogs in hunting (Gunnerson 1992).

Aboriginal Historic Period ca. 300-125 RCYBP

During the aboriginal Historic period European trade items appear in Native American site assemblages, the horse is introduced, and the Comanche peoples displace the Apache (Johnson 2008). The period closes in 1875 following the Red River War when native peoples were effectively removed as the dominant culture group in the area (Carlson 1998). A programme of Anglo-American settlement followed, which ultimately paved the way for the cultural and political structures that dominate the region today.

The Ten-Mile Buffer

The archaeological record of the Southern High Plains indicates that the regional concept has been essential to understanding the archaeological record of native peoples subsisting through hunter-gatherer lifeways. Theoretical models appear to be most usefully employed against a detailed background developed from theoretical propositions based in cultural and behavioural ecology (e.g. Kelly 1995; Winterhalder 2001; Bird and O'Connell 2006). Examination of the available cultural chronologies indicates that authors have rarely stuck within the region when discussing the Southern High Plains. Important sites are often located in the ecologically diverse areas at the margins of the Southern High Plains (e.g. Boyd 1995). The reasons for the apparent proliferation of sites at the boundaries of the region are diverse but are again likely to be the result of the predictable location of critical resources in these areas.

A useful example is the distribution of high-quality tool-stone suitable for manufacturing lithic tools and weaponry. Holliday observes that for the most part, all the lithic resources are only available at locations off of the Southern High Plains region itself (Holliday 1997:14). Therefore, in order to obtain these resources, the indigenous population would be forced to periodically travel to or trade with groups at the periphery of the region. The occurrence of distinctive types of tool-stone at archaeological sites in the interior of the Southern High Plains with well-known geographically discrete source locations confirms that at least the lithic material and probably the indigenous people, were highly mobile, often selecting high-quality source locations hundreds of kilometres distant (Holliday and Welty 1981; Holliday 1997).

For the purposes of the current project then, the regional view is supplemented by an arbitrary ten-mile boundary extending around the circumference of the area (Figure 4.12). The selection of a ten-mile extension was chosen to contextualise the Southern High Plains area without encroaching significantly into other distinct archaeological regions. This extension comprises much of the southern margin of the Lower Canadian river or 'Canadian Breaks' in the north, Boyd's (1995) Caprock Canyonlands to the east, a greater portion of the Edwards Plateau region to the south, the Mescalero dunes in the southwest, and the eastern margin of the Pecos river valley in the west. The formal delineation of this boundary is an attempt to explicitly contextualise the Southern High Plains within a dynamic cultural and ecological setting, which even today characterises it as a distinct landform setting between eastern and western portions of the North American continent.

Conclusion

The Southern High Plains of North America provide an ideal region within which to examine the archaeological evidence for domestic hunter-gatherer fire technology. Background research indicates that significant datasets have been developed over almost a century of research. These studies have resulted in a rich understanding of the development of the late Quaternary landscape and dominant grassland ecosystem. While numerous technological studies have been conducted within the region, they have largely focussed on aspects of lithic technology and in particular that of projectile point weaponry. In contrast, no previous regional scale examination of thermal-feature technology has been undertaken. Over 75 years of archaeological site data including basic information on the thermal features encountered, has been compiled and stored in state-level databases. These data are now available for electronic querying. The cultural chronology of the region suggests that hunter-gatherer lifeways have been the dominant subsistence strategy throughout much of the region's prehistory and it is assumed that these peoples have consistently employed fire as a primary technological strategy. The potential for the identification of LTFs during archaeological research is therefore high. The ten-mile extension to the periphery of the region is devised in an effort to contextualise the hunter-gatherer peoples that variously occupied the Southern High Plains within a sphere of resources that were critical to their survival. It is expected that these border areas will yield the strongest evidence for hunter-gatherer activity, as they are also the locations at which critical resources are located.

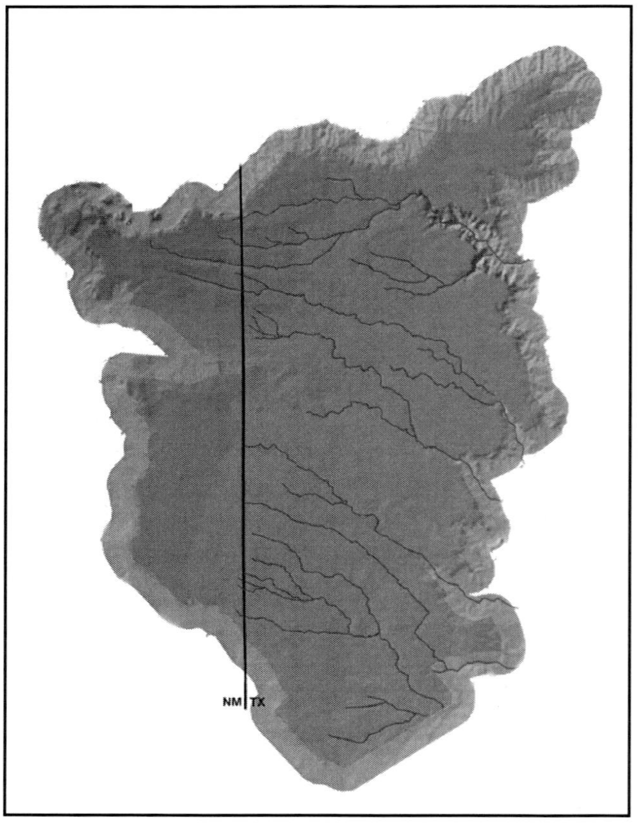

Figure 4.12. Southern High Plains region showing the arbitrary ten-mile extension used in the current program of research.

Several lines of evidence (archaeological, ethnographic, and experimental) can be drawn upon in order to understand the archaeological record for fire technology on the Southern High Plains. The first step in this is the development of a classificatory system with which to unify the accumulated regional dataset by means of a common descriptive vocabulary. A theoretical framework for the investigation of LTFs and the design of a hierarchical taxonomic key for their classification is set out in Chapter 5.

CHAPTER 5 - THEORETICAL FRAMEWORK

Abstract

The first approach in the study of LTFs is the development of a theoretical framework for their analysis. Historically, the archaeological investigation of hunter-gatherer groups has relied on artefacts rather than features as the primary evidence for constructing interpretative arguments. This emphasis, in spite of a growing corpus of feature-orientated description and quantitative data, has arguably resulted in the uncritical interpretation of feature-related technologies such as localised thermal features (LTFs). An interpretative project is devised in response, with which to explore hunter-gatherer fire features as a potentially meaningful technology requiring serious theoretical consideration. The project engages multiple lines of evidence to assess the variability expressed within the archaeological record. These lines of evidence are open-ended and potentially recursive as new lines of enquiry emerge. A taxonomic key for the classification of individual features is developed. This key facilitates the normalisation and incorporation of previously recorded state-level datasets by generating a common vocabulary for fire features. The classificatory system will allow future fieldworkers to more easily contextualise and therefore interpret the fire features they record. The assumptions and limitations of the project are considered and it is concluded that the theoretical prioritisation of hunter-gatherer fire technologies is a useful undertaking both as a means for understanding hunter-gatherer behaviour and as a tool for planning modern research strategies.

The archaeological investigation of prehistoric hunter-gatherer groups necessarily engages with restricted datasets, poorly preserved ephemeral locations of activity, and deep time scales (Kroll and Price 1991). Historically, the archaeological process has been an examination of discarded portable material culture (Davies 2000) and resultant synthetic works tend toward an emphasis on economic materiality focusing the analytical lens on individual artefact classes and fluctuations over long time scales. In contrast, the features constructed and utilised by hunter-gatherer groups for the controlled manipulation of heat through the medium of fire, here termed LTFs, have received very little theoretical consideration.

Critical examination of relevant literature identifies the largely implicit role LTFs assume when interpreted at a site level (Chapter 2). This situation is in part due to a lack of common terminology or rigorous theoretical framework with which to assess these feature types. In response, an interpretative project is framed which centres LTF construction and use as a technologically meaningful hunter-gatherer behaviour worthy of detailed investigation.

The proposed model examines variability in these feature types by exploring multiple lines of evidence derived primarily from fieldwork, heuristic investigations, and qualitative analysis. The relationship between these lines of evidence is open-ended and potentially recursive as new lines of enquiry emerge throughout the project. The construction of a simple hierarchical key for field identification is proposed as an initial classificatory tool for analysing the existing regional dataset. Lastly, assumptions and limitations inherent to the proposed research project are identified and discussed.

An Interpretative Model for the Investigation of Localised Thermal Features.

Theoretical approaches to thermal features as built, used, and abandoned by hunter-gatherer societies are necessarily studies of technology. It is proposed here, that recent trends in archaeological theory (e.g. Vanpool and Vanpool 2003, Pauketat and Alt 2005) can be positively utilised in formulating dynamic open-ended research questions that seek to explore the mosaic of technological activities associated directly and indirectly with the construction of LTFs by hunter-gatherer groups. The proposed research structure then, draws critically on multiple lines of evidence to provide an interpretative framework for assessing hunter-gatherer fire technology.

Archaeologically, it is hypothesized that distinct signatures will be observable at intra-site scales and that these will translate to trends observable at inter-site scales and furthermore, that these observations should be useful in the interpretation of prehistoric fire technologies as a function of time, space, and cultural behaviour. Ethnographic and experimental datasets are crucial components in the structure of this research model, as they provide two key lines of evidence in the ongoing assessment of the archaeological record.

The construction of an LTF is the basic unit of analysis in this research. It is also arguably an extremely useful structure with which to explore the archaeological evidence for hunter-gatherers within a particular region. Considered from a landscape perspective, "human mobility relative to other organisms tends toward saltatory movements" (Stafford and Hajic 1992:139), although hunter-gatherer mobility strategies vary over time and across space. Regardless of technological variation, it can be posited that the physical act of constructing an LTF always necessitates at least a temporary cessation in the ongoing ambulatory movement of hunter-gatherers (e.g. Bender 2001). The physical imposition of these features on the local landscape can therefore always be construed as a dynamic locus for human activity (e.g. Schlanger 1992). This activity may involve one or many people and may be as short as to be measured in minutes or as long as to be measured in seasons. Thermal features then contrast with the majority of hunter-gatherer technology in that the technological mechanism is not transportable and therefore can often be studied *in situ*. Their presence can be directly associated within a qualitative framework as the basis for the definition of the term campsite, underscoring a high potential for research.

Contrary to much thermal feature research (but see Petraglia 2002), recent research directions within the anthropology of technology have sought to emphasise the culturally embedded nature of technology (Pfaffenberger 1988; 1992) and focus on the changing relationships in an object or feature life cycle (e.g. Edmonds 1990; Lemmonier 1993; Karlin and Julien 1994; Dobres 1995; Greene 2004; Pauketat and Alt 2005, Bleed 2006). This view of technology has dramatically altered research directions in lithic analysis (e.g. Pigeot 1990; Andrefsky 2001) but has rarely been applied to the investigation of fire features (but see Lowell 1999). The perceived dominance of short-term artefact orientated material narratives encountered in archaeology (Lucas 2001) may be unravelled through the development of technological operational sequences in the life of features such as LTFs and their surrounding artefact scatters.

It can be assumend that social relationships were structured by technological organisation within the relatively confined space available around hunter-gatherer thermal features (Gravina 2004). The holistic investigation of technological features and surrounding cultural residues may therefore provide useful data for answering questions of social and technological organisation (Pauketat and Alt 2005). For instance, the creation of an LTF can be reduced down to a number of events or processes e.g. collecting fuel to burn, digging the pit, and processing the food to be cooked (see also, Dufraisse 2006 for an approach that considers the role of firewood in domestic thermal features during the Neolithic). These activities need not occur at the feature itself and most probably occur at some distance (e.g. food may be transported in, fuel gathered at the periphery of the camp), while other activities must have occurred in close proximity to the feature. Thermal-feature technology engages then, with a range of activities (technological sequences) parts of which are available in the form of the physical evidence presented by the archaeological record. Within this framework, decision processes are dynamically explored in terms of actions and events.

The implication is not that these actions and events are linear, simple, or predetermined. They are, in fact, sinuous, complex, and fluid actions recreated by social actors (e.g. Dobres 2000). A focus on the technological creation, maintenance, and abandonment of LTFs empowers human behaviour to be observed that is far from static. Each activity concerning the choice of type, construction method, and utilisation of an LTF in this sequence requires thought and decision making. This type of technological sequence has been described by Bleed (2006) as "wide" as opposed to the "narrow" sequential models often engaged in analysis of lithic reduction technologies. Assuming these wide decision sequences are technologically meaningful, then the biological and material residues present in the archaeological record are ideally situated for their investigation.

Research Model

A research model sets out a framework that can be used to reference and integrate the results from the four analytical strands of evidence (Chapters 5-8). This model is recursive in that the regional dataset is reinterpreted as the results of individual investigations are fed back into the regional model (Figure 5.1). This conceptual framework allows for multi-scalar diachronic research strategies to be framed, which examine the technological variability manifest in the regional LTF dataset through the detailed observation, recording, and testing of specific processes and research propositions. The programme is facilitated by exploring multiple lines of evidence and is inherently empirical in nature, as the baseline of inquiry is the archaeological identification of a single abandoned physical structure.

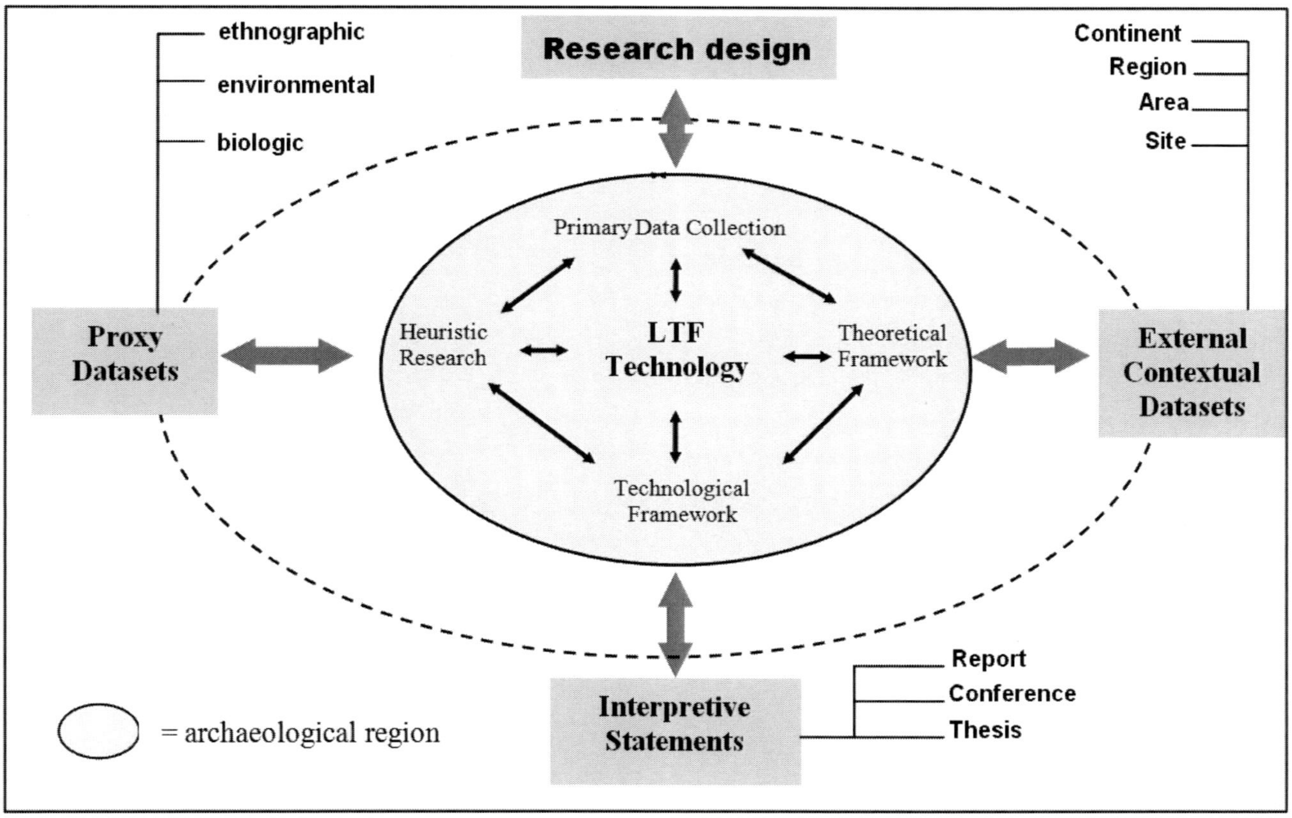

Figure 5.1. Research model for the investigation of localised thermal features.

Working up from a single feature then, larger frameworks and broader-range analogy are subsumed as the analytical focus is enlarged from feature to site to region. As Wylie observes, the "tension and revision of source-sided and subject-sided positions" act to effectively drive forward this dynamic process (Wylie 2002:125). New lines of enquiry can be commenced after the initiation of the project; similarly, old lines of enquiry can be discounted and written up as appropriate. The output does not, however, accept arguments of extreme relativism, as the basis for this construct are data derived from archaeological phenomena and experimental research. Attempts to find a closest fit for the data are acceptable and privileged over lower orders of likelihood. The product of the relationship between these active elements is in the form of original observations that frame a viewpoint through which to empower a humanistic ontology of past technological human – fire relationships.

The core of this framework comprises a continuing reassessment of the archaeological record in relation to four epistemological areas of inquiry (primary data collection, the theoretical framework, heuristic research, and the technological framework). The conceptual core of the model is not closed to external stimuli but merely acts as a focus to constrain inference at a regional scale. The more general interchange of information occurs at a wider scale and in less formal terms than the core regional research elements. The conceptual model is clearly weighted toward an inductive or interpretative approach to the investigation of LTFs.

A Hierarchical Key as a Means of Feature Classification

In order to practically implement the proposed preliminary research model, it is necessary to construct a classificatory device with which to assess the variability in the archaeological signatures for LTFs from a regional perspective. It has already been demonstrated that the identification of LTFs during the archaeological investigation of prehistoric hunter-gatherers is a relatively commonplace occurrence and that the operative terminology used by archaeologists to describe these feature types has been shown, to be highly idiosyncratic (Chapter 2). It is therefore not surprising that technological research trends toward a focus on the description rather than analysis. In contrast, Lewis Binford (1992) argues for the need to make explicit our claims for encountering a particular type of feature:

> "In short, we must claim that we have recognized a structured and, importantly complementary pattern among different things that has reference to past organizational dynamics. This claim is as good as

the arguments presented or alluded to in assertions that structure (a patterned arrangement of different things resulting from organized interaction among variables) is distributed in a complementary spatial pattern so as to implicate past dynamics of an understood form – in this case, the controlled burning of fuel in a single place" (Binford 1992:45-46).

The potential structure and utility of the identification device is directly related to the type of research being undertaken. The investigation of LTFs is dependant on the re-examination of secondary sources, which are mainly in the form of archaeological site archives and written reports. Therefore, in order to maximise the utility of the identification device, it is necessary to select variables that can be determined with reference to the available literature. More often than not, this evidence comprises very brief site-nomination forms, which include only the most basic information (Figure 5.2). Practically, this requirement necessitates the identification of commonly recorded simple variables, which can be arranged into a hierarchical structure.

The development of a hierarchical key (e.g. Dunell 1971) is utilised as the primary identification device for classifying the variability in the existing regional LTF dataset. The advantage of this approach is in the generation of interpretation-free categories, which can be used to discriminate variation and thus provide a qualitative structure for subsequent research development (Sullivan and Rozen 1985). This classificatory structure focuses on the material remains of a feature or artefact and seeks to assign a class based on the inclusion or exclusion of a series of key attributes. Classification is achieved by means of binary oppositions, which allow for the identification of objects within arbitrarily developed taxonomic classes (Dunell 1971). A potential drawback of the construction of such identification devices is that they necessarily rely on detailed a posteriori knowledge of the potential classes, which will be generated by the available dataset (Dunell 1971:106) and therefore are ideally suited to the investigation of archaeologically identified LTFs. The proposed identification device has five key variables; with each variable defined by the result of a dichotomous attribute (Figure 5.3).

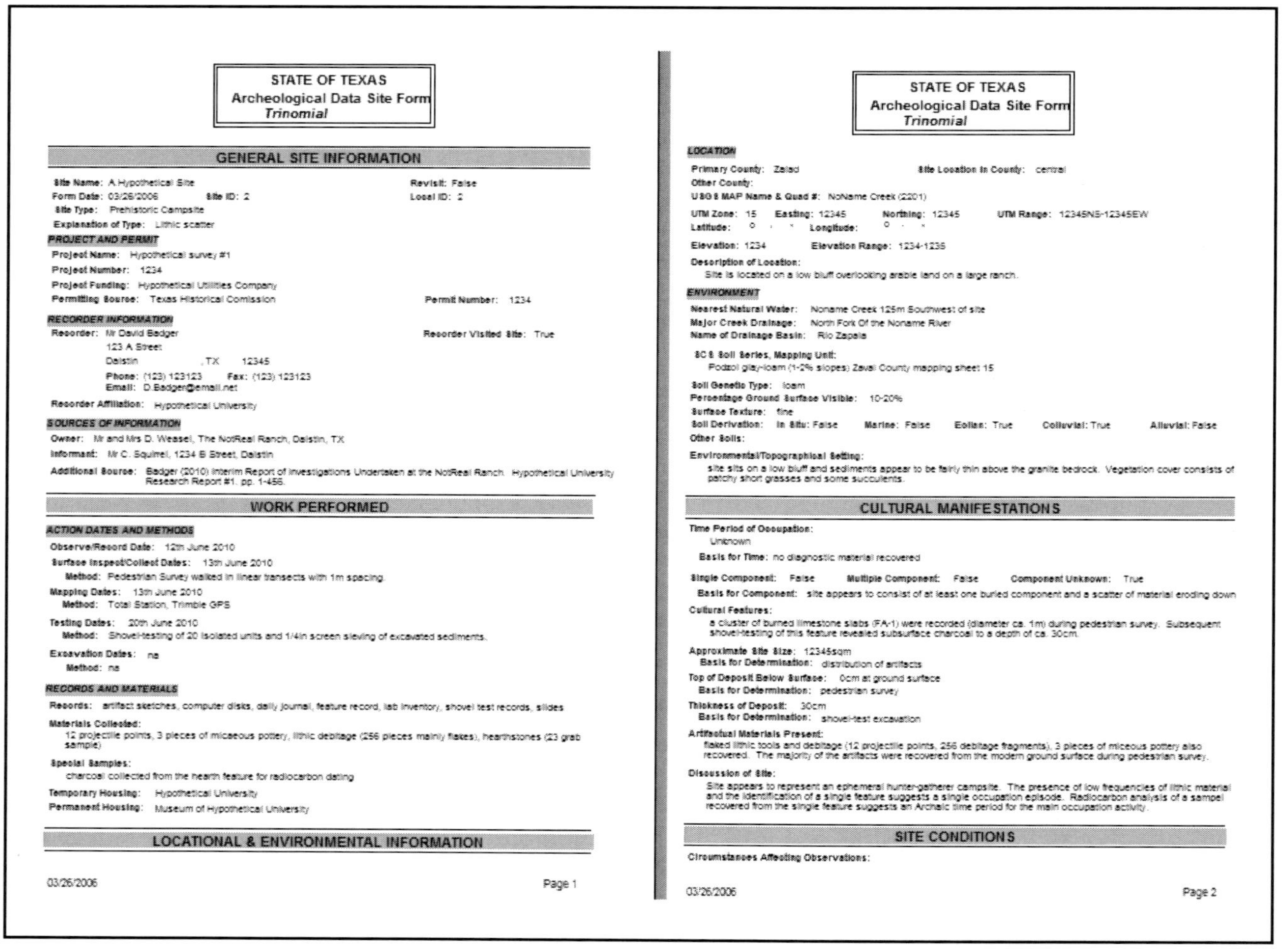
STATE OF TEXAS
Archeological Data Site Form
Trinomial

GENERAL SITE INFORMATION

Site Name: A Hypothetical Site — Revisit: False
Form Date: 03/26/2006 — Site ID: 2 — Local ID: 2
Site Type: Prehistoric Campsite
Explanation of Type: Lithic scatter

PROJECT AND PERMIT
Project Name: Hypothetical survey #1
Project Number: 1234
Project Funding: Hypothetical Utilities Company
Permitting Source: Texas Historical Comission — Permit Number: 1234

RECORDER INFORMATION
Recorder: Mr David Badger — Recorder Visited Site: True
123 A Street
Daistin, TX 12345
Phone: (123) 123123 Fax: (123) 123123
Email: D.Badger@email.net
Recorder Affiliation: Hypothetical University

SOURCES OF INFORMATION
Owner: Mr and Mrs D. Weasel, The NotReal Ranch, Daistin, TX
Informant: Mr C. Squirrel, 1234 B Street, Daistin
Additional Source: Badger (2010) Interim Report of Investigations Undertaken at the NotReal Ranch. Hypothetical University Research Report #1. pp. 1-456.

WORK PERFORMED

ACTION DATES AND METHODS
Observe/Record Date: 12th June 2010
Surface Inspect/Collect Dates: 13th June 2010
Method: Pedestrian Survey walked in linear transects with 1m spacing.
Mapping Dates: 13th June 2010
Method: Total Station, Trimble GPS
Testing Dates: 20th June 2010
Method: Shovel-testing of 20 isolated units and 1/4in screen sieving of excavated sediments.
Excavation Dates: na
Method: na

RECORDS AND MATERIALS
Records: artifact sketches, computer disks, daily journal, feature record, lab inventory, shovel test records, slides
Materials Collected:
12 projectile points, 3 pieces of micaeous pottery, lithic debitage (256 pieces mainly flakes), hearthstones (23 grab sample)
Special Samples:
charcoal collected from the hearth feature for radiocarbon dating
Temporary Housing: Hypothetical University
Permanent Housing: Museum of Hypothetical University

LOCATIONAL & ENVIRONMENTAL INFORMATION

03/26/2006 — Page 1

STATE OF TEXAS
Archeological Data Site Form
Trinomial

LOCATION
Primary County: Zalad — Site Location in County: central
Other County:
USGS MAP Name & Quad #: NoName Creek (2201)
UTM Zone: 15 Easting: 12345 Northing: 12345 UTM Range: 12345NS-12345EW
Latitude: ° ' " Longitude: ° ' "
Elevation: 1234 Elevation Range: 1234-1236
Description of Location:
Site is located on a low bluff overlooking arable land on a large ranch.

ENVIRONMENT
Nearest Natural Water: Noname Creek 125m Southwest of site
Major Creek Drainage: North Fork Of the Noname River
Name of Drainage Basin: Rio Zapala
SCS Soil Series, Mapping Unit:
Podzol glay-loam (1-2% slopes) Zavai County mapping sheet 15
Soil Genetic Type: loam
Percentage Ground Surface Visible: 10-20%
Surface Texture: fine
Soil Derivation: In Situ: False Marine: False Eolian: True Colluvial: True Alluvial: False
Other Soils:
Environmental/Topographical Setting:
site sits on a low bluff and sediments appear to be fairly thin above the granite bedrock. Vegetation cover consists of patchy short grasses and some succulents.

CULTURAL MANIFESTATIONS

Time Period of Occupation:
Unknown
Basis for Time: no diagnostic material recovered
Single Component: False Multiple Component: False Component Unkown: True
Basis for Component: site appears to consist of at least one buried component and a scatter of material eroding down
Cultural Features:
a cluster of burned limestone slabs (FA-1) were recorded (diameter ca. 1m) during pedestrian survey. Subsequent shovel-testing of this feature revealed subsurface charcoal to a depth of ca. 30cm.
Approximate Site Size: 12345sqm
Basis for Determination: distribution of artifacts
Top of Deposit Below Surface: 0cm at ground surface
Basis for Determination: pedestrian survey
Thickness of Deposit: 30cm
Basis for Determination: shovel-test excavation
Artifactual Materials Present:
flaked lithic tools and debitage (12 projectile points, 256 debitage fragments), 3 pieces of miceous pottery also recovered. The majority of the artifacts were recovered from the modern ground surface during pedestrian survey.
Discussion of Site:
Site appears to represent an ephemeral hunter-gatherer campsite. The presence of low frequencies of lithic material and the identification of a single feature suggests a single occupation episode. Radiocarbon analysis of a sampel recovered from the single feature suggests an Archaic time period for the main occupation activity.

SITE CONDITIONS

Circumstances Affecting Observations:

03/26/2006 — Page 2

Figure 5.2. Example of the types of information typically recorded for a hypothetical site in Texas, USA. and on file at the Texas Archaeological Research Laboratory and the Texas Historical Commission Sites Database.

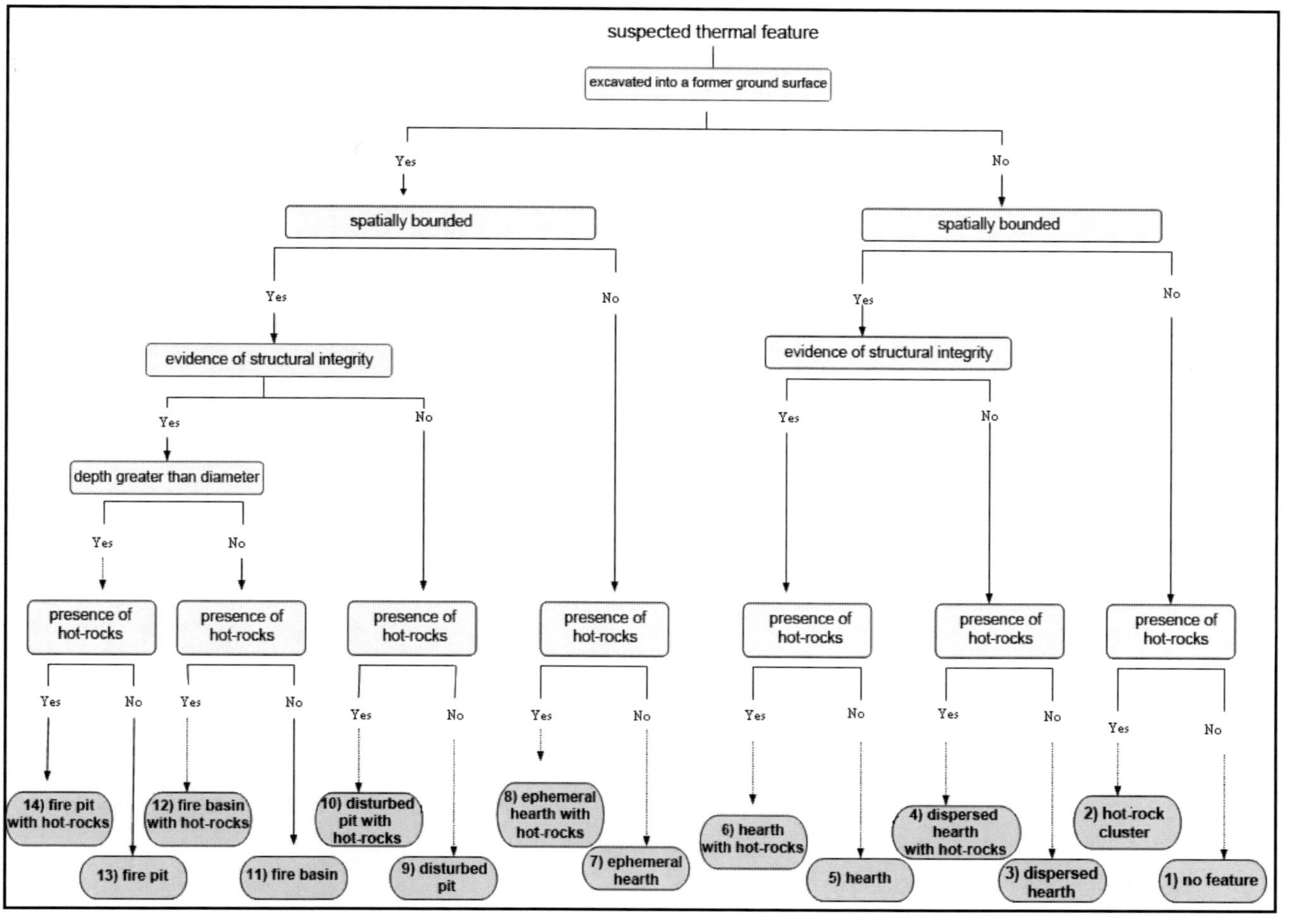

Figure 5.3. Hierarchical key for the classification of localised thermal features.

The realization that site-formation processes are particularly important in actively structuring the physical evidence for LTFs (Petraglia 2002) is explicitly acknowledged in the selection and construction of this classificatory system. Contrary to the taphonomic transformations documented on single artefact classes, the complex taphonomic processes that actively structure the physical remains for a particular feature are inherently more complex, as they are simultaneously acting on multiple elements of differing material types. Consider as an example of this situation a shallow basin structure excavated on the shore of a beach. Inside the excavated basin, pine logs are burned and granite stones are placed on the resulting hot coals. Later, as the rocks heat, fine strips of deer meat are cooked on the heated rocks. After only a single use, the feature is abandoned and the incoming tide washes away the burned wood and charcoal that lined the inside of the basin. Fresh sand is washed into the basin burying the granite rocks in a tight cluster, which are encountered after considerable time has elapsed by archaeologists excavating test-units along this beach.

This simple example illustrates the complex taphonomic relationships that are necessarily involved in the investigation of LTFs and therefore the consideration of these processes must be integrated into any scheme that attempts the classification of structures over entire archaeological regions. In the example, the only components of the hypothetical LTF that endured to be recovered by archaeologists were the granite hearthstones, which being rocks, are extremely difficult to break down and therefore obscure. Perhaps several years of wave action and erosion would have scattered these rocks rendering the interpretation of an LTF at a particular location impossible. The importance of the foregoing example is that the goal of the classificatory structure is the characterization and not the interpretation of the archaeological record in a particular area. This is a similar position to the widespread use of the hierarchical key concept in lithic debitage analysis (cf. Sullivan and Rozen 1985). The expected outcome of the research exercise is therefore a structure with which to qualitatively examine the research potential of LTFs, highlighting the strengths and deficiencies of the current database and thus providing a quantitative resource assessment that can be queried prior to the development of behaviourally meaningful and therefore interpretative research.

The Five Classificatory Keys

The framework developed for this project has five simple classificatory keys (Figure 5.3). As such a framework has not previously been attempted for the classification of LTFs, then it must be regarded as experimental. It is anticipated that the results of the overall project will suggest the need for additional keys, removal of unnecessary keys, creation of different keys, or restructuring of the hierarchical relationships. This realisation arguably strengthens the epistemological position of the overall project as a tool for the investigation of the regional archaeological record. The potential for reconfiguration maximises the utility of the framework to assess the regional record and therefore identify the areas that would most benefit from focussed qualitative research.

If the response to one or more of the keys is unknown, then the negative path is followed by default. For instance, if the subsurface integrity of a particular feature is unknown because it has been recorded during pedestrian survey and is exposed only from a plan perspective, then the feature should be recorded as having no subsurface expression. If the majority of features in a particular region are therefore recorded in a similar manner, then it raises an important research goal for that region i.e. the subsurface investigation of one or more of these features to determine if a subsurface expression is present or absent. Even if this information had previously been recorded implicitly, perhaps relying on excavator 'know how', the classificatory scheme will still 'flag' it as an area requiring research. This is a positive development in developing datasets that have potential to be useful for multiple researchers and agencies and that will be able to be queried long after both of which have long departed.

The first key is evidence for the excavation of the structure into a former ground surface. Evidence from diverse ethnographic and archaeological contexts has documented excavation as means of physically containing thermal features (e.g. Binford 1978; Ellis 1997; Galanidou 1997; Smith *et al.* 2001). The identification of prehistoric excavation is largely determined by the archaeological method deployed (Figure 5.4). The presence of spatially discrete concentrations of burned artefacts and high fractions of carbonized material within negative features are usually indicative of subsurface expressions of LTFs. Identification is not, however, mutually dependant on intrusive modern excavation and LTFs are often naturally exposed in section by erosional processes such as those acting on coastlines, or incised drainage systems (e.g. Basford 1980; Backhouse 2003; Figure 2.5).

Figure 5.4. Example of an intentionally excavated LTF structure encountered during archaeological research. Note the stained sediments encountered in the wall of an excavated trench. (photograph courtesy of the Anthropology Department, Museum of Texas Tech University).

Site records commonly provide information as to whether a particular feature has a subsurface expression or not and furthermore, this information can often be confidently inferred from the mode of investigation reported i.e. a pedestrian survey is less likely to encounter a feature with a subsurface expression than is an invasive method such as hand excavation. Non-intrusive methods such as some types of geophysical survey may also identify negative features. If an accurate method for the discrimination of these signals is devised and proven in the field through a systematic programme of groundtruthing, then LTFs with subsurface expressions may be characterised without any excavation, or the feature ever being exposed. This situation however is relatively unlikely given the wide range of archaeological signatures identified for LTF in Chapter 1. Nevertheless, future remote-sensing methodologies may not require excavation to occur to determine whether or not an LTF was excavated into a former ground surface.

The second key is the identification of a spatial boundary. Defined here as the identification of increased fractions of ash and charcoal within a spatially discrete area (Vaquero and Pastó 2001; Figure 5.5), the patterned arrangement of rocks or other material to demark the periphery of the structure, or any other physical manifestation by which a structural boundary can be clearly identified. Thermal staining is often the most visible physical evidence for a boundary. Stains result from the carbonisation of organic material through thermal alteration followed by physical degradation and inclusion into the archaeological record (e.g. Wright 2003). It is recognised that thermal soil stains can be created by a number of processes (cf. Sullivan *et al.* 2001) and need not necessarily result from *in situ* burning. Visibility of this phenomenon is not mutually dependant on excavation but it is considerably less likely to be observed during survey in well-vegetated or highly eroded environments. Site records again, commonly record the presence or absence of soil staining as it is a highly suitable context for radiocarbon assay.

Figure 5.5. North American example of a thermally altered soil stain defining the spatial boundary of a feature recorded during excavation at 41LU1, area 10 (Photograph courtesy of the Anthropology Department, Museum of Texas Tech University).

The third key is the presence or absence of structural integrity. Defined here as an observable relationship between burned material and residues within the internal matrix of the observed feature (Figure 5.6). Criteria include the arrangement of hearthstones to one another, the identification of multiple oxidised layers (e.g. Johnson 1987b), the recovery of burned faunal or vegetal material, and the inclusion of unburned material in the internal feature matrix. Inclusion or exclusion of a specific feature is therefore based on multiple and complex criteria and are themselves reliant on field experience and a posteriori knowledge of the regional LTF dataset. Features encountered during excavation are more likely to be positively identified as structurally intact. Site records can yield wildly different levels of reporting and description concerning the integrity of individual features as was described in Chapter 1. The purpose of the third key is to distil this information and make a judgement as to the integrity of the feature. The application of the third key will most likely be applied to only select features, which have been well excavated. It is hoped therefore the determination as to the integrity of these features will be supported by adequate supplementary documentation. If this is not the case for an individual feature, then a negative response should be recorded.

Figure 5.6. Example of a structurally intact feature; note the presence of fine ash around the exposed hearthstones and the obvious visual difference between the carbonized material within the feature and the surrounding sediments. (P.N. Backhouse)

The fourth key is a basic qualification of the internal geometry of the feature, requiring an assessment of the depth in relation to the diameter (Figure 5.7). Thermal feature analysis has previously sought to explore this relationship in terms of function (Shockey 1997) and as a means of classification (Ellis 1997:60). Identification is again, limited to excavation or the observation of features in natural exposures. Basic measurements such as these are often recorded for excavated features and therefore should be available in the site documentation.

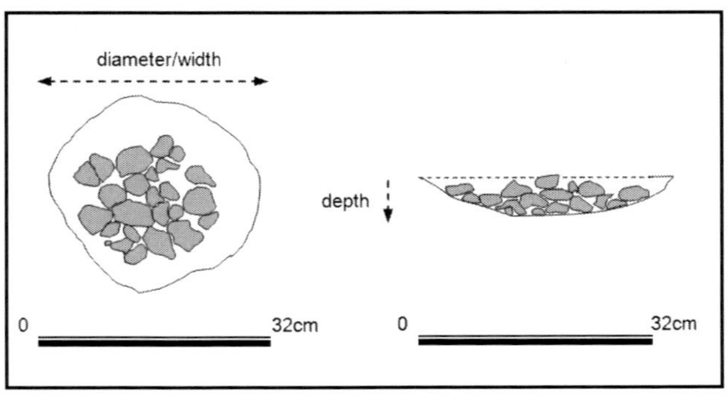

Figure 5.7. Example of the two measurements for the determination of diameter and depth of an LTF. In this example maximum width is greater than the maximum depth resulting in a basin-shaped structure.

The fifth key is the presence or absence of hot-rocks. This variable is mutually exclusive and as such is not determined by any of the preceding variables. Hot-rocks (hearthstones) are typically capable of preservation long after the eradication of the internal structure of the original thermal feature (Holdaway *et al.* 2002; Figure 5.8). Hot-rocks are not dependant on research strategy and are equally likely to be encountered during excavation (e.g. Brink and Dawe 2003) as surface survey (e.g. Hawkins 1998). Although this artefact class has long been looked upon with a certain derision by the archaeologist (cf. Brink and Dawe 2003) and is less than frequently recovered, perhaps for fear of filling up precious museum storage space with tons of burned rocks. Nevertheless, the presence or absence of hot-rocks is commonly documented during archaeological site nominations.

Figure 5.8. Caliche hot-rocks within an experimental LTF at Lubbock Lake Landmark, Texas (photograph courtesy of the Anthropology Department, Museum of Texas Tech University).

The proposed classification scheme results in the assignment of 14 mutually exclusive categories that serve to describe the range of variability observed in the regional archaeological record. The advantages of this schema are two-fold: firstly, the identified types are not technologically or functionally deterministic, and secondly the influence of different variables depends on the context of the original investigation (Sullivan and Rozen 1985:759). This result allows the researcher to actively assess the potential of the dataset for particular types of analysis. For instance, if the result of a survey of all the documented LTFs in a particular region indicated that 90% comprised hot-rock scatters (formally, class 2) then research focussed on the examination of variation in the internal morphology of individual features would be constrained by a small unrepresentative dataset, thus limiting the potential for meaningful inference at wider scales.

Assumptions and Limitations

The programme of research depends on basic assumptions and is constrained by inherent limitations:

Assumptions

1. The construction of localised thermal features for the purpose of manipulating fire is a technologically significant activity representative of hunter-gatherer groups and therefore is an archaeologically significant phenomenon.

2. LTFs, as constructed by hunter-gatherer groups, can result in physical signatures observable in the archaeological record.

2. Activities involving, and in proximity to, LTFs (such as cooking) can and often do leave physical, chemical, and biological residues, which are sometimes observable in the archaeological record.

3. The ability to routinely generate and manipulate fire is an inherent component of the technological skill-set possessed by all hunter-gatherer groups identified in the archaeological record since at least the last glacial maximum.

Limitations

1. The archaeological excavation of hunter-gatherer campsites often encounters a palimpsest of cultural material resultant of multiple activities, and/or occupations, and/or post depositional processes. These processes may obscure or destroy the potential for detailed spatial analysis.

2. Cleaning and maintaining activities undertaken by prehistoric populations, around the LTF area may obscure or destroy spatial patterns of cultural debris.

3. The proposed project inherently privileges archaeologically identified LTF structures and therefore may be influenced by the visibility of particular types of LTF and by the physical construction materials that have a tendency to survive for long periods and be successfully identified by archaeologists (such as hearthstones).

4. The construction and use of LTFs are not culturally specific to hunter-gatherer groups and therefore the potential exists to confuse structures created by sedentary

societies as having been created by hunter-gatherer groups.

5. Not all spatially confined fire technologies leave physical evidence that is identifiable in the archaeological record.

6. Natural processes capable of the production of spatially confined areas of burning (particularly lightning strikes) can leave physical evidence that is morphologically identical to the human-built structures termed localised thermal features.

Conclusion

The practical technological analysis of LTFs in a particular region or a time-period may go some way to providing some much-needed critical context on which to base our theoretical assumptions. This analysis will benefit the development of hunter-gatherer research by pointing toward the expected variability evident in any regional dataset and therefore moves toward a richer humanistic understanding of hunter-gatherer activity, critically drawing from the available data rather than grasping across continents for answers in far-removed and often problematic examples.

A novel research design for the holistic investigation of LTFs is devised in an attempt to resituate their study within a conceptual framework that emphasises LTFs as interpretatively meaningful components of hunter-gatherer technological strategies. Practical assessment and classification of a regional archaeological LTF dataset is the first step in this process. Assessment of research potential is facilitated by the application of a taxonomic key, which groups features based on the qualification of five commonly recorded variables. Application of this key allows for the rapid assessment of large regional LTF datasets and therefore creates a regional framework within which the results of the other analytical strands of research (Chapters 6-8) can be assessed. Questions arising from the results of the application of the taxonomic key are examined through experimental research (Chapter 7) and the regional archaeological record (Chapter 8) of the Southern High Plains. Lastly, the success of the application of the four analytical strands of evidence that comprise the interpretative project are drawn together and critically discussed in Chapter 9.

CHAPTER 6 - ETHNOGRAPHIC CONTEXT

Abstract

The second approach in the study of LTFs employs ethnographically recorded thermal features as a proxy for the types of construction, use, and abandonment processes occurring in the archaeological record. Geographically, the physical manifestation of thermal features is to some extent conditioned by the local environment and for this reason the North American continent and in particular the Plains region provide the basis of the ethnographic inquiry. Ethnographic data are explored across three research strands and the results indicate that North American native fire technologies were well developed by the time of sustained European contact in the sixteenth century. The realisation that the domestic structures that dominate the archaeological record are the only trace of a wider technology, capable of altering entire ecosystems, is profound. The variation in the form and function of domestic fire technologies, as documented by early explorers and ethnographers, is significant and should be an important caveat to simplistic site scale interpretations. At a wider focus, the data exhibit some culturally meaningful patterning, suggesting that at least in the historic period, technological trajectories of different thermal feature types may be usefully studied. Ultimately, the immense range and complexity of physical structures and processes identified in the ethnographic literature reinforce the need for research strategies that critically examine and engage with these features through integrated programmes of field and experimentally based investigation.

> "..documentation of the Indian use of fire is fragmentary at best. Historically documented incidents are rare; photography was invented after most tribes had disappeared or surrendered their traditional ways" (Gerald W. Williams 2000:8)

It is important to note that ethnographic documents consistently point to fire technologies as "the most ecologically effective and technologically powerful tool available to foragers, farmers, and pastoralists" (Lewis 2003:27). The critical use of ethnographic research therefore has the potential to provide much of the central contextual detail necessary for exploring the range in variation present when confronting hunter-gatherer fire technology from an archaeological perspective. It is likely that the largely domestic features that dominate hunter-gatherer archaeological sites and that are the subject of this study, mask the true complexity of aboriginal fire technology. Furthermore, it is clear that native fire technologies often extended as the primary human tool for managing whole ecological regions through active brush control and the encouragement of new growth (Westbroek *et al.* 1993; Stewart 2002). Freed from an intellectually subordinate position in relation to other hunter-gatherer technologies such as lithic reduction, then, serious consideration should be afforded to the fragmentary ethnographic descriptions of the physical operation sequences associated with the construction, use, and abandonment of localised thermal features.

Three complementary strands of ethnographic research are examined for their potential to contribute to development of the wider research model (Chapter 5). These strands are historic documentary sources written during the colonial explorations and expansion between the fifteenth and nineteenth centuries, anthropological fieldwork undertaken in the late nineteenth and twentieth centuries, and an informant interview conducted during an earth oven technology demonstration in Texas during 2005. All three research strands offer different perspectives on the range of aboriginal fire technology from the physical characteristics and construction sequences of individual features, through the cultural specificity of construction and use practices, to activities associated with LTFs including their location of and role within the wider site setting. In parallel with this three-tiered approach, I have attempted to continuously narrow the focus from a broad perspective of evidence for aboriginal fire technology presented in colonial texts, through the technological variation present in individual features recognised by ethnographers, to the discussion of an individual feature class (an earth oven) during an informant interview.

The ethnographic context considered here is not restricted to the North American Southern High Plains study area. This decision is in part due to the paucity of technologically meaningful ethnographic information available for the region and also the applicability of the present investigation to draw technologically meaningful statements from a wider geographic context. I seek therefore, to examine the potential range of thermal-feature technology, rather than use the types of facilities documented during the sixteenth-century European *entradas* onto the Southern High Plains, as a direct analogue for those occurring in prehistory. An emphasis is, however, maintained on the North American continent and on the Southern Plains region in particular as this environment effectively conditions the physical range of possible features, by limiting the availability of fuel and other resources, and therefore the physical remains that can be expected to be encountered in the archaeological record of the present study area.

Two important themes recur across the examined foci and throughout the literature: the realisation that wide-ranging and diverse fire technologies, most of which leave little or no detectable archaeological residue, are deeply embedded in all North American native societies; and that fire technology is typically multipurpose and dynamic, so consequently archaeologists should expect LTFs to have complex construction sequences, and use histories.

Historic Documentary Sources

Between the fifteenth and eighteenth centuries, European expansionist objectives facilitated the intrusion of a variety of explorers into geographic areas already occupied by hunter-gatherer groups. Chroniclers of these expeditions were not specifically interested in documenting hunter-gatherer fire technologies; often, however, some information was noted. These records comprise a useful line of investigation in understanding hunter-gatherer use of fire at the time of contact between the two cultures.

An example that underscores the colonial perspective also serves to illustrate (in this case dramatically) the consequences of underestimating aboriginal fire technology. The example concerns the early observations of the Tasmanian hunter-gatherer groups. Erroneously, the European observers suggested that these groups had lost the technology necessary for fire-making and therefore relied on fires kindled by lightning strikes (Gott 2002:654). Accounts of fire being carried from place to place by the indigenous people were interpreted as an inability to generate fire and therefore reliance on keeping portable torches lit at all times. This assumption was clearly incorrect, relying as it did on a colonial, impressionistic interpretation of the native Tasmanian populace. The interpretation, that aboriginal Tasmanians lacked the technology to create fire, had a direct impact on their lives, for in many ways it legitimated their dispossession by 'civilized' European colonizers (Gott 2002:655). It is perhaps most surprising that this opinion persisted until late in the twentieth century.

The foregoing example illustrates the need for critical evaluation of historic documents prior to acceptance and incorporation within a framework of fire technologies for any geographic region. The documentary evidence for aboriginal fire technology as recorded during the incursions of Cabeza de Vaca (1528), Coronado (1541), De Soto (1542), and the seventeenth-century Spanish mission settlements on to the Southern Plains of North America is not extensive.

Álvar Núñez Cabeza de Vaca's unplanned and often haphazard journey through what is now southern Texas does contain the earliest evidence for the practice of native groups (identified as Iguaces or Yguaces) deliberately setting fire to large tracts of grass and lumber (cited in Stewart 2002:139; and see Hester 1999 for an archaeological context). In addition, De Vaca notes several technologies at various points along his journey, in which fire is used to: provide warmth in open camps and within domestic structures (Ch.5.); dispose of the dead (in this case medicine men) (Ch.5); cauterise wounds (again medicine men) (Ch.5); cook roots in open features stirred by sticks (an overnight task undertaken by women) (Ch.6); produce smoke at the periphery of camps in order to ward off mosquitoes (Ch.6); hunt deer by encircling them with fire (Ch.6); trap enemies in warfare (Ch.8); create intoxication through smoke inhalation (Ch.8); boil liquids to be consumed in ceremonial customs (Ch.8.); cook game (again a female gender role) by broiling in oven features (Ch.9); heat hearthstones, which then are placed in a middle-sized gourd (filled with water) with rocks being cycled into and out of the gourd to keep it boiling (Ch.10) (all translations of Cabeza de Vaca's *La Relación* [1555] in Kreiger 2002, appendix 1 and original chapter in parenthesis).

A wide range of fire technologies are therefore recorded by De Vaca and it is interesting to note that in many cases there appears to be a clear sexual division of labour in that women are much more likely to be in charge of the production and use of these thermal facilities. It is also likely that De

Vaca offers a unique insight into these technologies, as he was a participant compelled to intimately understand them for his own survival, all the while able to operate outside the accepted aboriginal social spheres for both male and female gender roles (Wade 1999).

Not long after Cabeza de Vaca had struggled along the very southern boundaries of modern Texas and Mexico, a much more deliberate incursion crossed into the Plains region from the southwest. The expedition led by Don Francisco Vázques de Coronado is thought to have travelled across the Southern High Plains, pulled eastward by the intoxicating lure of non-existent riches (Morris 1997). In contrast to the De Vaca party, this much larger expedition largely failed to document the aboriginal fire technologies they witnessed during their crossing (see English translation in Winship 1922). The reasons for the discrepancy between these two early accounts is most likely to be utilitarian. De Vaca was forced to deal with aboriginal technologies because he and his party were often literally starving, whereas Coronado with a retinue of 1,800 marchers (Morris 1997) never had to deal directly with native technologies for survival.

A recurrent theme in these early descriptions of travel on the Southern Plains, and the Llano Estacado in particular is the use of buffalo chips (manure) as a fuel source for camp fires, which was apparently a direct result of the lack of trees. Much later, the smoky *Cibolero* [New Mexican buffalo hunters] camps of the early to mid-nineteenth century (Morris 1997) used the same fuel, also the result of the largely treeless environment of the interior of the High Plains. Three hundred years earlier, Pedro de Casteñada (chronicler of the Coronado expedition) confirms a lack of fuel, providing a useful account of the physiographic setting of the Southern High Plains:

> "It has no woods except in rivers that are in some *barrancas* that are so concealed that until you are standing at the edge of them, they are not seen" (Spanish translation in Morris 1997:126).

The ethnographic use of buffalo chips as a fuel source for LTFs by historic Plains Indian groups (Cree, Kutenai, Assiniboin, Crow, Pawnee, and Apache) has been previously noted and, examined experimentally, proved to be capable of only low-temperature fires (Holland 1984). The rampant mesquite (*Prosopis*) that covers the region today (providing an obvious source of high-energy fuel) is a late development, which appears to have, at least in part, been actively suppressed by human-set fires prior to the late nineteenth-century enclosure and settlement of the region by Anglos (see examples in Stewart 2002). Fabiola Cabeza de Baca's memoirs of the Hispanic sheep-herding *Pastores* who inhabited the Southern High Plains during the late nineteenth century confirm this important environmental change:

> "There is little similarity between the Llano of today and that of last century. The Llano, then, was an endless territory of grass and desert plants, with nothing to break the monotoy except the horizon and the sky" (Cabeza de Baca 1994:3).

It is clear from the foregoing texts, that at least by the sixteenth century, fire technology had become a fundamental component of aboriginal life-ways throughout the southern portion of North America and in all probability this situation was typical of the entire continent (Driver and Massey 1957; see Brown 2000 for a similar perspective from the east coast of North America). This is perhaps unsurprising given that it is very likely that the earliest North American colonizers brought fire with them and therefore its use has considerable time depth prior to European contact (for a Paleoindian context see LaBelle 2005). Regardless of the antiquity for the first anthropogenic use of fire in North America, accounts of the earliest encounters between Native American and European explorers clearly demonstrate the complexity of indigenous fire-based technologies by the sixteenth century. These statements add important context to the LTFs observed in the archaeological record and therefore are an important caveat for any attempt to interpret thermal features in simplistic survivalist or calorific models.

Early explorers to the Southern High Plains add indirect but no less useful evidence for the availability of different fuel resources, a vital consideration in the construction of an LTF. It appears from the European evidence, particularly Casteñada's descriptions of the juniper canyonlands of the eastern escarpment, that the most abundant fuel (and water) resources were to be found at the periphery of the region. These observations have obvious implications for archaeological expectations, suggesting that longer-term, perhaps winter, campsites and consequently an increased visibility of LTFs, might be expected in these resource-rich areas.

Anthropological Fieldwork

The proliferation of early to mid-twentieth century ethnographic accounts does not correlate with the later development of ethnoarchaeological research (e.g. Gould

1978), and often appears of limited direct use to the goals of material culture research (Rathje 1978). Nevertheless, the ethnographic record highlights important contextual, variation not easily recognisable from the archaeological record. It is important to note that the Plains region at the time of European contact was one of rapid and dynamic change and the cultural groups identified occupying the south of the region all relatively recent arrivals (Figure 6.1).

Ethnographic accounts for the use of LTFs are dominated by their role in subsistence activities and of these, cooking is the most widely documented (e.g. Hodge 1912; Steward 1933; Kelly 1932; Du Bois 1940). Although subsistence-based technologies are likely to be represented by the vast majority of the features documented in the archaeological record of the Plains, the role of fire in ceremonial activities should not be underestimated, as the eclectic examples drawn together by Winifred Blackman (1916) clearly demonstrate. It should be noted that the Native American production of fire in the later half of the twentieth century was dominated by the frictional fire drill method (rotating a stick between the palms of the hand; Driver and Massey 1957) and this would probably leave no physical evidence in the resulting archaeological record.

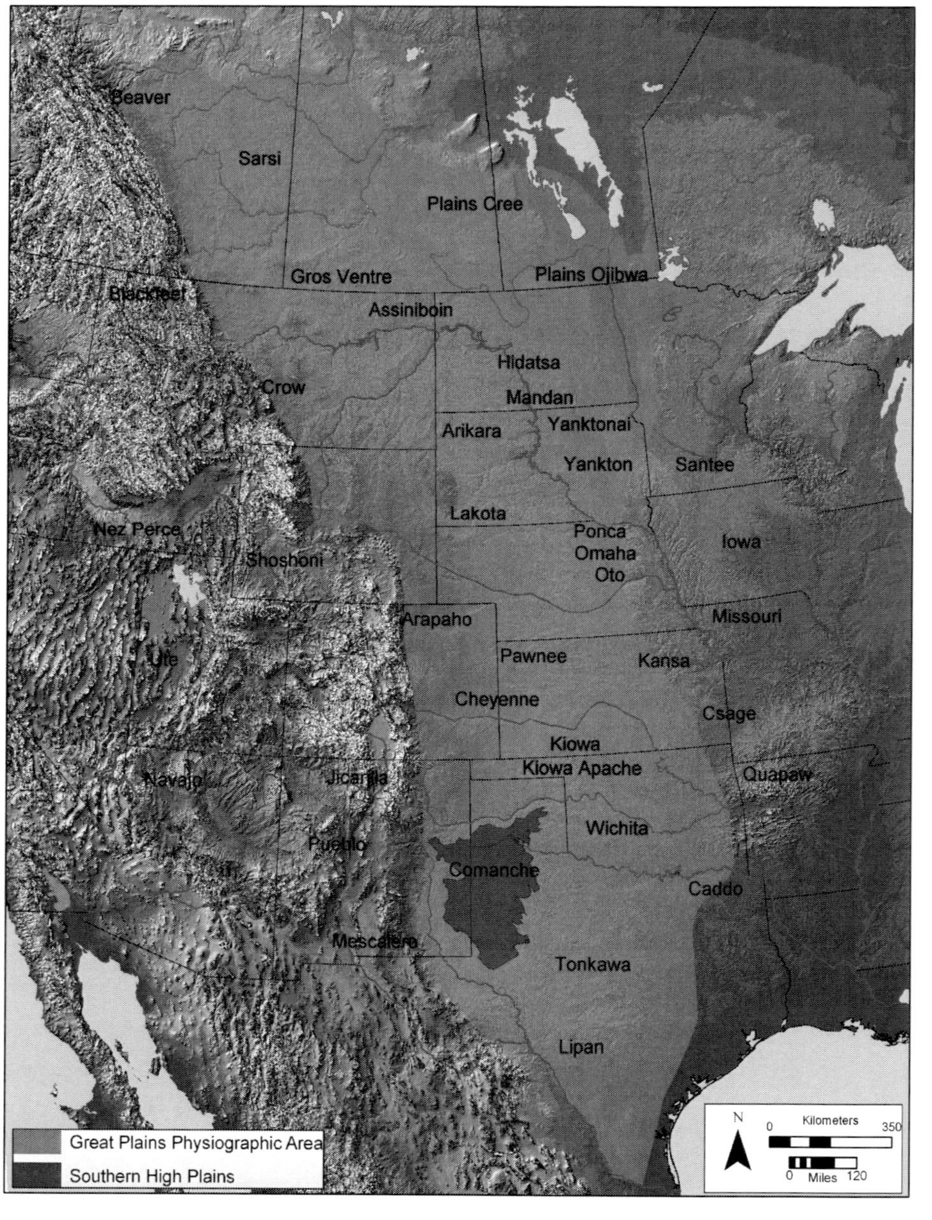

Figure 6.1. Locations of the major Native American ethnic groups on the Great Plains by the early 19th Century (modified from Carlson 1998:3).

The use of hearthstones within LTFs for 'hot-rock' technology, appears to be a late Upper Palaeolithic human adaptation (Petraglia 2002) and also has been recorded ethnographically from a diverse range of historic and modern groups (e.g. Akins 1988). The identification of hearthstones (also known as Fire-Cracked Rock [FCR] and hot-rocks) within a feature often is linked to functional interpretations involving cooking. This situation again oversimplifies the ethnographic evidence for their use. Steven Lovick identifies at least five additional uses for hearthstones from ethnographies recorded in the Northern Plains area; these include sweat lodge activity, heating of structures, the production of ceramic temper, as a means to enclose hearths, and in structure burning (Lovick 1983:41).

Domestic Facilities

Thermal features are integral to providing warmth and light to hunter-gatherer groups and therefore domestic facilities are crucial to the survival and well-being of any one group. Domestic thermal features (typically built within residences) contrast with specialised (and largely exterior) features (e.g. Binford 1987). Among the Comanche, for instance, domestic (interior) hearths were rarely used for cooking (except in cold winter months). Instead, large external features were constructed for food preparation (Rollings 1989). Domestic features appear to typically fulfil a wide range of different functions during their life cycle (e.g. heat, light, smoke, protection) and are therefore crucial technological components of any hunter-gatherer campsite. Cooking is a large component of these domestic features and the variability in this process is explored separately.

Ethnographic data examined by Galanidou (2000) suggest some correlation between cultural group and LTF technology. Her data indicate that open-basin hearths were the only type of domestic LTF used by many southern hemisphere hunter-gatherer groups (including South African, Australian, and Paupa New Guinea) (Galanidou 2000:247-248). Among Kalahari hunter-gatherer groups, the hearth emodies the social identity of the family unit (Guenther 1996:80). From the foregoing, it can be assumed that the production and control of fire often has profound and important cultural meaning. We also might expect that domestic thermal features would be represented by physically similar signatures in the resultant archaeological record. The quantity of LTFs constructed and used during the occupation of a site also appears to be another culturally dependent variable, as does the reuse of previously utilised hearths left over from earlier occupations (Galanidou 2000).

The location of domestic LTFs is often symbolically important, as was noted by Francis La Flesche in a more sedentary context among the Plains, Osage:

> "the house of the chief should be held as sacred as it represents two life-giving powers – the Earth and the Sun. The house stands for the earth and must have two doors, one opening toward the rising sun and the other toward the setting sun. The fire that is placed midway between the two doors represents the sun, whose pathway symbolizes endless life, and thus passes through the middle of the house that stands for the earth. The fireplace was also consecrated and the fire taken therefrom by the people to start their home fires was thought of as holy and as having power to give life and health to those that use it" (La Flesche 1921: 68-69).

On the Plains, the conically shaped tipi is emblematic of historic period Native dwelling structures and was first documented by the Coronado expedition in 1540-1542 (Winship 1922). Although, many other lodge, pit house, and wickiup designs were also in use, the tipi was historically used all year round by more than 15 seperate Plains groups (including the Arapaho, Cheyenne, Commanche, Kiowa, and Kiowa Apache); and during seasonal hunting excursions by the Mescaleros, Jicaillas, Nez Perces, Utes, and eastern Shoshonis (Carlson 1998:60). It should be noted that these groups are believed to represent relatively late arrivals to the region and it is largely unknown what types of structures were in use prehistorically. Nevertheless, the placement of the domestic hearth within tipi structures is necessarily towards the door rather than at the exact centre of the structure, thus allowing for a larger living area in the rear of the structure (Laubin and Laubin 1977).

The Wickiup was commonly used by some Apache groups and ethnographic evidence indicates the hearth was also located toward the centre of these structures (Melody 1989). The central location of the interior hearth in Pawnee lodges is inscribed in mythology, wherein the Council of Chiefs ordered the sun to send down a ball of fire through the centre of a lodge and in doing so burning down a tree that was growing there and marking the traditional central location for the domestic hearth within Pawnee living spaces (Hall 1997:170).

The physical form of ethnographically documented domestic LTF generally comprises a circular shallowly excavated basin, although there is some evidence that Cheyennes and Arapahoes used a rectangular type (Laubin and Laubin 1977). All groups were obviously more than capable of making a square or even triangular feature had they chosen to. The prevalence of circular morphologies is perhaps functionally conditioned by native digging technologies (commonly a digging stick, or bone digging tool such as a bison scapula or horn core) which use small rounded blades that diffuse the downward force required to excavate the hole throughout the active end of the tool. In contrast, modern metal shovels with large blades, act as force multipliers that can easily excavate vertically into sediments. Native implements are therefore more suitable to a horizontal scraping motion rather than vertical excavation and consequently much more effort is required to excavate a straight-walled feature. A psychological element may also well be present in the shape of these features as Black Elk (Lakota) states: "The power of the world is always in circles, and everything tries to be round. There is no power in a square" (Black Elk quoted in Carlson 1998:114).

Cooking Facilities

The reliance of archaeologists on the identification and correlation of LTF with cooking facilities is an old problem (e.g. Lovick 1983). An early example critical of this implicit relationship can be found in Gilmore's (1917) use of ethnographic data in a rebuttal to an article published in American Anthropologist (Sheldon 1905) identifying ancient fireplaces in the South Dakota badlands. Gilmore questioned a group of Teton elders then living on the land where the fireplaces had been observed. Their response was that the structures were not ancient fireplaces at all but rather that they were underground caches of food stored away for future use (Gilmore 1917:584). The credibility of this interpretation was reinforced by the recollection of one member, Red Hail, that he was there when the caches were first excavated over 60 years previously (Gilmore 1917:585).

Nevertheless, fire is vital for culinary processes as it significantly increases the range of foodstuffs available to hunter-gatherer groups. Cooking is essential for storage (through drying) and improves nutritional returns (dependent on cooking technique) (Stahl 1989). Ellis's recent review of 100 North American ethnographic and ethnohistoric documents summarises information relating to the types of foods exploited by various groups and the ways those foods were processed and cooked (Ellis 1997:54). Although the study focussed on hot-rock technology and a wide range of thermal-feature types, the results exhibit variability within all aspects of the dataset. At least 15 cooking techniques were recognized including roasting, grilling, smoking/drying, container frying, container boiling, baking, moist baking, roasting, broiling, searing/charring, and smoking/drying (Ellis 1997:81). In addition, the results underscore cultural variability in the ways the same foods were cooked, the types of features built, and how they were used (Ellis 1997:50-51).

Variation in the type of cooking facility constructed is clearly conditioned by the type of food selected to be cooked and the time-frame within which it is going to be consumed. Driver and Massey note that:

> "Over a thousand species of plant were eaten, and if we add to this the lists of mammals, birds, fishes, and invertebrates consumed by Indians, the total might approach 2,000 species. Now if we combine these by twos, threes, fours, and sometimes more, into food recipes, we get an enormous number of dishes (Driver and Massey" 1957:228).

Cora Du Bois (1940) observed the butchery and processing of hunted deer carcasses by *Wintu* peoples in western North America. The description highlights the intensely personal and often specialist nature of cooking, the complexity of the operation sequence, and the importance of human decision making when considering behaviour at a fine scale:

> "Meat roasted in strips on hot coals. Slices from hams pounded with a small pestle (satak), dampened with water, wrapped around a clean hot rock. Resulting bundle laid in hot coals with folded edges of mat underneath; then covered with coals. Meat, when "Dry and nice" taken from coals, a little water sprinkled on the edges to make them unfold, and rock removed. Roasted meat usually eaten with acorn soup. Men hunting in the hills might simply roast whole side or a quarter over fire. Meat often only partly cooked. This described by feminine informant with considerable disdain for lack of cultural nicety as "hunter's way of cooking" (Du Bois 1940:10).

Patterns of variability in the composition of ethnographically documented cooking facilities highlight the difficulty in detecting specific technological processes (e.g. Gifford-Gonzalez 1989; Kent 1993; Jones 1993). The variability in the data, must, however, be consistently weighed against the minimum amount of heating required to make a particular food digestible. For instance, strips of meat cook rapidly over a hot ember-bed whereas root foods are often ethnographically recorded as being cooked within the ember-bed itself (e.g. Thoms 2003). Bushfoods from arid and semiarid areas, such as the Southern High

Plains, often require intensive processing and longer cooking times to be edible (Wandsnider 1997) and may be poisonous to human consumption if they are not thoroughly cooked.

Cooking remains a well-documented function for hearthstone technology. Akins' (1988) review of ethnographic sources reveals the variability in cooking processes and activities associated with hot-rock usage and shallow-basin hearths (Akins 1988:Table 22.1). Culturally, these ethnographic accounts highlight the use of hot-rocks, and by extension LTFs, as a highly adaptable technology. For instance, Kelly's (1932) account of the Surprise Valley Paiute lists at least four separate activities (meat boiling, bone grease rendering, seed meal mixing, and carcass bag boiling) that involved adapting the respective technologies of hearth pits and hot-rocks to a given situation. For the Surprise Valley Paiute then, the types of thermal structure and cooking technique appears to be largely a matter of individual preference (Kelly 1932:97). The utilisation of hearthstones has ethnographically been largely associated with water boiling as a cooking method in pre-ceramic societies (Thoms 2003, 2008, 2009).

Boiling Technologies

Ethnographic evidence for the Plains region indicates that boiling was a common method for processing food. Two methods of boiling are recorded in this region (Figure 6.2). Firstly, direct boiling in which a liquid filled container is placed on a fire to heat. The second method, indirect boiling, involves the removal of heated hearthstones from a feature and their placement in a seperate liquid filled container (Driver and Massey 1957). Dakota Indians have used a cow stomach for this activity (Spencer *et al.* 1977). Direct boiling generally required more manufactured paraphernalia than indirect boiling and was probably not available to hunter-gatherer groups living on the Southern Plains until after ca. A.D. 400 when pottery containers slowly became available through trade with the Puebloan groups living in aggregated communities in New Mexico (Johnson and Holliday 1995, 2004). These containers were used in the direct boiling method and would have needed to be traded for and consequently carried from camp to camp. It is therefore very unlikely that any pottery would enter the archaeological record unless the vessel became broken or unusable. In contrast, containers used for indirect boiling on the Plains were more readily available and required no transport between camp locations. Examples included the paunch, hides, and thoraxes of a dead animal or specially constructed baskets. The distribution of boiling technologies presented by Driver and Massey (1957) suggests that there was some evidence for the direct method of heating, apparently using largely non-ceramic apparatus (Figure 6.3). The same authors identify the indirect hearthstone heating method as the dominant boiling technology on the Llano Estacado and throughout much of the Southern Plains region (Figure 6.4) Furthermore, the preferred container for holding the liquid to be boiled is recognised as being mainly animal based and therefore organic. It is important to note that an organic container fashioned from existing animal parts obviously implies a meat-based economy. This is not a startling realisation for considering the well documented bison economies of the contact period communities on the Southern Plains (Carlson 1998). It is however significant that an organic cooking container, as compared to an inorganic one such as a pottery vessel, must always be prepared as and when the need arises (an exception may be the use of rawhide containers that could be transported between camp locations). It therefore requires additional time, energy, and skill to post-process the animal prior to commencing cooking.

The indirect cooking method is well described by Laubin and Laubin and is worth quoting at length:

> "After One Bull [Sioux] had gathered eight or ten satisfactory stones about as big as his fist or a little larger, he cut four green poles about five feet long and tied them together into a quadripod. To this he fastened the paunch, which had been thoroughly washed. Although the paunch is like a skin bag, he handled the opening as if it were square. He thrust four small skewers through the "corners" and with thongs, tied each of these skewers to a pole. To one side he laid a big fire (crisscross), with stones on top. Water was poured in the paunch until it was about half-full, and then small pieces of meat were placed in it. When the stones were hot, they were picked up with green forked sticks, also prepared in advance, and placed in the "kettle". The very first stone added to the water brought it to a violent boil...As the boiling died down, another stone was added, and so on. It took less than half an hour to cook the little pieces of meat. The broth made a rich soup. Because the stones had been carefully selected there was no grit or sand at the bottom" (Laubin and Laubin 1977:147-148).

In regions (such as the Southern High Plains) with few trees for the construction of the suspension method, it seems a hole excavated into the ground often used instead.

(Laubin and Laubin 1977).

It is clear that the archaeological record should be favourably biased toward the identification of cooking facilities in which indirect boiling occurred. This process requires significant quantities of hearthstones, which may exhibit physical evidence for their utilisation (e.g. discoloration or angular fracture) in the indirect boiling operation sequence. In the case of both technologies, it is extremely unlikely that the container used to hold the liquid to be boiled will be preserved in the archaeological record. Possible exceptions would be the accidental breakage of a ceramic vessel during the direct method. Finally, Driver and Massey (1957) note that despite clear differences in costs, groups often use both boiling methods at different times to cook the same food. This realisation reinforces the importance of agency when choosing between competing cooking technologies.

Stone boiling technology is perhaps ideally suited to foraging groups, which require high residential mobility, and frequent moves in order to acquire food (Binford 1980). The spatial distribution presented in figure 6.4 confirms a positive correlation between stone boiling technology and the Plains region, demonstrating its utility as an expedient means with which to process bison while maintaining a minimum amount of material to be transported between camp locations.

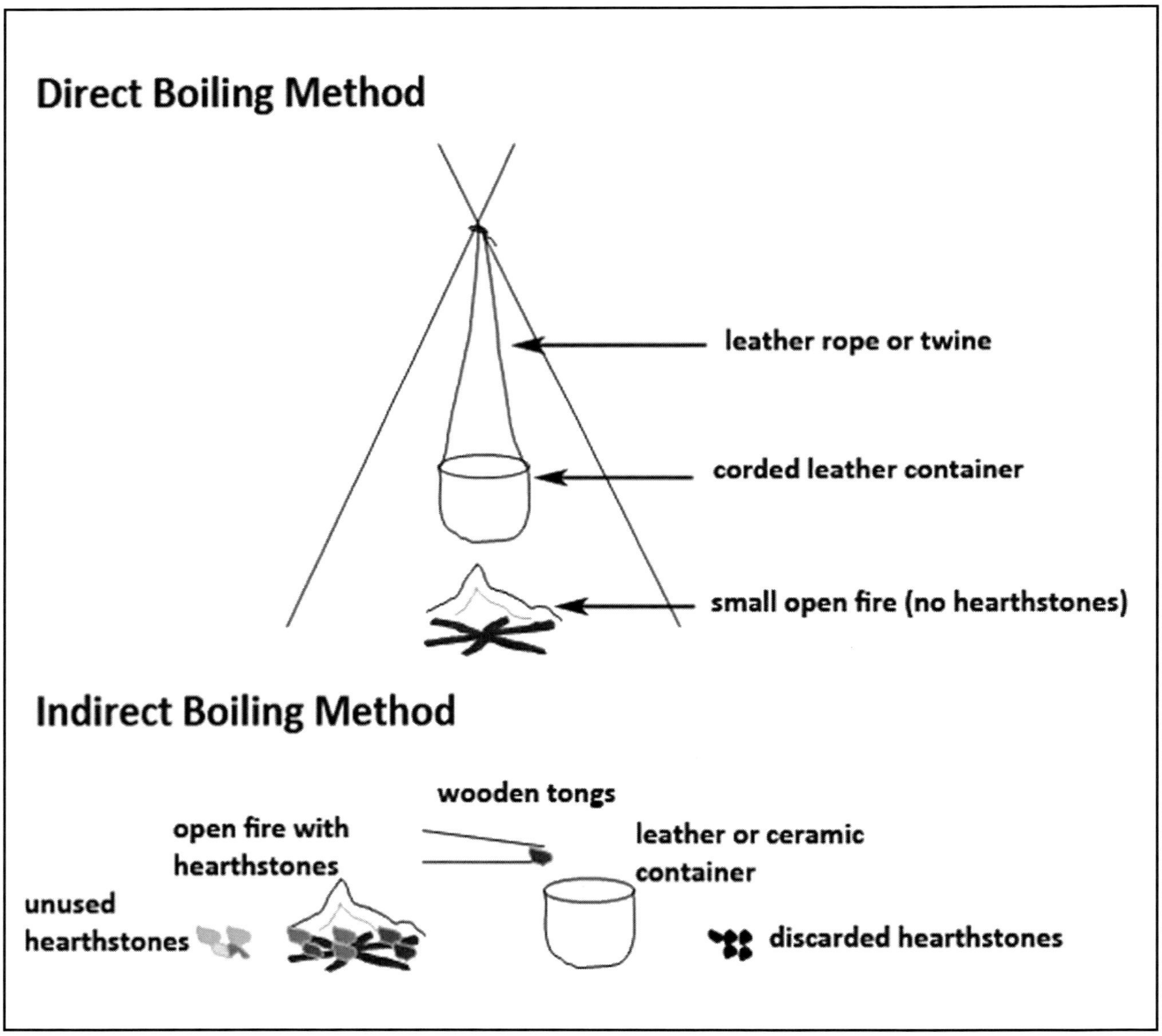

Figure 6.2. Direct and indirect container boiling methods (after Holmes 1903, fig. 2).

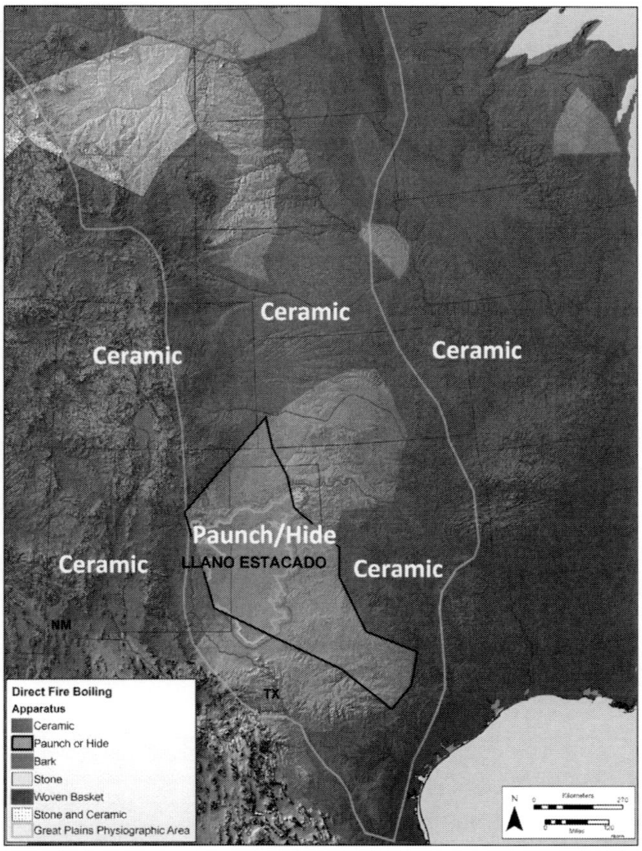

Figure 6.3. Ethnographic spatial distribution of direct fire boiling technology in relation to the Southern Plains and the Llano Estacado Research Area (data modified from Driver and Massey 1957, maps 41 and 42).

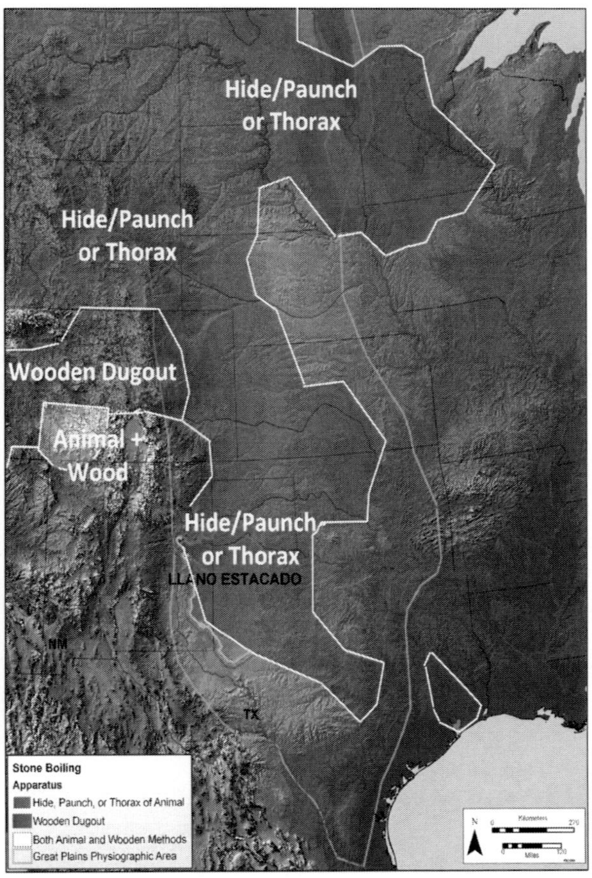

Figure 6.4. Ethnographic spatial distribution of indirect fire boiling technology in relation to the Southern Plains and the Llano Estacado Research Area (data modified from Driver and Massey 1957, maps 43 and 44).

Earth Oven Technologies

Earth oven technologies as used primarily for plant processing are not generally associated with the Southern Plains region, but are ethnographically common technologies of the peoples inhabiting the arid landscape to the southwest of the study area (Figure 6.5). Driver and Massey describe a typical feature used to cook the bud of the Agave plant:

> "After the pit was thoroughly heated with fire and hot stones, the agave was thrown in, covered with earth, and kept sufficiently hot by a fire on top. Often a number of families or a whole community cooked their agave in a single huge pit, which might be used year after year" (Driver and Massey 1957:233).

Plains groups often roasted bison meat in closed cooking facilities and Newcomb notes the Lipan Apache often roasted appendicular elements in pits, before cracking open the cooked bone for the fatty marrow contained within (Newcomb 1961:113).

The utilisation of earth oven technology clearly requires significant participation and energy input from members of the group or community in order to amass the necessary resources. Because these large facilities operate at longer time-scales (both in preparation and cook time) then significant planning depth is also required. Edward Curtis detailed early twentieth-century account of the Mescalero Apache Mescal gathering and cooking reveals a very useful technological sequence for earth oven cooking:

> "As fast as the plants are cut the women place them in burden baskets and carry them to the pit, load after load. To make it possible for each woman to identify her mescal after the cooking, each piece is branded with a distinguishing device – a property mark. The gathering of mescal continues for several days, an area covering the radius of perhaps two miles being stripped of its budding plants, for such only are harvested" (Curtis 1907:18).

He continues to describe the cooking process itself:

> "Just at daylight the old woman in charge takes her place at the rim of the pit and prays that the cooking maybe successful and that the people may be in condition to partake in the food. In igniting the fuel the old-fashioned fire sticks must be employed; to use matches would bring ill fortune. When the fuel in the pit becomes a blazing mass the women go to prepare breakfast, but are soon at work again gathering brush and grass to cover the mescal. Within four hours the fuel is entirely consumed and the red-hot stones settle to the bottom of the pit. When it is certain no fuel remain unburned, as even a small quantity would spoil the quality of the mescal, the head-woman says "it is good", and with great eagerness the followers begin to fill the pit. There is need for haste in throwing in and covering the mescal, as the steam must be confined to prevent the hot stones from scorching it. The covering consists of alternate layers of green brush, grass, dry leaves, and finally a layer of earth, about six inches in thickness. After forty-eight hours of steaming the seething mass is uncovered and each woman removes her portion" (Curtis 1907:18).

The production of a surplus is inherently necessary in order to sustain the group between harvesting and production episodes. Additionally, a strong sexual division of labour is apparent in Curtis's text. The scale of earth oven technology is perhaps best suited to the collector groups with high logistical mobility and less frequent moves to resources (Binford 1980). The spatial distribution presented in Figure 6.5 appears to largely confirm this situation, as the greatest use of earth oven technology appears to be at the resource-rich periphery of the plains and in more sedentary communities of the southwest.

Other Cooking Facilities

Grilling appears to have been one of the most important ethnographically documented methods of cooking in the Plains region. Driver and Massey (1957) attribute this to the high frequency of meat in the diet of the region. They go on to list some of the grilling methods, which include; skewers excavated into the sediment at angles around the main feature; on racks above the fire; on a preheated hearthstone removed from the fire; or lastly in the ashes of the fire itself (Massey 1957:233). The skewer method is described below and was apparently commonly used among the Kiowa and Commanche groups:

> "Sometimes fresh meat would be roasted on a green stick that was sharpened at both ends. This stick would be stuck in the ground so that when the meat was put on the other end it would be over the fire, and this would be moved several times so that the meat would be cooked thoroughly on both sides." (From an interview with Frannie Hudson Crowell, White Resident of Caddo Territory, in La Vere 1998:85).

Small mammals were often roasted without any prior processing in the ashes of the feature (Newcomb 1961). An open fire feature was also necessary in order to process and soften meat, prior to being pulverized and dried in the production of the long-lasting and durable pemmican (Newcomb 1961). Interestingly, the ethnographic data compiled by Driver and Massey indicate that practices of meat being sun or air dried dominated much of the Southern Plains and the Llano Estacado (Figure 6.6). Based only on these data, LTFs constructed for the purposes of meat drying would, for the most part, not be expected in the archaeological record of the study area.

Symbolic Facilities

LTFs are often directly utilised for ritual activity in which the physical transformations of heat, light, and smell are woven into ceremonial activities. Ethnographic accounts often record the link between the hearth and control of the production of fire. For instance, Steward's ethnography of the Owens Valley Paiute records that spiritually powerful individuals fashioned the sticks used to light the hearth, without these special people then fire would not exist (Steward 1933:276).

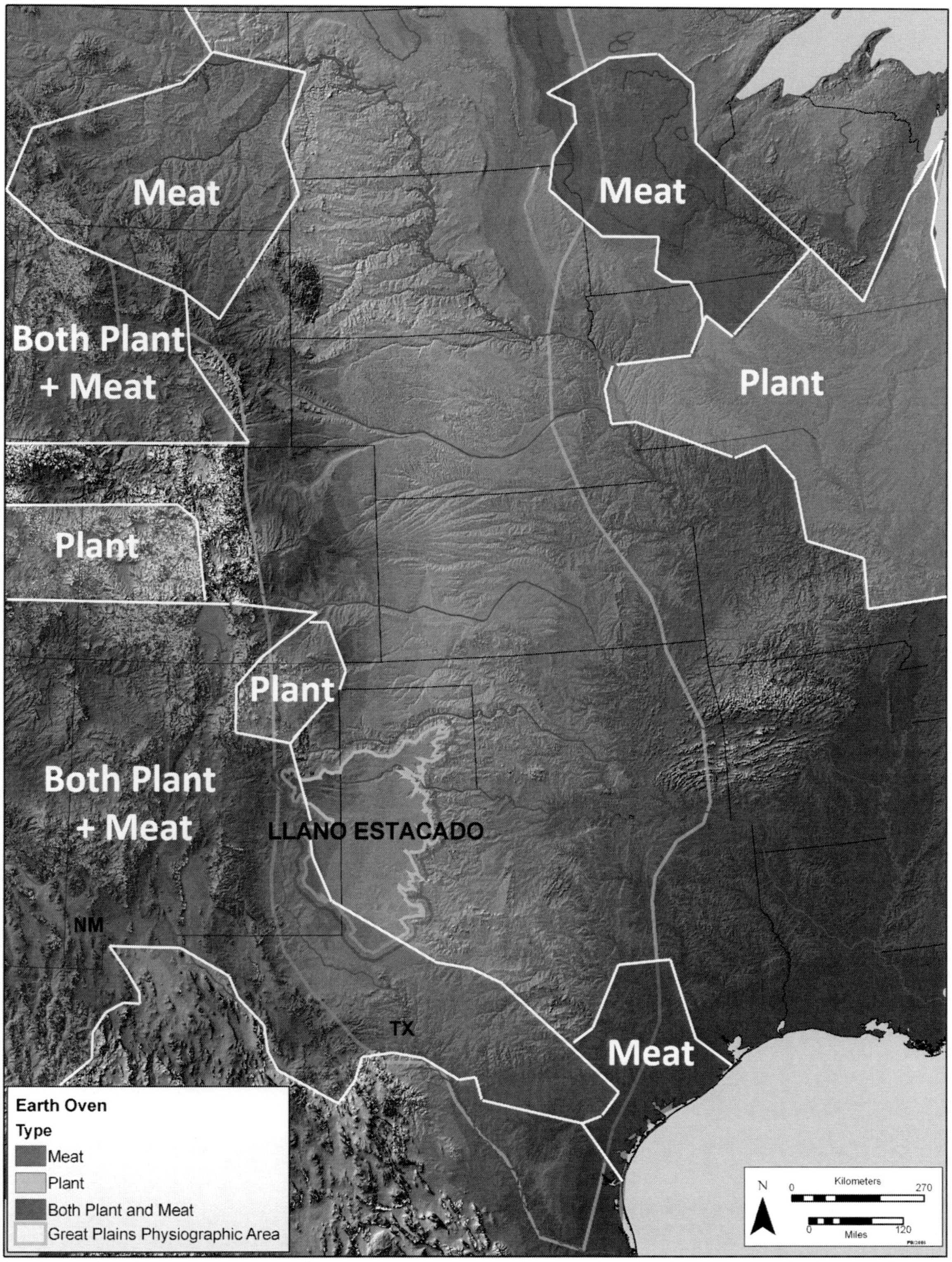

Figure 6.5. Ethnographic spatial distribution of earth oven technology in relation to the Southern Plains and the Llano Estacado Research Area (data modified from Driver and Massey 1957, map 45).

Francise La Flesche (1921) recorded some important examples among the Osage of the ceremonial usage of fireplaces, which extended to the actual naming and linguistic characterisation of individual family groups:

> "The three groups of seven gentes each are spoken of as: The Wa-zha´-zhe, who possess seven fireplaces; The Ho^n´-ga, who possess seven fireplaces; The Tsi´-zhu, who possess seven fireplaces. All of these 21 fireplaces are war fireplaces, for the people of these three groups were organized as military bodies for defensive purposes…these war fireplaces are kept separate when speaking of the gentile order for the purpose of commemorating certain portions of the story of the tribe" (La Flesche1921:53-54).

Particular LTF often have a documented spiritual effect on the food being cooked on it. For example, Klamath peoples of the North American northwest coast follow a strict procedure for cooking sucker fish (*Catostomus rimiculus*). The first fish is roasted and allowed to disintegrate to ashes within the hearth. Subsequent fishes must all be roasted in this facility for the fear that otherwise no more will come. If the rite is observed, "suckers will be plentiful" (Spier, 1930:149).

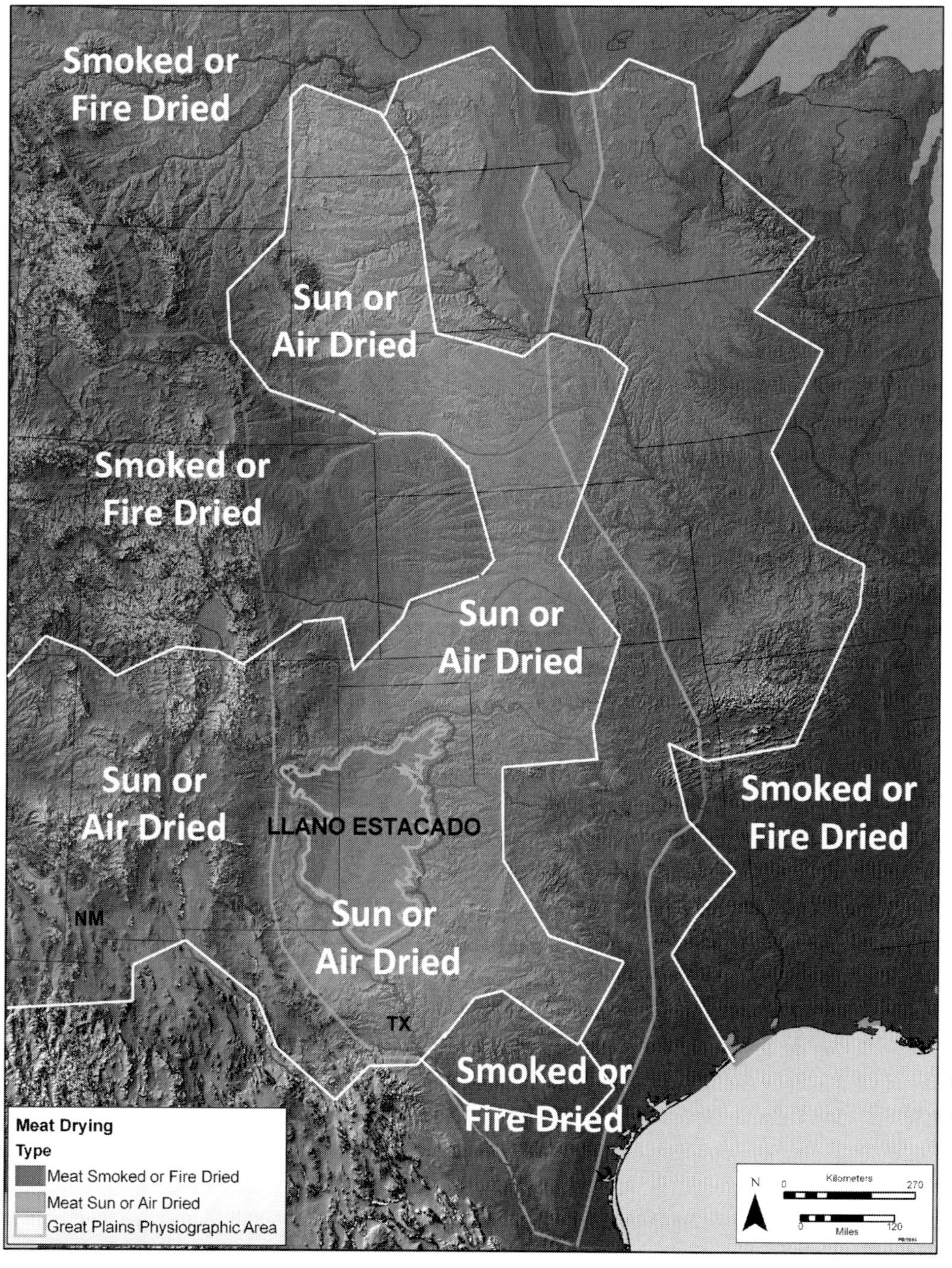

Figure 6.6. Ethnographic spatial distribution of meat drying technology in relation to the Southern Plains and the Llano Estacado Research Area (data from Driver and Massey 1957, map 53) .

Native American sweat lodges are another example (Barrett 1909 in Ellis 1997:49; Akins 1988). A small thermal feature is often built in the center of a temporary structure. The feature is used to heat hearthstones onto which herbs can be added in order to generate fumes (Ellis 1997). The need for continuous hearthstone cycling meant that these structures often generated large quantities of discarded hot-rock waste. Native American medicine lodges are another similar example in which extremely specialised procedures, often blending medicinal expertise and shamanistic ritual, centred on the hearth (e.g. Opler 1941). Hearthstones used in LTFs are referenced among the North American Hupa groups for whom there are many "venerated stones" that result from different ritual activities. These special stones are often only used (heated) once (Goddard 1903-1904:80-81 in Ellis 1997). The purpose of these thermal feature-oriented ritual functions would obviously be extremely difficult to detect in the archaeological record.

Plains Examples

Water vapour sweat lodges were typical of the contact period Plains tribes (Driver and Massey 1957) and are characteristically small domed structures with round ground plans. These temporary buildings were constructed for the use of single individuals. The function of the structures involved the heating of hearthstones in one or more features that could be internal or external to the sweat lodge. Once heated, water was poured onto the rocks, causing steam (Driver and Massey 1957: 314).

The use of LTFs in ritual performance and medicine is well documented on the Plains. The Cheyenne fire dance required that the performers danced over hot coals until they had been cooled (Hodge 1912). Also among the Cheyenne, Edward Curtis described a medicine ritual in which blood was sucked from a sick infant and spat into a nearby fire (Gidley 2001). A central fire was an important element of the Wachita Deer Dance, or Dance of the Medicine Men, in which the fire was centeral to performance elements that lasted for a four day period (Curtis 1930:70). Again, these uses of LTFs would be extremely hard to identify, although a sweat lodge may leave a very distinct archaeological signature.

Other LTF Technologies

From the preceding discussion the conclusion might be drawn that LTFs are ethnographically only associated with cooking, ceremony, or both. This is far from the case and a wide range of alternate uses have also been documented. The ability to dye organic material by (heating material/ in water with roots) is one such example (Martin 1793). Although overplayed in the popular media, communication could be achieved by means of smoke signals or light beacons (Hodge 1912; Swanton 1928). Evening campfires were often used as a teaching aid by Plains groups who used them to gather children together in order to receive instruction (Carlson 1998). Other important but under-researched activities that involved the construction of an LTF include the production of light and heat, protection, insect collection, tree felling, lithic manufacturing, and the production of salt.

The foregoing ethnographic data clearly indicate that Native American culture and fire technology are deeply intertwined with one another. Fire creation myths are common among individual groups and a good example is the crafty fox of the Jicarilla Apache fire myth who, is punished for stealing fire. The fox is forced to transport fire but not allowed to use it for himself (Russell 1898:262). It is clear that late nineteenth and early twentieth-century ethnographic records are a rich source for examining the technological variation in the construction and use of LTFs and in addition add a vital humanistic element to the enquiry that is too often painfully missing in the archaeological record.

Informant Interview

An informant interview was conducted with Ray Olachia (hereafter referred to as Ray) during an educational demonstration at the Lubbock Lake Landmark. The demonstration involved (ca. 15) local high school teachers in the construction of a caliche-lined earth-oven with the purpose of cooking some native (prickly pear) and some non-native foods (beef brisket). The results are interpreted with the caveat that they are based on the knowledge, attitudes, and opinions of one person who himself has not needed to subsist on traditional technologies such as earth-oven cooking since early childhood. The interview is clearly therefore not rigorous in presenting Apachean or Yaki earth-oven technology. The technological construction and practices described by Ray do however have potential to be observed in the archaeological record and offer useful insights as to the construction stages (and their resultant archaeological signatures) involved in earth-oven technologies. Viewed from a heuristic perspective, the interview offers useful insights as to the particularity of cultural practice as manifest in native fire technologies, and thereby directly challenges the Euro-American assumptions that often implicitly underlie archaeological interpretation of these technologies.

A useful example occurred early on in the demonstration when Ray asked the teachers to collect rocks with which to line the base of the excavated oven pit. Although a large pile of previously burned caliche rocks were within easy distance (less than 5 meters) for collection by the participants, Ray instructed them to instead obtain unburned rocks from farther away (greater than 50m and dismantling a wall in the process). When questioned about this practice, Ray stated that when the rock is burnt it is 'used up' and 'won't work'. Caliche hearthstones are fairly easily identified as burned or not burned by their characteristic transformation in colour from white to blue-grey hues (Lintz 1989, Backhouse *et al.* 2005). There is, however, very little apparent structural integrity loss between heating episodes and this would appear to indicate that selection of 'fresh' hearthstones is largely a culturally motivated practice. Interestingly and in contrast to the practice observed with the hearthstones, Ray had no preference as to the type of wood used for fuel.

Once the small pit was lined with 'fresh' caliche hearthstones, Ray instructed fuel to be added and a fire lit on top of the hearthstones. Although the sexual division in the Euro-American participants was roughly equal, five male teachers stepped forward from the group to undertake the fire-lighting task (Figure 6.7). Of the five, two of the older men quickly took charge, dominating the activities of this task group. Females and the younger male participants either helped by collecting the tinder or were observers in the fire-lighting process. The gender roles taken on by the Euro-American participants are in direct contrast to Ray's suggestion that building fire was 'women's work' (which also correlates with the vast majority of the foregoing ethnographic research). The foregoing gender roles of the participants were observed to be reversed later in the activity in the preparation of food for placement within the oven. During this activity, the women and their younger male colleagues worked together, the older men maintaining close observation on the completion of this task (Figure 6.8). The example highlights the potential complexity in age and gender relations associated with fire technology when it is situated within any cultural context. Furthermore, none of the observed complexity would have left any physical evidence in the archaeological record.

Figure 6.7. Men of European American descent dominate the race to light an educational earth oven , an event in direct contrast to Ray Olachia's stated, Mescalero Apache sexual task divisions (Note: the peripheral positions of European American females in this activity). (P.N. Backhouse)

Figure 6.8. Women of European American descent dominate the preparation of food for placement in the earth oven (Note: the peripheral positions of the older European American men in this activity). (P.N. Backhouse)

Although Ray remembers using earth ovens to cook various meats such as (cow and pig) he suggests that their traditional usage is for plant processing and in particular for the preparation of Agave (*Agave lechuguilla*) and Sotol (*Dasylirion wheeleri*) for consumption. Agave is an indicator species of the Chihuahuan Desert. The modern distribution extends to the southern and western periphery of the Southern High Plains study area and has been documented in Crane County, Texas and Eddy County, New Mexico (source USDA, 2006). The range of Sotol is found farther north in New Mexico and extends along the western boundary of the Southern High Plains study area with communities documented in Chaves, DeBaca, and Curry Counties (source USDA, 2006). No communities of Sotol have been documented in the Texas portion of the study area. Both species are poisonous if consumed raw and must be cooked for considerable amounts of time to render them edible. (Ray states 36 hours as a typical cooking time). Because these species potentially occur within the current study area, earth-oven technology, such as that described by Ray, has potential to be observed in the regional archaeological record. Furthermore, the size of the features described by Ray (over 2m in diameter) are hearthstone intensive, especially if new rocks are required for each new use.

The basic construction sequence of an earth oven for processing vegetal material, described during the interview and constructed by the demonstration participants, exhibits a distinct archaeological signature that may help to differentiate it from other types of LTF technology when observed in the archaeological record. The knowledge that these features were constructed to process foodstuffs over long periods of time implies significant investment in construction and at least a 2-day stay at a particular camp location. The technological sequence, in the case of the demonstration, took 2 hours before the second fire could be ignited and the food left to cook. The pit was mechanically excavated prior to the demonstration and is not included in this time estimate. The technological sequence described by Ray is very similar to the one already recorded by Curtis among the Mescalero Apache some 100 years previously (Curtis 1907):-

1) excavate a large basin-shaped hole ca. 2m in diameter;

2) line the interior of the hole with cobble-sized hearthstones (in this case caliche);

3) build a fire on top of the hearthstone lining;

4) wait for the fire to die down and an ember bed to develop;

5) add food to be cooked on top of the ember bed;

6) cover food with more cobble-sized hearthstones (again caliche was used);

7) build a second fire on top of the new layer of hearthstones;

8) leave the whole facility until the food is considered to be cooked.

9) excavate and consume food.

Archaeologically, the signature generated by this technological construction sequence should be large scatters of hearthstones and burnt sediments around a central feature. Because the lower hearthstone layer is not required to be removed to retrieve the processed food, it is more likely that this will remain in place following the final use of the feature. Obviously, if the cultural practice described by Ray of using hearthstones only once is being followed, then features would need to be completely emptied before any additional uses could occur. This process would also generate large numbers of hearthstones.

Conclusion

The ethnographic evidence presented here underlines the realisation that LTFs were constructed by Native American groups for a wide variety of purposes, not all of which would leave any significant trace in the archaeological record. Furthermore, LTFs are but a small component of a wider fire technology which was itself extremely sophisticated by the time of European contact. Variability in the data appear to suggest that cultural/behavioural attributes are locally the prime factor in determining what types of thermal feature/s were constructed, the activities for which the feature/s were used, and the appropriate use of space around the feature at different times in its life-history. The high degree of cultural variation identified in the ethnographic sources is a positive outcome of this review and highlights the cultural as well as economic importance of the LTF for hunter-gatherer groups.

In actuality the type of facility was often conditioned by the local environment and the availability of necessary resources. This predictable relationship is particularly true of features that need to be consistently constructed, such as those subsistence-based facilities used to process various foodstuffs, often on a daily basis. Because the ethnographic data suggest that the technological components of these facilities are to some extent environmentally conditioned, for instance, the fuel type and food processed, then distinct and culturally meaningful archaeological signatures should be expected to be encountered when features are examined at a wider regional focus. Ultimately, the immense range and complexity of physical features and processes identified in the ethnographic literature reinforces the need for experimentally based investigation.

CHAPTER 7 - EXPERIMENTAL RESEARCH

Abstract

The third approach in the study of LTFs is the application of experimental research to target questions generated during the classification of the regional dataset for the Southern High Plains. In total, four separate experiments were carried out to address specific research hypotheses. The identification of hearthstone technology at more than three-quarters of the sites in the regional dataset underscored the need to prioritise the investigation of this technology. Two experiments assessed the taphonomic changes and performance characteristics of various local rock sources that have been commonly recovered and interpreted as hearthstones. The discovery that regionally the majority of LTF sites were encountered at the modern ground surface, initiated experiments that attempted to gauge the role of site-formation processes on the structural integrity and displacement of individual assemblages. Preliminary experimentation was also undertaken in order to examine the functional dynamics of a commonly encountered feature type, a shallow basin hearth filled with caliche hearthstones. The results added much-needed quantitative datasets useful in explaining the physical performance characteristics and potential for such features to survive in the archaeological record. The overall outcome of the programme strongly suggested that experimentation could be positively applied in understanding hunter-gatherer domestic fire technology at both site and regional levels. Additionally, new models and hypothesis generated by this programme of research feed back into the wider interpretative project.

> "an archaeology of fire…would insist on the importance of the role of experiment in analyzing the phenomenon; therefore the study of the complex control of fire in prehistoric societies could be inferred through replication of the design (i.e., proportions, scale, functioning, materials) of the pyroobjects found in archaeological remains. Experimentation is a process that involves, besides the ethical approach of the observer-archaeologist, an emic approach of the participant-archaeologist too, whose accounts should be united in one single narrative." (Gheorghiu 2002:93).

A programme of experimental research was initiated in order to actively examine the physical evidence for the types of LTF recorded most frequently on the Southern High Plains. In total, four formal experiments were designed and executed. Each experiment was designed to answer a specific research question developed from the wider perspective drawn from the regional dataset. The results of this modular approach are reported in full in the resultant publications (Backhouse 2004; Backhouse *et al.* 2005b; Backhouse and Johnson 2007a,c).

The chronologic development of the experimental programme mirrored the development of the wider programme of research. Individual experiments were devised and carried out throughout the programme's data-collection phase (roughly 2002-2006). The sequence of experimentation can be summarised as:

- Experiment 1. Measurement of the performance characteristics of the geologic rock-type caliche, in order to make inferences as to its usefulness for hunter-gatherer hot-rock technologies (Backhouse *et al.* 2005b);
- Experiment 2. Measurement of the effects of site-formation processes acting on LTFs and surrounding artefact scatters during exposure to environmental conditions for the purposes of developing a model that can more accurately predict the relative site condition for open-air hunter-gatherer sites (Backhouse and Johnson 2007a);
- Experiment 3. Measurement of the physical outputs of a simple basin-type LTF qualified in terms of heat, light, and smoke in order to determine if non-economic humanistic lines of questioning can be developed by the experimental method (Backhouse 2004);
- Experiment 4. Measurement of the performance characteristics of the geologic rock types found in the Ogallala Formation, in order to make inferences as to their usefulness for hunter-gatherer hot-rock technologies and as an aid for their identification during fieldwork (Backhouse and Johnson 2007c).

Prior to the current investigations, very little experimental research has focused on these processes on the Southern High Plains (an exception is Lintz 1989). A brief examination of experimental archaeology and its relationship to LTF research provides the background and situates the context for the current programme of research.

Background

Experimentation in order to examine archaeological questions has a considerable history spanning some 150 years of practice (Coles 1973). The epistemological basis for experimental research, however, only began to coalesce in the later half of the twentieth century (e.g. Ascher 1961; Tringham 1978) and Ascher (1961) formally set out the objectives of experimental archaeology:

> "…each imitative experiment is an attempt to test a belief about cultural behaviour, relying implicitly on the first proposition: all cultural behaviour is patterned. The statement of the hypothesis describing the particular pattern involves artefact classes and has implicit within its second proposition: artefacts produced from the same scheme, or used according to the same scheme, exhibit similarities which permit their division into groups which reflect those schemes. Taken together, the two propositions form the implicit broad working hypothesis of the imitative experiment" (Ascher 1961:806-807).

Succeeding processual applications increasingly emphasised empirical data and hypothesis-driven research as a component of middle-range theory building (Trigger 1989). Experimental or actualistic studies conducted from this perspective increasingly have become oriented toward the measurement and analysis of by-products, which include both economic and technological variants (e.g. Tringham *et al.* 1974; Tringham 1978; Bleed and Meier 1980). The primary criticism of this approach has centered on the difficulties involved in directly associating modern experiences of the archaeologists conducting the experiment with those of the people being studied (Trigger 1989; Thomas 2000).

Despite this criticism, experimental archaeology appears to be a fundamental, rather than paradigmatic, component of the contemporary archaeologist's interpretative toolkit. Archaeological research continues to utilise experimental projects in largely positivist frameworks, although inferences derived from this work often are guarded or self limiting in terms of their interpretative potential (e.g. Schiffer *et al.* 1994). The primary objectives of the experimental research undertaken as part of the current project fit within the normative model of hypothesis testing.

In addition to the strong tradition of 'normal science' mode of experimentation, a small body of more radical approaches have highlighted a humanistic or pedagogic perspective. An excellent example is Stuart Townend's recent examination of the reconstruction of a later prehistoric roundhouse in Britain:

> "building is considered to be primarily and for the most part about interpretation and negotiation of relations between people and things; the reconstruction or 'original' construction is therefore, while not incidental, a secondary phenomenon. It also means that experimental archaeology is not an inherently scientific exercise and does not consist primarily in neutral, de-personalised deduction" (Townend 2002:73).

Using the above example as a blueprint for an alternative experimental approach, a secondary humanistic objective was deliberately factored into the current research programme by conducting all the experiments in an actualistic setting i.e. non laboratory. The rationale for this decision was to facilitate a continuing dialogue between the author and other researchers, and members of the public that encountered the experiments.

The broader historical development of experimental studies closely correlates to the development of experimental thermal-feature research. These studies are characteristically empirical and partitive, focussed on emphasising economic value, couched in terms of energy, calories, efficiency, and return rates. Within this broadly economic schema, a range of processes have been systematically examined. These studies include the assessment of the functionality of different LTF morphologies in terms of internal geometry (Shockey 1997; Score and Mithen 2000), the experimental management of different fuel types (Théry-Parisot 2002), the application of hot-rocks to boiling food (Williams 1990), and to cooking strategies (Wandsnider 1997; Dering 1999; Speth 2000; Smith et al. 2001; Roberts et al. 2002; Church and Lyman 2003; Pagoulatos 2006). Common to all these approaches is a research agenda that seeks to investigate a specific output or primary product of the LTFs quantifiable in terms of energy. Economic models are particularly dominant in the North American literature; a useful example being Smith et al.'s experimental roasting of Sego Lilies in earth-oven features:

> "The 62h of effort produced only 12,830 calories a fairly low return. At 62h of collection to fill the oven it would take four people 15.5h or approximately 2 days to obtain enough bulbs. The production from one full oven would supply only 6.4 person days of food, assuming 2000 kcal per day for daily caloric requirements" (Smith *et al.* 2001:174).

The measurable output of these experiments facilitates their potential to wider quantitative approaches, such as behavioural ecology with its emphasis on time, energy, and reproduction (e.g. Smith and Winterhalder 1992;

Kelly 1995; Winterhalder 2001; Bird and O'Connell 2006). The current programme of research differs from the preceding examples because the focus of experimentation is the mechanism of fire technology itself i.e. the LTF rather than its by-products (e.g. fire-processed foods). In contrast, the design of experiments 1, 3, and 4 (Backhouse 2004; Backhouse and Johnson 2007a,c) are all concerned with measuring the direct performance characteristics of components of the feature (i.e. the performance of particular geologic rock types as hearthstones in experiments 1 and 4 (Backhouse *et al.* 2005b; Backhouse and Johnson 2007c)), or the feature itself (Backhouse 2004).

Technological approaches to thermal-feature experimentation are uncommon in the archaeological literature. A recent study by Joris Sergant and colleagues is a useful example of the application of an alternative methodological perspective (Sergant *et al.* 2006). This research investigated the hypothesis that surface hearths were a commonly constructed, but minimally archaeologically visible, component of Mesolithic sites on the northwest European plain. Research was undertaken in order to examine the active role of hearths in site-formation processes as demonstrated by the distributions of burnt artefacts in relation to an experimental feature. Experimentation completed during this research was largely heuristic, by building various surface fires and examining the dynamics between the feature and simulated material culture. For instance, the researchers introduced various lithic materials into a hearth and plotted the resultant distribution of material, as individual rocks were super-heated and ejected from the feature (Sergant *et al.* 2006:1001). The results indicate that goal-oriented experiments and detailed spatial analysis can be effectively combined to generate models for accurately locating hearths based only on the spatial distribution of burned artefacts. This case study clearly illustrates the inductive (or heuristic) use of experimentation as a powerful research tool, with the potential to be used in parallel with more traditional problem-orientated deductive approaches. The design and implementation of Experiment 2 (Backhouse and Johnson 2007a) follows a similar inductive position as the preceding example, but applies a more rigorous spatial methodology to understanding the site formation processes affecting hunter-gatherer sites.

Much of the previously stated experimental research is essentially atomistic, tending to examine an aspect of hunter-gatherer fire technology in relative isolation. Integrated and complex examinations of domestic fire technology combining ethnographic and archaeological evidence, such as that attempted by the current research programme, have only rarely been undertaken. An example of an integrated approach is LuAnn Wandsnider's (1997) pioneering research on hunter-gatherer cooking systems and underscoring the potential of situating the experimental examination of LTFs within a wider contextual network of research. The inferences gained from this perspective can be extended to answer questions beyond the simple testing of material properties. Despite this example, holistic programmes of experimental research have arguably yet to be fully undertaken by archaeologists.

Experimental Objectives in Light of the Regional Dataset

Examination of the archaeological record compiled for the Southern High Plains study area indicates that hearthstone technology was recognised at over 75% of sites with evidence for domestic fire technology (Figure 7.1). Unfortunately, the majority (n=450) of the sites with identified hearthstone clusters have no temporal association (Figure 7.2). Of those sites that have age estimates, there appears to be an increase in hot-rock use during the Archaic through Ceramic/Mogollon cultural periods (i.e. mid through late Holocene). This pattern is, however, most likely to be the result of the high sample sizes for these two periods in comparison with Paleoindian and Protohistoric contexts (Figure 7.2).

Geographically, the identified sites are largely located at the periphery of the region (Figure 7.3) with significant clusters of sites with evidence for hearthstone technology located in the southwest of the area along the Mescalero Dunes, along the rough canyonlands off of the eastern escarpment, and along the Canadian river valley. In the interior of the region, sites with evidence of hearthstone technology largely follow the incised draw systems. This pattern is particularly noticeable along Midland Draw and Mustang Draw in the south; Yellowhouse Draw and Runningwater Draw in the east, and Tierra Blanca Creek in the north of the region (Figure 7.3). Based on the frequency of sites identified in the archaeological record, the age of these sites, and their geographic distribution, hearthstone technology appears to be significant.

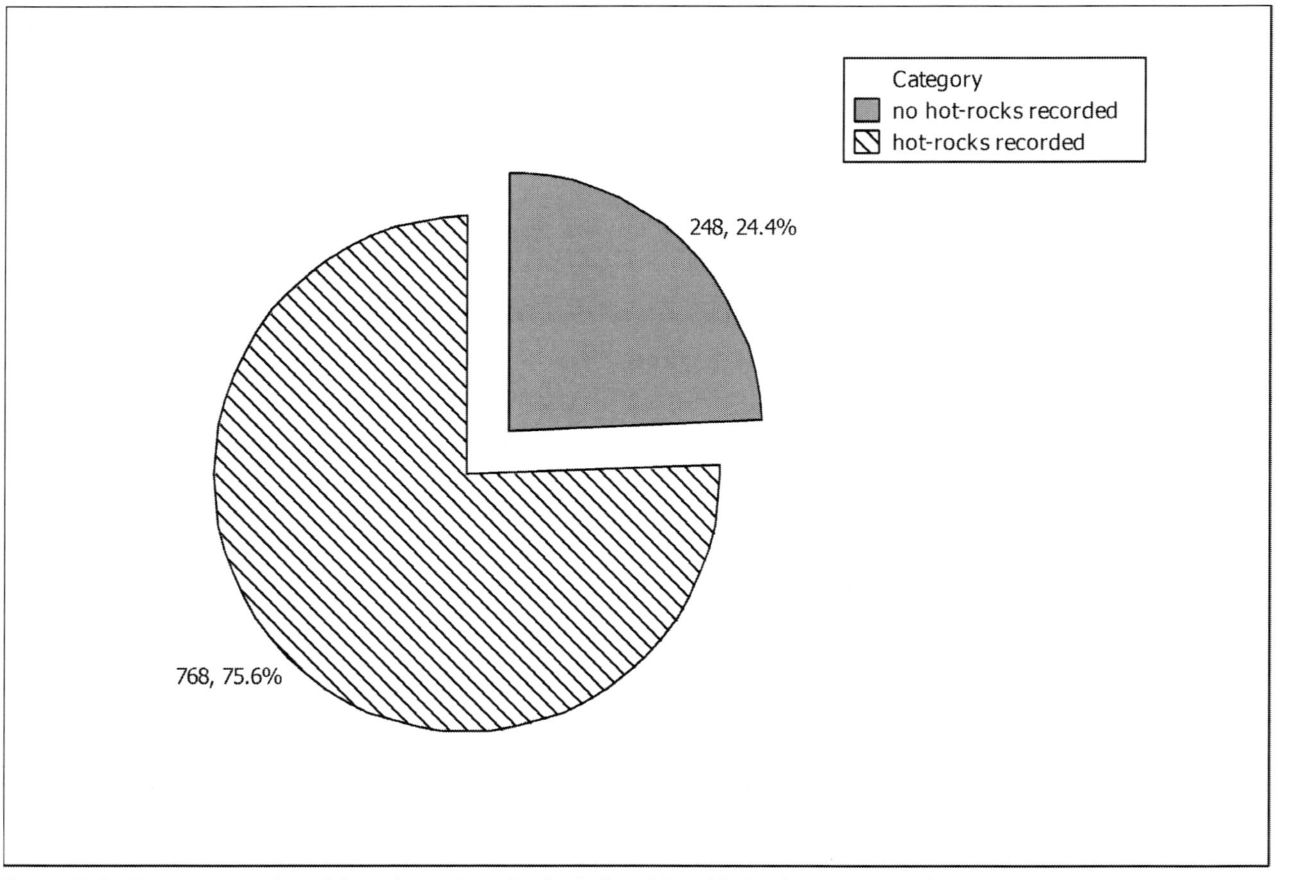

Figure 7.1. Frequency of prehistoric archaeological sites identified with evidence for hot-rock use.

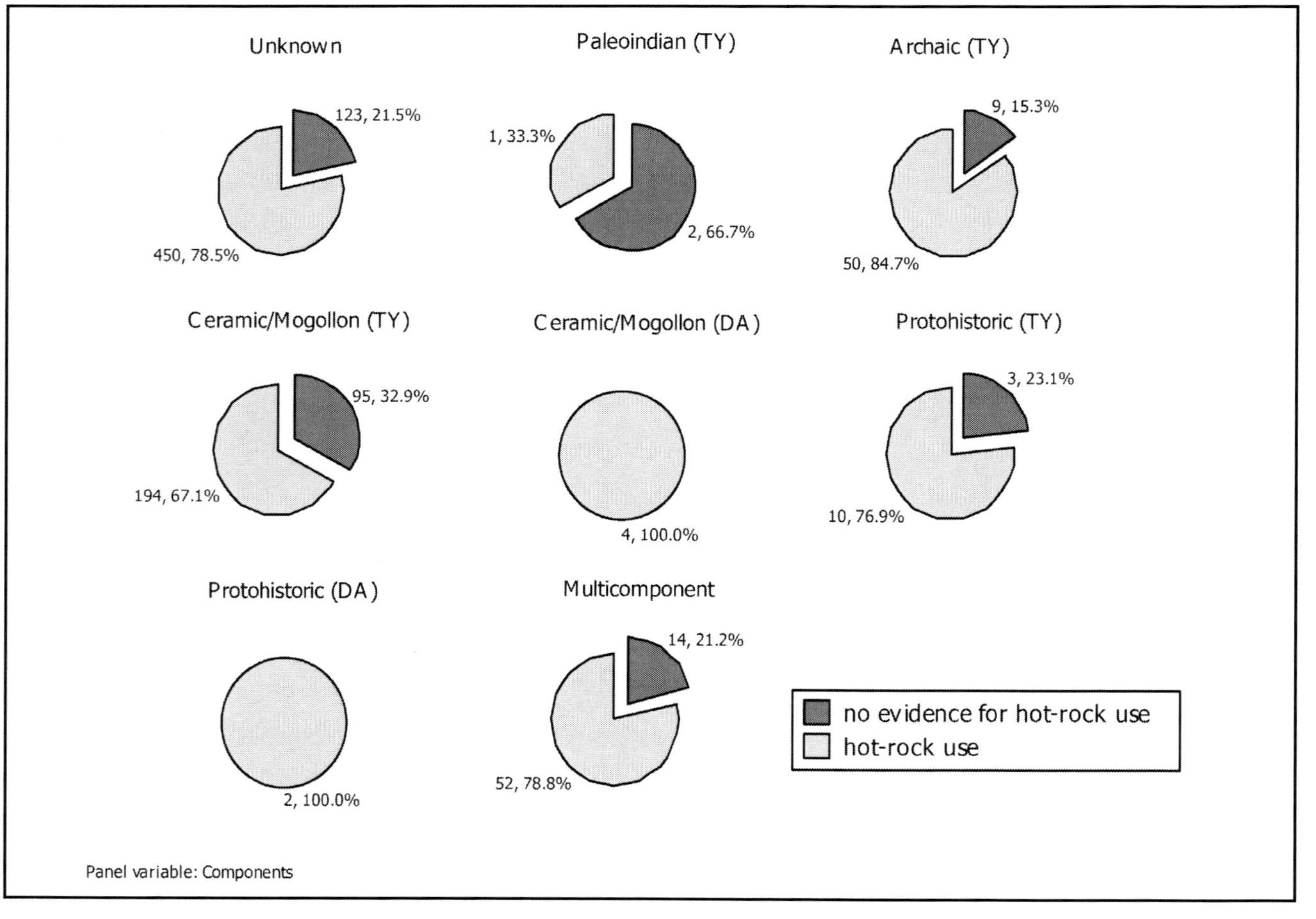

Figure 7.2. Temporal distribution of evidence for hot-rock use at prehistoric sites with LTF components within the study area. Note (TY) = age of site estimated from typological analysis of artefacts, (DA) = age of site secured from radiometric assay e.g., radiocarbon dated.

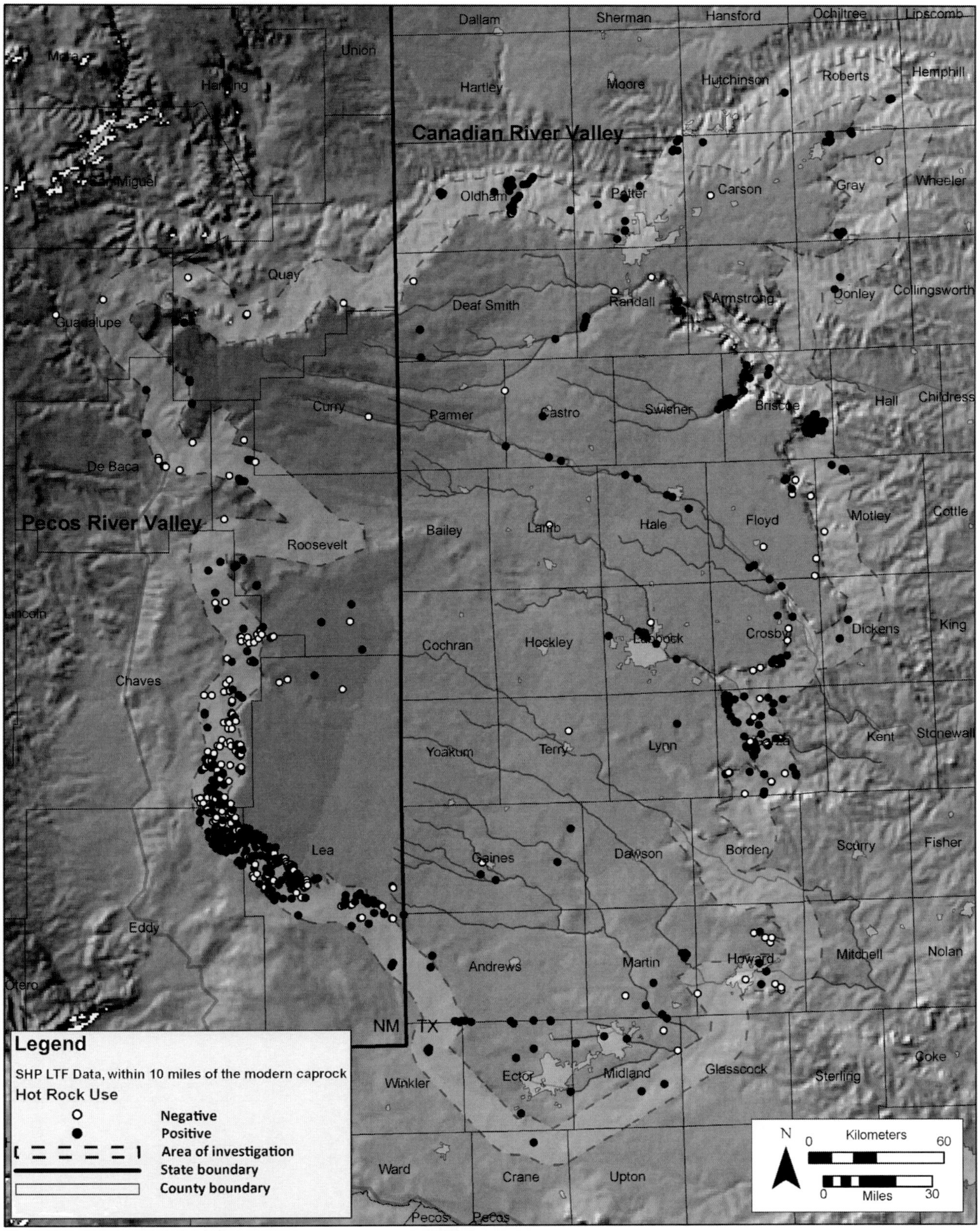

Figure 7.3. Geographic distribution of prehistoric archaeological sites within the LTF database with evidence for hot-rock use.

Examination of the application of the hierarchical key classification scheme (see Chapter 5) to the LTF dataset indicates that at the majority of sites on the Southern High Plains, insufficient information exists to make even a basic identification of the primary type of LTF facility (Figure 7.4). Nevertheless, at sites with identifiable LTF components, the vast majority consist of hot-rock clusters (n= 348). These clusters are typically exposed on the modern ground surface (e.g. Figure 2.6) and are therefore often encountered during pedestrian survey research methodologies. The original structure of these features is not readily apparent from the archaeological evidence and it is reasonable to posit that site-formation processes have often actively removed any structural evidence for thermal features. This would result in a regional record of LTFs dominated by the identification of sites that resulted from hearthstone technologies.

Hearthstones are the most ubiquitous artefact type on the Southern High Plains. Experimentation carried out to further examine the dynamics of hearthstones is therefore highly appropriate in this geographical setting. An abundance of hearthstones is not uncommon in other North American contexts and has led to some useful, although largely piecemeal, experimentation to be completed. A good early example of such research was undertaken by Richard Zurel in the Great Lakes area (Zurel 1979). The research highlights three clear benefits of a focus on hearthstones: firstly, the quantity of the arefact type available and therefore a potential for large datasets; secondly, hearthstones are an excellent additional line of evidence with which to test interpretative statements generated from more traditional research enquiries (ceramics and lithics); and thirdly, hearthstones are unlikely to have been previously collected or otherwise disturbed by collectors or agricultural activity (Zurel, 1979:1). Nearly thirty years later, this rationale for the intensive investigation of what were once hearthstones is still appropriate.

Experimental Research Goals

The regional database identified hearthstone technology, site formation processes, and the basic physical mechanics of a typically encountered feature type as regionally significant phenomena that might be usefully explored from an experimental perspective. Three broad goals of the programme of experiments therefore were:

1. to empirically examine hearthstone (or hot-rock) technologies in terms of both performance characteristics and materials analysis;

2. to examine the extent to which site-formation processes affect the regional record, both in terms of site-scale and regional-scale patterning; and

3. to examine the performance characteristics of localised thermal-feature technology through the study of inputs, outputs, and materials-based research.

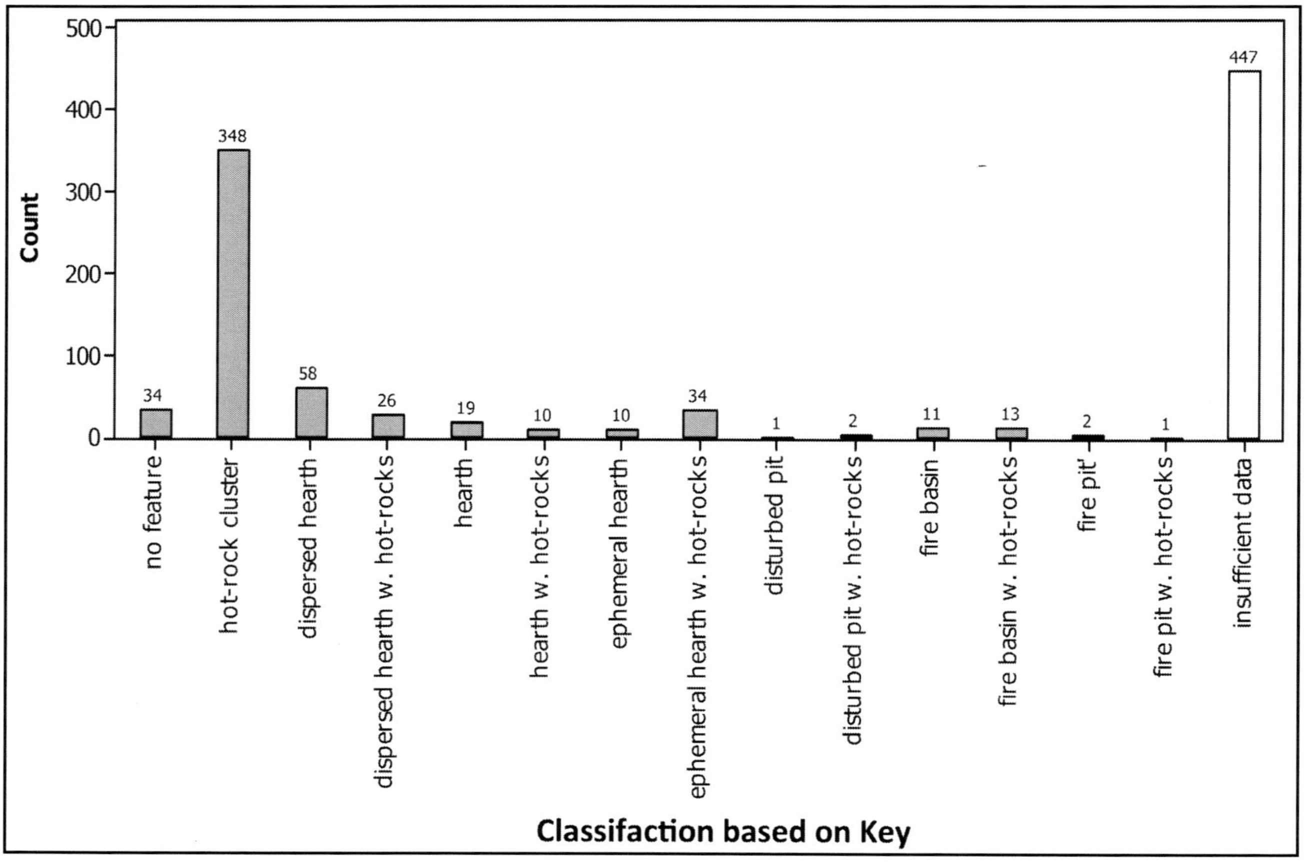

Figure 7.4. Classification of the primary LTF types at prehistoric archaeological sites within the study area.

The Experiments

Within the three wider research goals, four individual experiments were designed and tailored to address particular research questions. These experiments were undertaken separately over a period of five years and are presented as self-contained reports, numbered in the sequence in which they were completed (Table 7.1).

In complement to the research goals identified for the present study, the experimental work also sought to investigate site-specific problems generated during an ongoing programme of regional research. This fieldwork component was undertaken under the auspices of the Lubbock Lake Landmark Regional Research Program, directed by Dr. Eileen Johnson (Museum of Texas Tech University). The goal of the regional research programme is the examination of grassland hunter-gatherers and their adaptation to ecological change. The programme of fieldwork was often beneficial to thermal-feature research as it directly engaged with the material evidence for hunter-gatherer fire technologies on a regular basis.

Table 7.1. Summary of the objectives of the experimental programme by experiment.

Description	Research Question	Reference location
Experiment 1	Measurement of the performance characteristics of caliche as hearthstones useful in hot-rock technology.	Backhouse *et al.* 2005b
Experiment 2	Measurement of the effects of site formation processes acting on LTF and surrounding artefact scatters during exposure to environmental conditions	Backhouse and Johnson 2007a
Experiment 3	Measurement of the physical outputs of a simple basin type LTF qualified in terms of heat, light, and smoke	Backhouse 2004
Experiment 4	Measurement of the performance characteristics of Ogallala Formation gravels as hearthstones useful in hot-rock technology	Backhouse and Johnson 2007c

All the experiments reported here were devised by the present author in consultation with Dr. Eileen Johnson. A specialist report for each experiment was systematically submitted to a peer-reviewed journal for evaluation and publication (n=3) or alternatively presented at an appropriate conference (n=1). The manuscripts generated by this process are the result of collaborative efforts by myself (as the primary author), Dr. Eileen Johnson, and other Lubbock Lake researchers (Table 7.1). It is my opinion that the active involvement of the research team was a positive one, which enabled us to actively understand different aspects of LTF technology. Any errors resulting from the experimental work are, however, the responsibility of the current author.

In one instance (Experiment 2, Backhouse and Johnson 2007a), an experiment was continued beyond its original design timeframe and therefore that which was submitted for peer review and publication. The rationale for the extra study period was necessary to answer questions that had arisen from the first phase of the experiment and to continue to track the physical changes in the observed material over a longer time span. This extension represents a major addition to the originally reported research and is discussed here for the first time.

Methods

Methods particular to each of these experiments are discussed in detail in their relevant publication (Backhouse 2004; Backhouse *et al.* 2005b; Backhouse and Johnson 2007a, c). The direct engagement with open flames and the active construction of features in 'real-world' (or actualistic) settings were common to all the experimental research. The degree of control exerted on the experiments varied depending on the specific research question(s) being examined. Quantification and replication were important to the experimental programme. Also important was the pedagogic element that introduced students and archaeologists accustomed to looking at the static result of fire (charcoal and burnt stones) to the dynamic processes that control their transformation.

All the experiments were undertaken in controlled conditions at the Lubbock Lake Landmark Experimental Research Area (Figure 7.5). The safe completion of all experiments was an important consideration due to the dangers inherent in working with flames in an outdoor setting (Wright and Bailey 1982). A standard set of safety equipment was on hand for each experiment and consisted of:

- 1 class-C type fire extinguisher
- safety clothing (Backdraft II, firefighting gloves)
- nearby water sprocket and attached hose
- protective glasses
- shovel and loose sediment

As previously stated, the experimental area is not in a public-accessible portion of the park and therefore safety considerations concerning the public were not appropriate. The hot dry climate on the Southern High Plains dictates that 'burn bans' are often in effect during the summer months. Experimentation did not take place during these times. The periodic mapping assays associated with the site-formation experiments (Backhouse and Johnson 2007a) were sometimes conducted in extreme temperature conditions (Figure 7.6). During these times, access to the nearby climate-controlled laboratory facilities was greatly appreciated by the field research team.

Figure 7.5. Experimental research area at Lubbock Lake Landmark (P.N. Backhouse).

Hot-Rock Technology

Two experiments (Experiments 1 and 4) were designed specifically to examine the first of the identified experimental research goals (Chapter 1), the role of hearthstones in hunter-gatherer fire technologies. The use of hearthstones at more than three-quarters of archaeological sites with identified LTFs on the Southern High Plains (Figure 7.1) is the primary justification for the attention on this artefact class. The first experiment (Experiment 1, Backhouse *et al.* 2005b) sought to examine the morphological transformation and performance characteristics of caliche hearthstones when heated in an experimental basin-hearth feature. The second experiment (Experiment 4, Backhouse and Johnson 2007c) examined the morphological characteristics and performance characteristics of other locally available (non-caliche) gravels as heat reservoirs in hearthstone technologies.

Research questions were examined by heating sets of different rock types in basin hearths and subsequently recording any changes in structural integrity, colour, weight, and temperature. The objectives of the two experiments differed, dependant on the rock type being tested. Caliche, already a well-known regional hearthstone resource, (e.g. Lintz 1989) was primarily tested to see if a predictable relationship between firing temperature and specimen morphology could be established, which could be used in the analysis of field-recovered specimens. Ogallala Formation gravels were tested to determine whether they were suitable for hearthstone technologies and if they were then to assess whether any predictable morphological changes were observable that would assist in field identification.

Figure 7.6. Alejandra Matarrese mapping the displacement of hearthstones during a bitterly cold winter's day on the Southern High Plains (photograph courtesy of the Anthropology Division, Museum of Texas Tech University).

The results of the caliche experiment (Experiment 1, Backhouse *et al.* 2005b) suggested that when this rock was subjected to intense heating during repeated actualistic experimentation, then the observable morphological transformations were highly variable when considered at a specimen scale. However, at an assemblage scale, several trends became apparent, indicating that fairly predictable structural transformations are likely to occur when temperatures above 204°C are achieved and sustained (Backhouse *et al.* 2005b:710). The results indicate that when caliche is heated above this threshold temperature then hearthstone colouration becomes darker, the rocks undergo weight loss, and begin to fracture (Backhouse *et al.* 2005b:712). Furthermore, the experiment illustrated that Honey Mesquite (*Prosopis glanulosa*) was more than capable of achieving temperatures within the basin hearth that were capable of altering the physical appearance of the experimental caliche. This previously assumed relationship had not previously been explicitly tested.

The results of the Ogallala Fm. gravel experiment (Experiment 4, Backhouse and Johnson 2007c) indicated that, in particular, Potter member quartzite underwent significant and predictable structural transformations when subjected to intense heating. Quantitative comparison of the performance characteristics of the gravels demonstrated that there was no heating advantage in the selection of these gravels versus caliche. The macroscopic transformations observed on the Potter member experimental hearthstone assemblage suggests that this material type often exhibits a reddened rind toward the cortical surface of individual specimens, and large curvilinear spalls are likely to detach from the exterior surface of heated cobbles (Backhouse and Johnson 2007c:1376). Both transformations have subsequently been observed archaeologically in structurally intact LTFs (such as the one pictured in Figure 7.7a) and have proved to be highly useful in identifying less well-preserved features during subsequent surveys (Figure 7.7b) .

Both hearthstone experiments have revealed dynamic aspects of hunter-gatherer LTF technology that are not directly observable from the existing site perspective (i.e. based only on the results of fieldwork). For instance, Experiment 4 suggests that there was no advantage to using one type of geologic material over another in terms of hearthstone efficiency (Backhouse and Johnson 2007:1376). It is therefore possible to predict that the hearthstones used in LTF construction should reflect the most readily available resource for any given site location. At locations where this model does not correspond to the archaeological record, then alternative behavioural scenarios can be explored by the active examination of the physical evidence.

The two hearthstone experiments also proved a fundamental, but previously assumed, relationship between temperature and alteration in the structure of commonly available local rock types. This is important because the temperature threshold (*cf.* Purdy 1975) for these physically characteristic morphological transformations is above that typically generated by other less direct processes (i.e. prairie fires). The physically transformative characteristics of different rock types that have been subjected to cultural burning, can practically and predictably be identified during fieldwork and interpreted by reference to the experimentally generated assemblage.

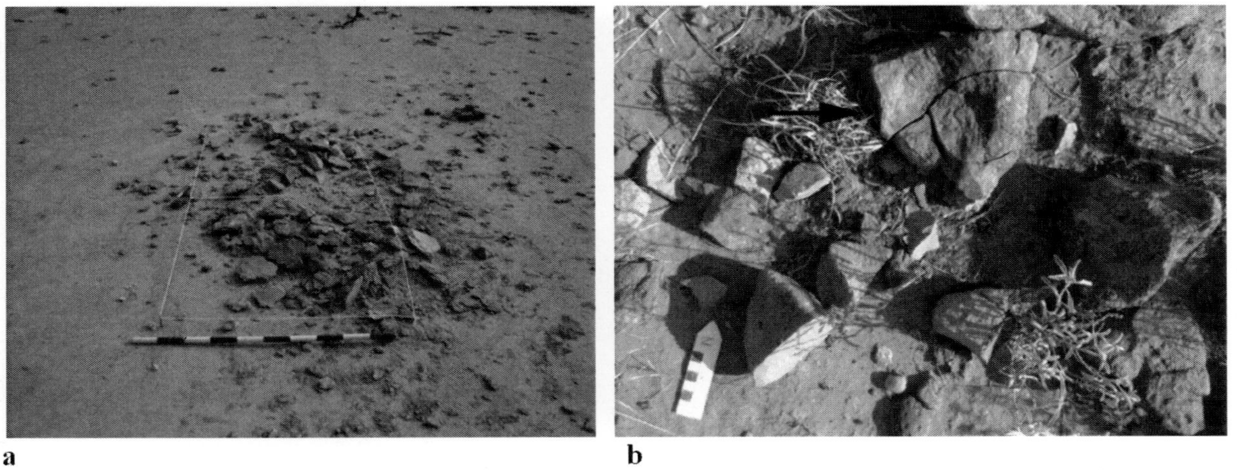

a b

Figure 7.7. Experiment 4 facilitated the field identification of Ogallala Fm. gravels as indicators for hunter-gatherer domestic fire technologies; a) large structurally intact LTF with Potter member hearthstones identified during pedestrian survey at PLK-Locality 4; b) isolated Ogallala Formation hearthstone (indicated by black arrow) from an eroded hearth identified as 41GR146 (P.N. Backhouse).

Site-Formation Processes

One experiment (Experiment 2, Backhouse and Johnson 2007a) undertook to examine the site-formation processes operating on open-air hunter-gatherer campsite assemblages in response to the high encounter rates of apparently highly disturbed sites identified in the regional dataset (Figure 7.4). The experiment was designed to test three specific research questions (Backhouse and Johnson 2007a:177):

1. To what extent is the spatial patterning of archaeological materials altered by site formation processes, in a low depositional environment, such as the modern ground surface of the Southern High Plains?

2. What transformative effect (if any) does the exposure to the elements have on artefacts commonly found in association with LTFs?

3. How does the structural integrity of an LTF alter after it falls into disuse?

These questions were examined through a multi-component experimental methodology that recorded observations on a simulated assemblage of cultural material over a one-year period. Equipment used to record specific processes included a total station to record any artefact displacement around a simulated basin hearth, artefact analysis to examine any changes to the morphology of hearthstones after being placed on the modern ground surface, and photo-documentation to examine the structural transformation of an expended basin hearth facility (Backhouse and Johnson 2007a).

The results of the spatial component of the experiment indicate that smaller specimens were less likely to be recovered than larger specimens. In fact, almost 17% of the experimental assemblage was not recovered after one year of exposure (Backhouse and Johnson 2007a:184). This result is significant in that it suggests that material was either not observed during data recovery (i.e. human error), was not visible during data recovery (i.e. had become buried or hidden by another specimen or a feature of the area such as vegetation), or had been removed from the boundaries of the experimental area altogether (Backhouse and Johnson 2007a:184). Subsequent analysis of the periodic (biweekly) mapping of the specimen locations suggested that unexpected cultural processes during mapping were most likely to be responsible for the loss of the smaller specimens.

Specifically, the detailed examination of the spatial displacement and loss of specimens indicated that larger specimens (such as hearthstones) were more likely to have undergone horizontal displacement than smaller specimens (such as lithic debitage) (Backhouse and Johnson 2007a:191). The horizontal displacement is most likely to be the result of researchers accidentally brushing or scuffing the larger objects during the regular data-collection assays. Similarly, vertical displacement is most likely to be the result of researchers accidentally standing on the small material and pressing it underfoot into the sandy sediments. Both these unexpected results provide useful proxies for the formation processes operating at sites during their initial aboriginal occupation and also during subsequent data collection by archaeologists. Put simply, we should perhaps expect that humans will tend to walk over smaller objects of cultural material either accidentally or deliberately and in this way these specimens appear more likely to be pressed in to the modern (or ancient) ground surface (vertical displacement). Conversely, humans might more easily observe and therefore directly avoid larger artefacts such as hearthstones. These objects are more likely therefore, to be accidentally scuffed (horizontally displaced) as humans move about within the relative spatial confines of a feature-orientated activity area. It should be noted that both displacement processes have been previously predicted, but not tested, from an archaeological perspective (Stevenson 1991).

The intensity to which both displacement processes should occur within an open-air camp would most likely be related to the number of people, and duration of initial occupation (cf. Petraglia 2002). The displacement model does not just apply to the original occupation of a site but extends to any subsequent human presence, whether in the form of another hunter-gatherer group occupying the same location at a later date, or a hunter-gatherer group stopping to use resources (such as to recycle previously discarded hearthstones or lithics). Furthermore, the presence of pedestrians during later historic and modern periods could also initiate a subsequent round of artefact displacement (Backhouse and and Johnson 2007a:189). Assuming that displacement is relative to the intensity of the human presence, then we might expect that in a modern setting, the sites which have been most frequently visited by local collectors and professional archaeologists should exhibit the highest rates of material displacements (both vertical and horizontal).

In order to test this hypothesis I ran a follow-up experiment, which again plotted the position of seeded specimens within the same experimental area for the period of one year. This time though, I actively encouraged heavy pedestrian traffic (through an arrangement with the Lubbock Lake Landmark park docents) of tour groups through the northern portion of the experimental area. The results of this follow-up experiment show significant displacement of specimens in the northern portion of the experimental area, along the route most frequently taken by the tour groups (Figure 7.8). The distribution again indicates that the larger hearthstones were most likely to be displaced, and several specimens were moved over 2 meters from their seeded locations

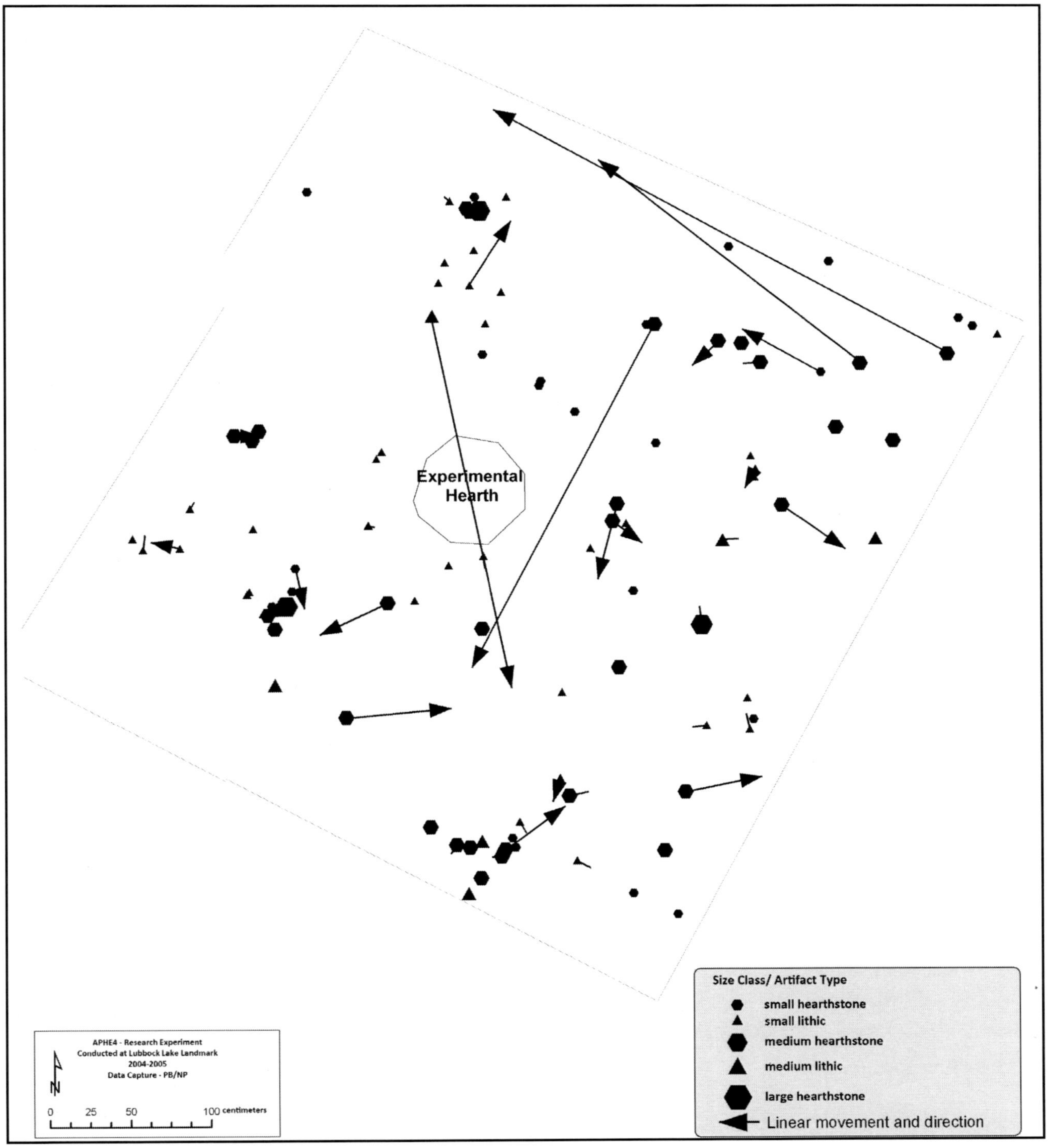

Figure 7.8. Spatial displacement of simulated cultural material around a hearth after one year; note the extreme horizontal displacement of specimens in the north of the experimental area along the route of the tour groups.

With regards to the regional dataset, the ramifications of the spatial displacement experiment are particularly significant, considering the high number of sites that are represented by hearthstone clusters at the level of the modern ground surface (Figure 7.4). If trails or paths pass through these sites, then we should expect high-intensity post-depositional spatial displacement of material from both human and animal traffic. The vast size of the Southern High Plains landscape, however, suggests that most sites are not bisected by human-worn trails. The effects of animal traffic are yet to be investigated, but it can reasonably be postulated that a similar-size dependant displacement effect may occur (cf. Gifford-Gonzalez *et al.* 1985). Nevertheless, at these sites, displacements of a lower intensity might be expected from both the activities of interested amateurs and professional archaeologists.

The spatial displacement suggests that at least three levels of artefact displacement are likely to be occurring within the Southern High Plains study area:

1. during the initial occupation of the campsite;
2. as a result of human traffic through sites post abandonment; and
3. as a result of the investigation of sites by amateurs and professionals alike.

It is likely that more than one of these factors is present in most sites recorded in the study area. A positive result of the spatial-displacement experiment predicts that small artefacts are less likely to be significantly horizontally displaced from their discard location by cultural processes. Therefore, the high-resolution survey and excavation techniques which target small artefact classes (such as lithic debitage) should have significant potential to infer behaviourally meaningful spatial distributions at a site level. It should also be remembered that human pedestrian traffic is but one of many site formation processes that can result in the spatial displacement of cultural material. Other processes include, but are not restricted to: animal traffic, agricultural practices, geomorphic processes such as erosion, weather events, climatological conditions, bioturbation, and the large earth movement necessary during construction projects.

The second objective of experiment 2 (Backhouse and Johnson 2007a) was the re-examination of typically recorded specimens following the year long exposure on the modern ground surface. This portion of the experiment recorded no significant alteration to flaked lithic material specimens seeded within the experimental area. However, caliche hearthstones appeared to undergo significant structural transformations during the prolonged exposure timescale of the experiment. The recorded initial variation in hearthstone coloration became less pronounced during the course of the experiment. In addition, significant fracturing was observed to occur on a small percentage of the seeded specimens (Backhouse and Johnson 2007a:185).

The hearthstone morphology results have significant implications for interpreting the regional archaeological record. Considering that hearthstones comprise the primary evidence for LTFs on the Southern High Plains (Figure 7.4), then the post-depositional processes that affect their physical appearance are clearly important to understand. The morphology experiment indicated that hearthstones are susceptible to physical alteration during prolonged periods of exposure. The apparent distributional preference identified in the regional dataset for hunter-gatherer campsites in low-deposition environments, i.e. at high points in the landscape, (Figure 7.3) then it can be assumed that most hearthstone assemblages have undergone at least some physical transformation. Interpretations that rely on an assessment of fracture frequency of a particular hearthstone assemblage to determine the function of a particular LTF or site must therefore be reassessed in light of this experimental perspective. For example, boiling technologies are widely acknowledged as resulting in a high frequency of fractured hearthstones (e.g. Brink and Dawe 2003). The relationship between hearthstone fracture and function of any particular feature is clearly more complicated than previous research might imply (e.g. Lintz 1989) and the susceptibility of caliche hearthstones to fracture while exposed on the ground surface (both ancient and modern) requires further investigation.

The results of the hearthstone morphology experiment suggest that hearthstone size is largely the result of initial size, function, and length of time exposed. It is therefore reasonable to hypothesize that as the age of caliche hearthstones increases, then specimen's size should become generally smaller. This hypothesis can be tested in the archaeological record of the Southern High Plains.

The results of monitoring the structural transformation of a basin hearth-type feature also provided a useful contextual dataset. Whilst the feature generally appeared to change very little (i.e. very low rates of sediment deposition occurred and the overall structure was largely the same), nevertheless several significant processes were recorded during the course of the experiment. Firstly, the organic evidence for burning (pieces of charcoal and wood) was quickly flushed from the feature during heavy rainfall (Backhouse and Johnson 2007:186). The larger organic pieces probably migrated toward the base of the feature as a result of simple gravitational dynamics. Secondly, the feature accumulated wind-blown seeds, which appear to have become trapped within the internal matrix. After several months, plants began to grow in the well-drained sediments (a result of the hearthstone fill) of the internal feature (Backhouse and Johnson 2007a:187). These results, suggest that basin-hearth features exposed in low deposition, or indeed eroding environments, are vulnerable to some structural transformation after abandonment. The internal arrangement of hearthstones was, however, largely unaltered after one year of experimentation.

In order to determine whether these results were applicable over longer time-scales, I continued to observe the same hearth used in this experiment. After 792 days (2.1 years) of exposure, the internal structure was again largely unchanged at an assemblage scale (Figure 7.9a). At a specimen scale, the hearthstones appear to have undergone continued physical transformations resulting in the additional fracturing of several of the large hearthstones (Figure 7.9b). Additionally, the colouration of several of the larger hearthstones appears to have become consistently lighter over time, with the result that several of the formerly blue-gray hearthstones (typically Gley 2 4/10B) were almost white by the end of the second year of exposure. The results of the additional year of study then appear to confirm the fracture and discoloration processes observed after one year on the caliche hearthstone assemblage (Backhouse and Johnson 2007a). The overall structure of the basin hearth, however, remained largely unchanged after the additional year.

In terms of archaeological analysis, these results provide a more detailed baseline for the upland site-formation processes that appear to dominate the Southern High Plains dataset. Importantly, the investigation reveals the complexity of attempting to model the processes that act to structure the archaeological record at site-specific scales. The experiment clearly demonstrates that hearthstones are easily horizontally displaced and are therefore most likely to be poor spatial markers of human behaviour. Hearthstone morphologies suggest that both the colouration and fracture mechanics of burned-caliche hearthstones are complex and continue to transform specimens well after their cultural lifespan. Lastly, basin-hearth features retain some structural integrity over relatively short time-spans. However, their ability to act as a trap for air-borne seed dispersals suggests that they are susceptible to destruction over longer time scales, especially if larger species seeds (such as mesquite) become trapped within their matrix.

Figure 7.9. Basin hearth taphonomy experiment after two years of experimentation: a) overall view of the feature; b) close-up of an internal hearthstone that has fractured (P.N. Backhouse)

Site-formation processes acting on LTF datasets clearly have significant implications for understanding the archaeological record of the Southern High Plains. The three components of the experimental programme all suggest that research designs and interpretative positions need to incorporate these processes in order to arrive at a more sophisticated understanding of the hunter-gatherer domestic fire technologies utilised by populations inhabiting the region. Furthermore, the validity of the conclusions of this experiment should be practically assessed against the archaeological record at site-specific and feature-specific scales.

Performance Characteristics

One experiment (Experiment 3, Backhouse 2004) examined the performance characteristics of a shallow basin hearth feature. Specifically, the output of the experimental facility was recorded in terms of heat, light, and smoke. Standardised observation and recording was undertaken at periodic intervals during the use-life of the facility. Furthermore, these observations were recorded at varying predetermined distances from the feature and at a set height (0.5 meters) above the modern ground surface, roughly equivalent to chest height of a seated individual.

The purpose of experiment 3 was to examine how an LTF might actively influence human behaviour at a micro-site scale. The assumption being tested was that the apparently meaningful patterns of human behaviour often identified in the archaeological record are interpretatively related to LTF placement at a site level (e.g. Binford 1978). Put simply, the strongest evidence for human activity around an LTF should logically be at the point where the maximum benefit is achieved i.e. at the point of most warmth and light, and where there is the least smoke. In order to test this hypothesis, I directly measured heat and light emission at one-meter increments from the centre of a typical basin hearth facility (up to 3.5 meters from the hearth). Smoke was only indirectly observed and any significant impact at each recording station was noted.

Predictably, the results of the experiment indicated that there was a clear relationship between heat and proximity to the fire (Figure 7.10). For example, at roughly 0.5m from the centre of the feature (station 1), the temperature was on average 8.6°C warmer than the ambient air temperature. While at 3.5 meters from the centre of the feature (station 4), the temperature was on average only 0.27°C warmer than the ambient air temperature. A variety of factors, are worth consideration in terms of placing this data in a human context. Huizenga *et al.* list several human factors which include temperature, air velocity, humidity, metabolic rates, physiological constraints, and clothing (Huizenga *et al.* 2001:692). These factors can greatly affect the potential impact the results may have on any one individual. In general, a temperature of around 25°C at chest level is considered ideal for modern humans (Huizenga *et al.* 2001:698). This statistic suggests that a location about 1.5 meters from the facility (station 2, average temperature 24.6°C) might be most suitable for humans not constantly involved with tending the facility.

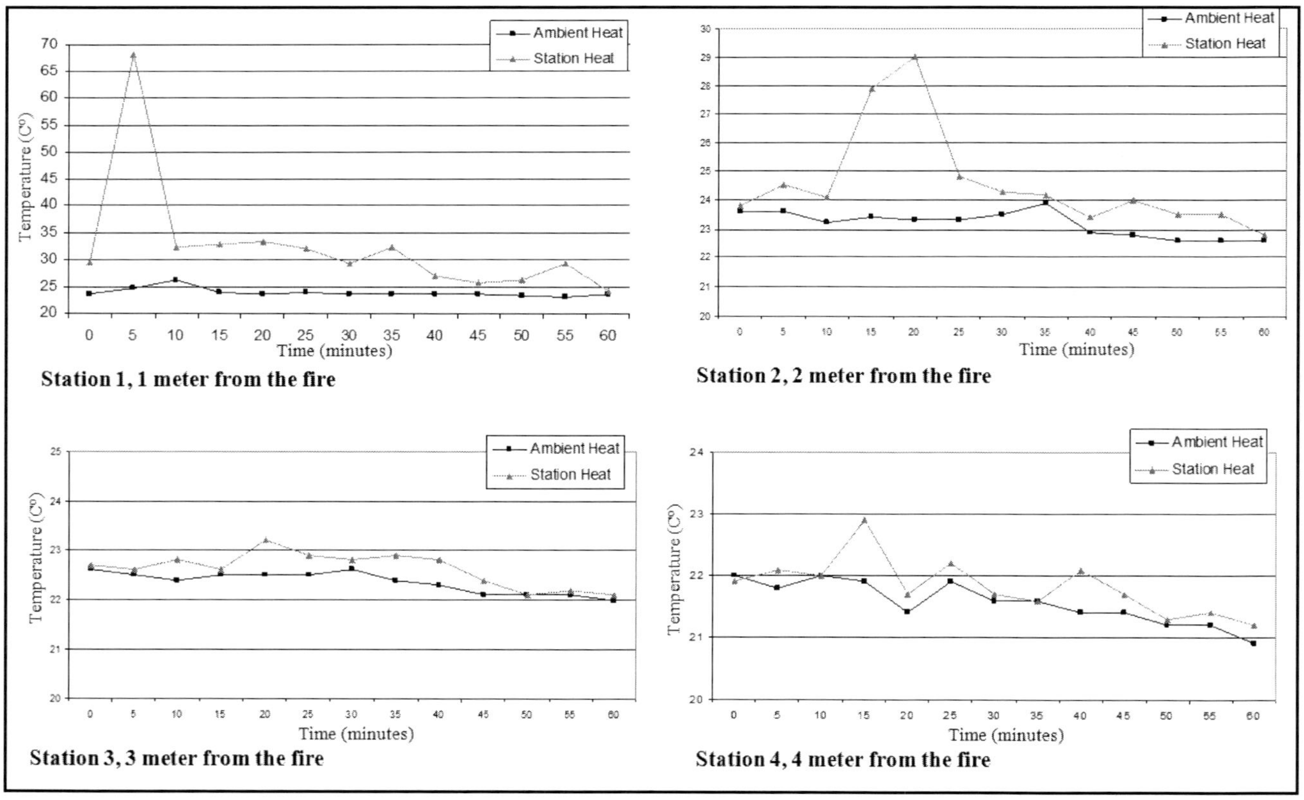

Figure 7.10. Heat dissipation from an experimental basin hearth plotted for a one-hour period: a) 0.5 meters from the hearth; b) 1.5 meters from the hearth; c) 2.5 meters from the hearth; and d) 3.5 meters from the hearth.

Few observations of this kind are available ethnographically and the results are presumably highly dependant on activities being performed, environmental conditions, and other factors such as clothing. Lewis Binford (1978) measured the distance of individuals sitting around an open hearth at the Mask site. The results indicated that distance between people was socially dependant on the number of people sitting around the feature, nevertheless an average distance between individuals of between 60 and 70cm was recorded (Binford 1978:349). Comparative data as to the social distance between individuals living in different ecological settings is not readily available but would provide an interesting comparison with which to examine the degree to which social space is a function of environmental stresses (i.e. extreme heat and cold).

Measurements of light emission from the facility were somewhat problematic. The low resolution of the plotted data (Figure 7.11) is likely to be the result of the low precision associated with using a standard camera photo light meter as the primary data collection device. Although the experiment was completely undertaken under the cover of darkness, artificial light sources most likely to be negatively affected data capture. Nevertheless, the mean ambient light during the experiment can be calculated as 1.55 footcandles. The ambient light level appears to be fairly consistent throughout the entire experiment (standard deviation of ambient light readings is 0.826 footcandles).

The results suggest that the area from 0.5 to 2.5 meters away from the LTF received the greatest amount of light (station 2 mean light reading = 2 footcandles; station 3 mean light reading = 2.23 footcandles). Assuming that the spatial patterning of human activities, such as craft production, may have been controlled by the need for light, then the area obtaining the highest light readings might be expected to exhibit the strongest archaeological evidence. This assumption must, however, be positioned within a framework of the minimum light levels required to carry out different tasks. As with heat, the amount of light necessary is dependant on a host of external variables, such as time of day, presence of other light-producing technologies such as torches, weather conditions (particularly phase of the moon but also local weather conditions such as fog), and individual physiology. Unfortunately, very little research has been conducted on this subject for human populations in general and hunter-gatherers in particular.

As already stated, the effect of smoke was not empirically recorded as part of the present experiment. Spatially

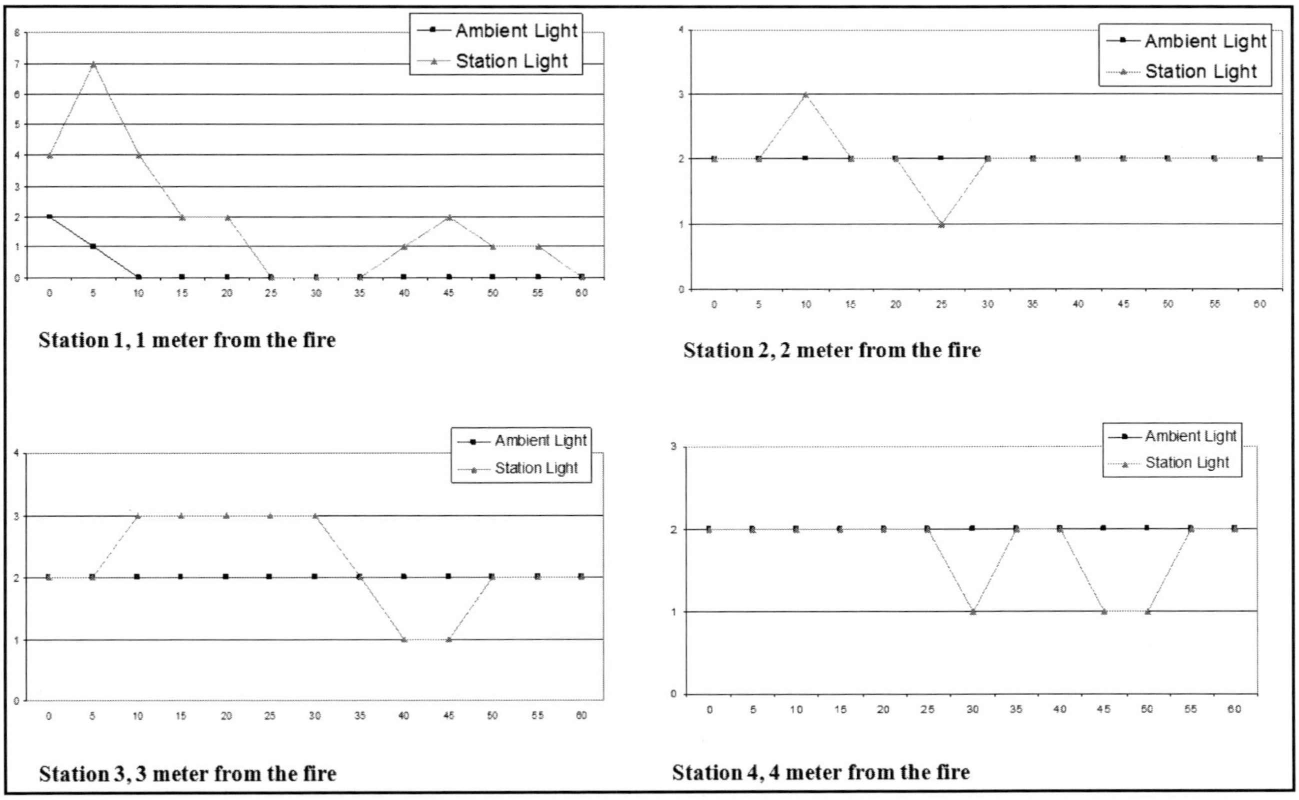

Figure 7.11. Light dissipation from an experimental basin hearth plotted for a one hour period: a) 0.5 meters from the hearth; b) 1.5 meters from the hearth; c) 2.5 meters from the hearth; and d) 3.5 meters from the hearth.

referenced observations during the course of the experiment confirmed that smoke was a decreasing hazard as distance increased away from the LTF. Specifically, smoke was most dense at distances between 0.5m and 2.5m from the fire. In these areas, the smoke often caused eye irritation, difficulty breathing, and impaired vision. These effects most likely would be less than desirable for any individual in close proximity to the feature. Furthermore, in the area closest to the fire, smoke swirled in random directions, which were not related to the general wind direction (wind direction was consistently from the east and ranged between 9 and 13mph throughout the duration of the experiment [data source: NOAA weather station located at Lubbock International Airport]).

The results of the experiment are useful in formulating a model for predicting human use of space around an LTF that appears to be common within the Southern High Plains study area (Figure 7.4). The results of heat and light measurements are complementary and would appear to indicate that an area around 1 to 2 meters from the feature would be expected to yield the maximum benefit to humans in terms of warmth and light necessary for craft production. Obviously, the radiation of both heat and light are variable over the lifespan of the fire and by the type and quantity of fuel utilised. A general distributional pattern toward the LTF might therefore be expected in situations where fuel resources are scarce or under severe weather conditions. Conversely, distributions well away from the thermal feature might be expected if fuel was plentiful or if temperature and light were not considered important characteristics by the inhabitants of any particular campsite.

Smoke also complements the results of both the heat and light data in that it is most prevalent within 2.5m of the LTF. The localised production of smoke is not beneficial to humans and therefore must be considered as a negative effect for humans occupying the physical space around a basin hearth. The observation that localised swirling of smoke is a constant phenomenon of such facilities appears to indicate that human activity would most likely be required to frequently shift position in relation to the feature, perhaps using the entire circumference of the area in an effort to maintain maximum heat and light without the discomfort of the smoke. Obviously, if the LTF were housed within a larger feature then the air flow (and therefore smoke dissipation) might be expected to be radically different than for exterior facilities. The experimental results, however, stand in contrast to some ethnographic sources, which suggest that individuals consistently occupied the apparently smoke-free areas upwind of open-air hearth facilities (e.g. Binford 1987).

Overall, the observation and measurement of three physical outputs of an LTF frequently constructed on the Southern High Plains has resulted in the formulation of a simple model for predicting the relative frequency of archaeological material that might be expected at various distances from the facility. Specifically, the model predicts that the area between 0.5 and 2.5m surrounding the circumference of the feature should indicate an increased presence of human activities. This model can be actively tested in the regional archaeological record by examining the physical evidence for human activity in the areas surrounding LTFs. Field research might also examine the types of activity (and spatial relationship to the LTF) represented in the archaeological record. One question to be investigated is whether evidence for lithic reduction is typically located close to or away from basin hearth features. Lastly, the experiment suggests that an individual working directly with the feature (such as a cook) most likely would be required to operate around the periphery of the facility, moving frequently to avoid smoke and to collect additional materials (such as fuel).

Summary and Conclusions

The three research questions, identified from the regional dataset and targeted for experimental research all provided significant insights as to the prehistoric domestic fire technologies of the Southern High Plains. Furthermore, all three have generated subsequent hypotheses that are testable by returning to the assembled regional archive or by initiating new programmes of archaeological fieldwork. A useful example is the realisation that in areas where caliche is not the dominant lithology, then hunter-gatherers appear to have selected various different and more easily available rock types for use as hearthstones. This scenario could be quantified by spatially examining the frequencies of different rock types identified as hearthstones throughout the study area.

Experimental research has also significantly increased our understanding of hearthstones in the wider technological tradition of domestic fire technology. The examination of different rock types has identified threshold temperature ranges above which fracturing occurs, the typical morphologies represented in the archaeological record, and the physical transformations in structure and appearance that they undergo following discard. This has allowed important distinctions, such as the difference between culturally and naturally burned rock, to be more accurately determined. Experimentation has emphasised the importance of modelling site-formation processes that previously were largely assumed. These processes together with the hearthstone data suggest that the assemblages that are encountered archaeologically are likely to have undergone considerable transformations as a result of a complex interaction between various site-formation processes that are continuously acting to structure the archaeological record. Simple behavioural interpretations, such as that burned, fractured rocks scattered across a prehistoric site are the direct result of stone boiling technologies, clearly require reassessment and greater empirical support in the light of these experiments.

The experimental construction and measurement of the physical outputs of a shallow basin-type feature reinforced the need to dynamically conceptualise the activities of people at site-specific scales. Construction of the feature required many trips to and from the experimental area to collect resources and once lit, the full circumference of the area around the feature was necessarily utilised in order to avoid the localised emission of dense smoke and heat. Although it is impossible to postulate the physiological levels of human comfort in the past, the LTF is clearly always a highly dynamic component of any site requiring significant construction preparation, and almost constant maintenance.

An indirect result of the experimental research programme was the active involvement of the local research community at Lubbock Lake in thinking about how thermal-feature technology was practically achieved. The engagement of researchers in building the experimental facilities, collecting fuel, and gathering hearthstones encouraged a discussion of the thermal-feature technology, which carried over into other aspects of the wider regional research programme. For instance, rather than considering hearthstones recovered during survey as abstracted objects largely devoid of potential for meaningful research, the laboratory research staff began to ponder the type of feature in which they were used, based on their size, morphology, and material type.

The three experimental research goals pursued here represent only a fraction of the potential for experimental archaeology to actively inform regional investigations. It is hoped that the establishment of an experimental research area at the Lubbock Lake will facilitate the continuation of fire-feature research.

In short, the experimental programme proved to be very useful for hypothesis testing, pedagogic engagement, and an excellent heuristic device, which generated as many research questions as it attempted to answer. The dynamic results of the experimental research therefore help to make sense of the static residues identified and excavated in the archaeological record and presented in the next chapter.

CHAPTER 8 - ARCHAEOLOGICAL RESULTS AND ANALYSIS

Abstract

The fourth and final approach in the study of LTFs is the examination of variation in hunter-gatherer domestic fire technology in the archaeological record of the Southern High Plains. The spatial and temporal distribution of hunter-gatherer campsites, derived from data recorded on site-survey forms, is plotted and a simple behavioural model advanced to explain the resultant patterning. The validity of this model is tested by examining the role that site-formation processes may have played in making some sites more visible than others. Lastly, the archaeological evidence for domestic fire technology is assessed at high-quality (determined by technique of excavation, preservation conditions, and integrity of features) sites. At these locations, high-resolution datasets, including the current programme of fieldwork, are utilised to qualitatively characterise the physical evidence for aboriginal domestic fire technologies. The archaeological record of the Southern High Plains demonstrates that the physical evidence for fire technology is well represented, and variation in time, spac,e and technology appears to be evident in the dataset. The causes of this variation are likely to be both cultural and taphonomic. The study of localised thermal features on the Southern High Plains appears therefore to be an excellent research avenue for understanding both hunter-gatherer technology and the broader utilisation of the Southern Plains landscape.

The archaeological evidence for hunter-gatherer domestic fire technology is well expressed within the Southern High Plains study area. Variation in this dataset is examined from spatial, temporal, and technologic perspectives. Lines of evidence are explored and the effects of non-archaeological events, such as political decisions and geomorphologic processes, are considered. The regional dataset extends over two modern state boundaries and the consequence of this political separation has arguably resulted in the construction of very different ways for thinking about, engaging with, and ultimately recording evidence for prehistoric hunter-gatherer campsites.

A scaled presentation of the archaeological results is undertaken. Firstly, a course-grained regional perspective derived from the application of the hierarchical key to state-level datasets is presented. At this scale, the hierarchical key identifies broad technological trends and site formation processes that can be usefully examined by site-scale investigations. Secondly, a necessarily coarse-grained spatiotemporal framework is applied to the regional distribution by identifying the age of the LTF-bearing site components. Lastly, fine-grained, individual site data are examined for their contribution to understanding variation in hunter-gatherer fire technology across the four broad cultural time-periods previously established for the Southern High Plains. A sample of the archaeological literature and the results of recent fieldwork, initiated as part of the Lubbock Lake Landmark Regional Research Program (LLLRRP), are utilised for the purposes of technological analysis. Two of the dominant themes, hearthstone technology and the role of site-formation processes in structuring the dataset identified from the archaeological record of the Southern High Plains study area, were briefly highlighted in Chapter 7 and practically investigated by experimental research. The complete presentation and critical examination of the regional results are, however, undertaken here.

Results of the Compilation and Characterisation of the Regional Dataset

The regional dataset was derived from site records on file with the SHPOs (State Historic Preservation Officers). A single exception to the use of SHPO database was at one site, Blackwater Draw Locality 1, where LTFs were not identified (perhaps due to its complex site history, see Hester 1972). Nevertheless, this site was included in the technological discussion because it potentially provides the earliest evidence for hunter-gatherer activity available in the region.

Site survey forms proved to be suitable sources of information in the establishment of a regional dataset for the identification of LTFs. The material evidence for cultural activities involving the spatially delimited use of fire was well represented in the states' record forms and a total of 1,016 archaeological sites exhibiting evidence for the physical remains of localised thermal features (LTFs) were identified within the Southern High Plains study area (Figure 8.1).

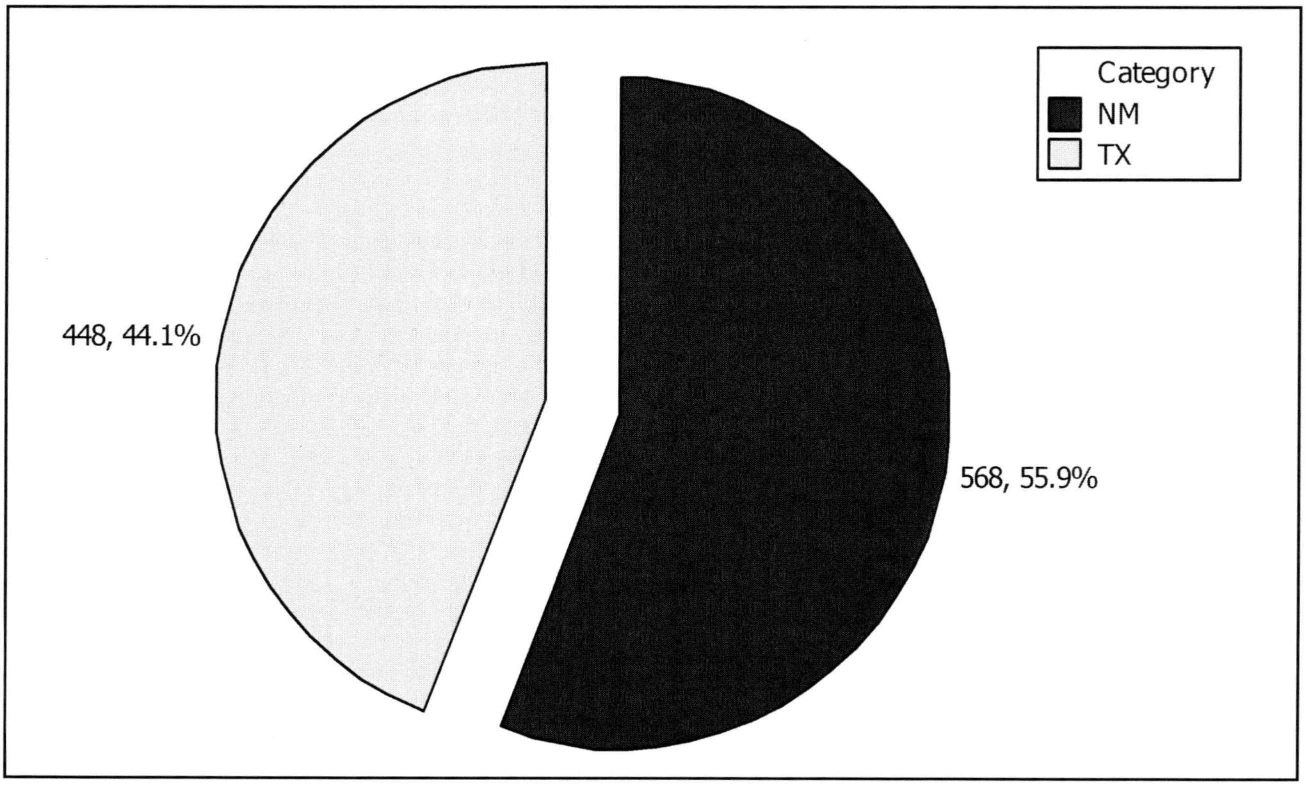

Figure 8.1. Archaeological sites with LTFs by state within the study area.

Geospatial

The distribution of sites is spread fairly evenly between the two represented states. However, since two-thirds of the study area is within the modern boundaries of Texas (Figure 3.3), the higher number of sites identified in the New Mexico portion is potentially significant (Figure 8.2). Specifically, the eastern flank of the Pecos river (the Mescalero Plains), in the southwest of the study area, demonstrates significant site densities (Figure 8.2). It is important to note that the distribution in this area is constrained by the arbitrary ten-mile buffer used in this study. The known distribution, in fact, extends throughout the southeastern portion of New Mexico (see Figure 3.4 for the unconstrained state dataset distribution prior to geoprocessing).

Inspection of the spatial distribution of sites by modern county reveals that sites with evidence for LTFs are unevenly distributed among the counties identified within the study area (Figure 7.3, 8.3). Counties in New Mexico are far larger than their Texan counterparts and it is therefore not possible to directly compare these two datasets. Three counties (Chaves, Eddy, and Lea) dominate the New Mexico distribution. In Texas, two counties (Briscoe and Garza) exhibit site densities which are several orders of magnitude more dense than the average (N = 14/county).

Perhaps the most notable feature of the spatial distribution is the lack of sites identified within the interior of the region (Figure 8.2). Proportionally, around 90% of the sites identified are located in the diverse ecotonal areas that surround the boundaries of the Southern High Plains. The 10% of sites identified within the interior of the region tend to be located along the length of the incised draw systems, and in particular are located along Runningwater, Yellowhouse, Mustang, and Tule Draws, as well as along Tierra Blanca Creek. Lesser distributions of sites are clustered in the Andrews Dunes to the south of the area and in the Lea-Yokum Dunes that run along the length of the Seminole valley in the central west section of the region (Figure 8.2). Very few sites within the interior of the region are not, then, located either along the draw systems or within the dune fields.

A handful of exceptions to this pattern are observable, such as 41LY52 in Lynn County, which is located at the periphery of a large saline lake (Tahoka Lake). For the most part, however, sites with physical evidence for hunter-gatherer fire technology are predictably located either in the draws or dune fields of the interior of the landscape or around the physiographic and ecotonal boundary that surrounds the region. Surprisingly, sites are almost completely absent adjacent to the episodic playa lakes on the surface of the region (Sabin and Holliday 1995).

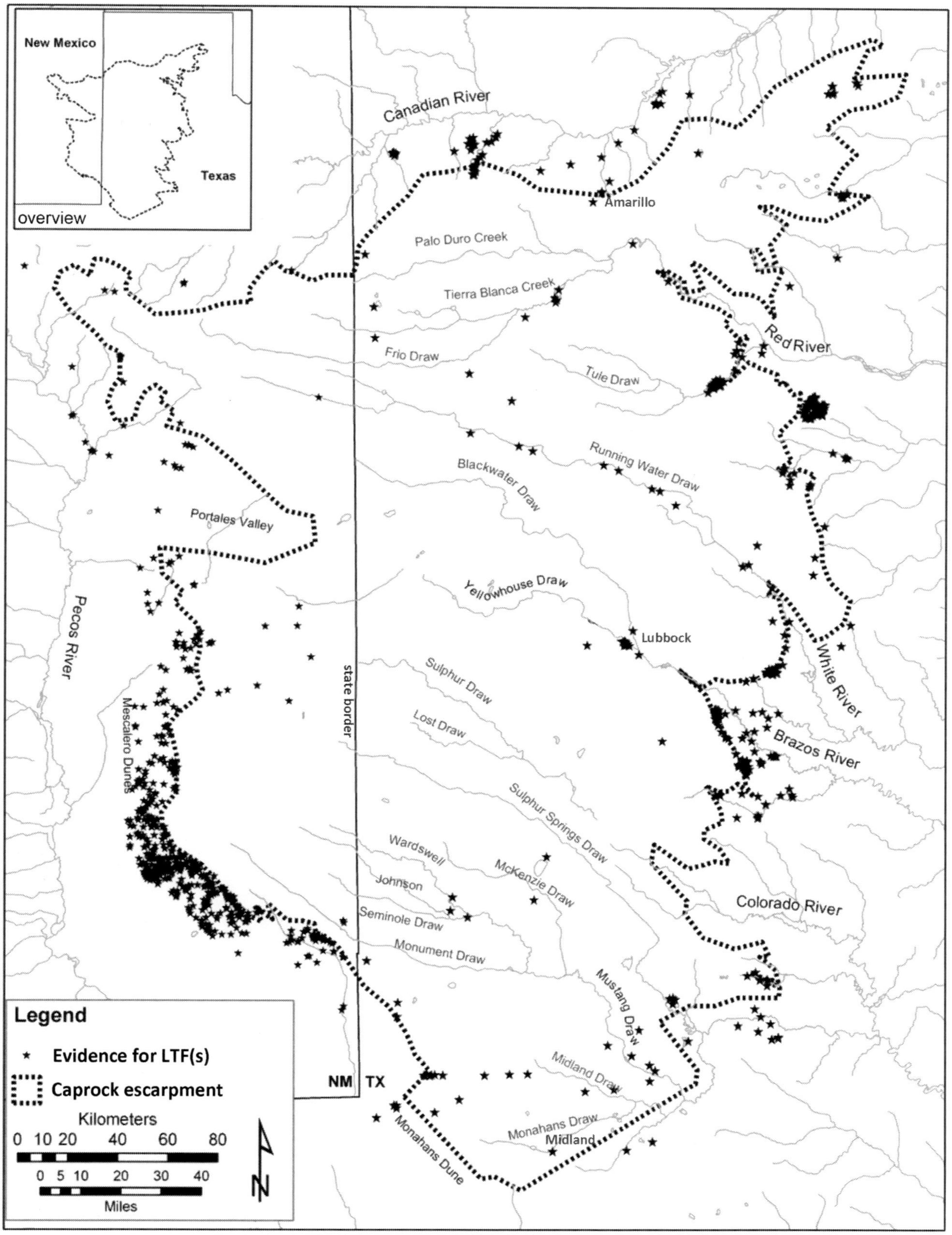

Figure 8.2. Spatial distribution of sites with LTFs identified in the research area.

The spatial distribution of archaeological sites with concomitant evidence for domestic hunter-gatherer fire technology thus far presented indicates a seemingly straightforward model for explaining the distribution of hunter-gatherer populations in this landscape. Specifically, and without introducing any time-depth, an interpretation can be advanced that the site patterning represents hunter-gatherer groups favouring the biologically more productive areas at the periphery of the region. Put simply, human populations appear to have been concentrated along the margins of the Pecos River to the west of the caprock escarpment and also strung out along the ecologically rich sections of the eastern canyonlands (Figure 8.4). These groups might have made periodic use of the draw systems to navigate into the interior of the region, presumably to kill and butcher bison, before returning to the predictable resources off of the escarpment.

This scenario makes intuitive sense; one of the most efficient subsistence strategies for human populations to utilise the energy locked in the vast grasslands of the interior of the region was to harvest the bison herds that consumed the grasses and transferred them into usable energy resrves (Hill 2006). As already stated, the diverse ecology and natural resources (particularly material for stone tools; e.g. Holliday 1997, Hurst *et al.* 2010) at the boundaries of the region would appear to be an ideal logistical base for hunting expeditions onto the Southern High Plains (Boyd 1995). The general lack of long-term habitation sites (i.e. villages) in the archaeological record of the region may be interpreted by considering that high residential mobility should be expected for foraging subsistence strategies reliant on the acquisition of herd animals (Kelly 1995). Populations may, therefore, have moved frequently along the boundaries of the region, tracking the bison herds on top of the Llano, whilst subsisting on a mixed diet of smaller game and plants while at temporary camps along the margin of the area.

In order to test the validity of this simple model to explain the distribution of sites with evidence for domestic fire technology (campsites) observed in the archaeological record, it is first necessary to explore relevant non-archaeological factors that may have resulted in over-representation or under-representation of the geospatial dataset. I expand the examination of these factors here, prior to the introduction of either temporal or technological considerations, because to assume that the density of sites at the periphery of the Llano Estacado is purely a result of the distribution of past human populations would most likely be a mistake.

Agricultural

The immense interior of the region historically has always been the most intensively farmed (see Chapter 4). The socio-political foundation for the agricultural development of the Southern High Plains was effectively facilitated by the removal of the aboriginal population following the Red River war of 1875 (Carlson 1998). However, it was not until the early twentieth century that mechanised agriculture, primarily cotton farming, began to radically transform the landscape of the Southern High Plains (U.S. Census of Agriculture, 1900, 1930).

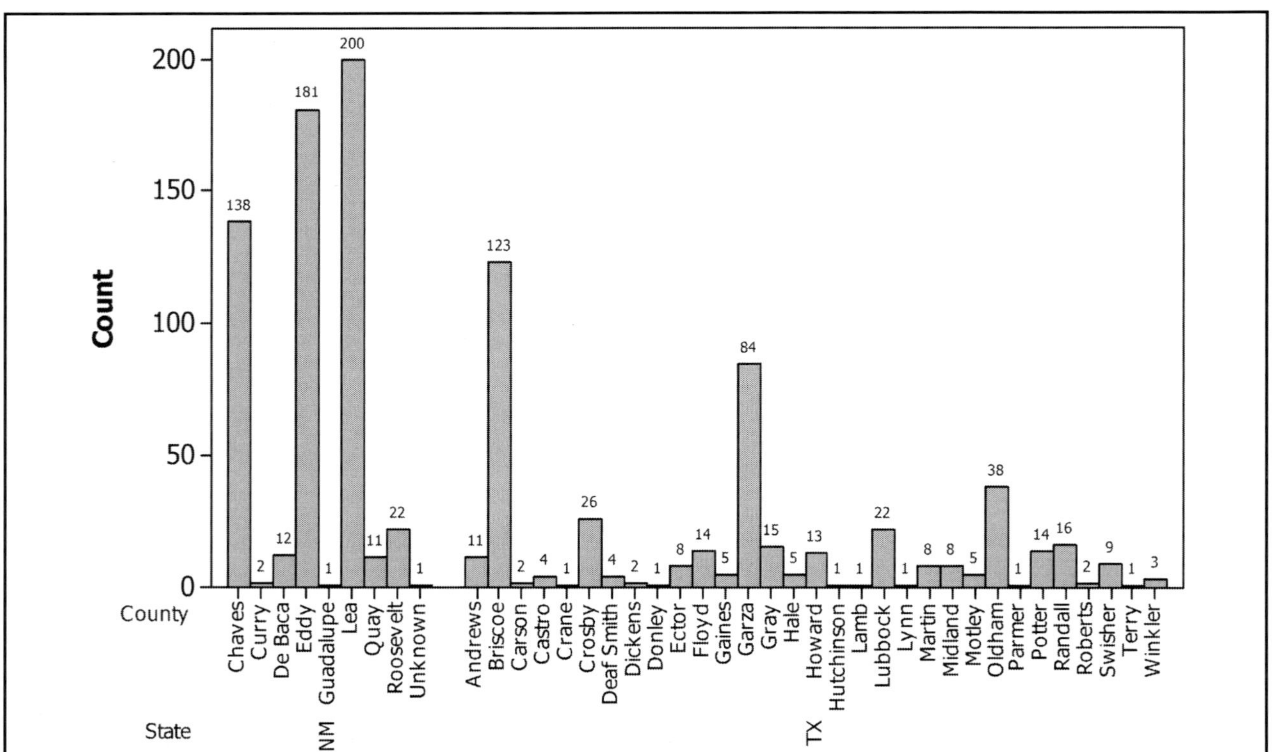

Figure 8.3. Number of sites with evidence for domestic fire technology by State and County.

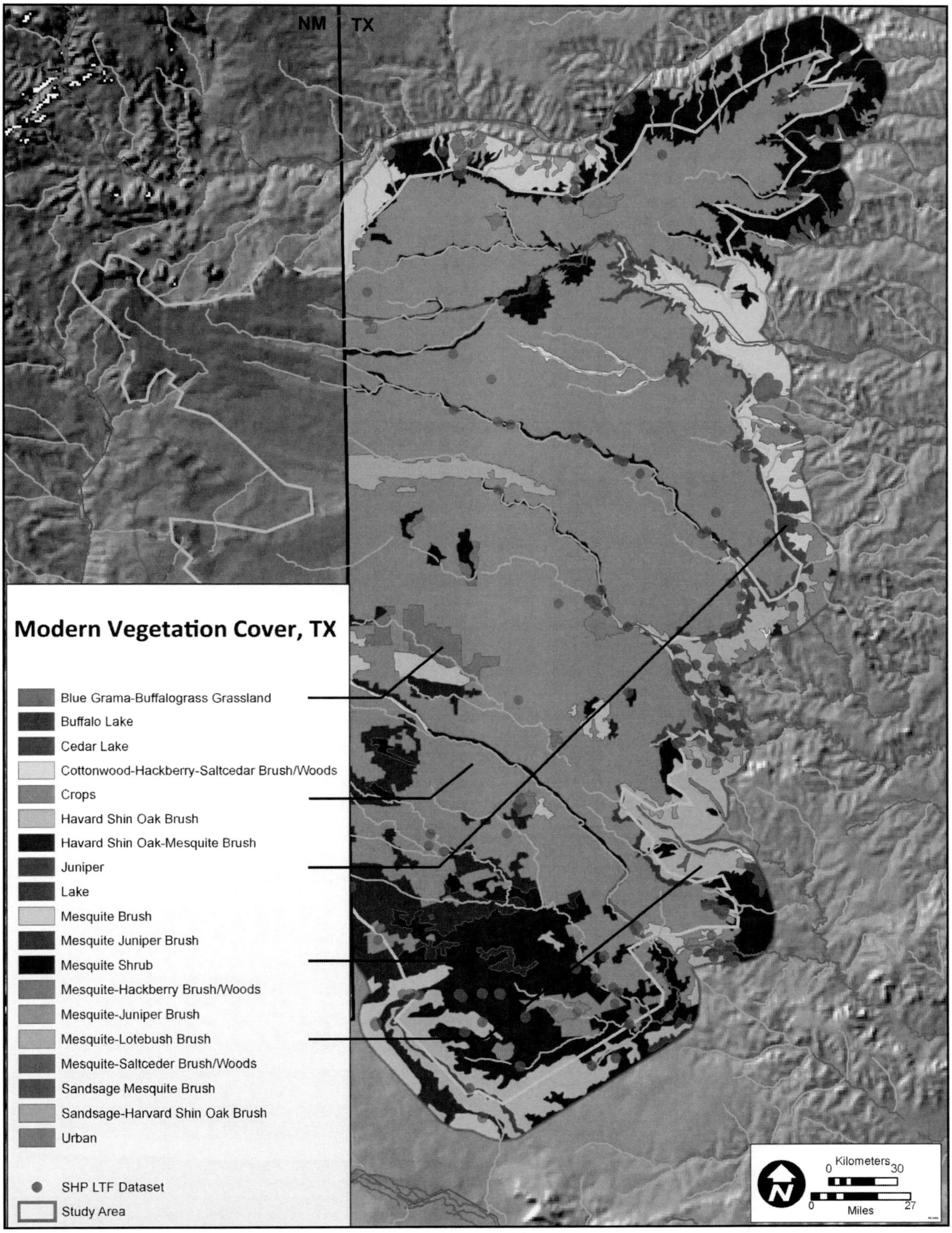

Figure 8.4. Spatial distribution of sites with LTFs identified within the research area overlaid on the modern vegetation communities along the Texas potion of the area. Note: important vegetaion communities indicated (data: 1:100,000 scale vegetation community dataset, Texas Park and Wildlife).

The vast flat terrain is ideally suited to crop production whereas the rugged canyonlands at the periphery clearly are not. Spatial examination of the intensity of farmland by county, compared to the distribution of archaeological sites identified in the regional dataset, indicates that far fewer sites are located in the most intensively farmed counties (Figure 8.5). The most intensive farming practices have occurred in a northwest through southeast trending band that bisects the region from Quay County (NM) in the northwest through to Howard County (TX) in the southeast. The distribution, therefore, demonstrates clearly that farming has not evenly affected the preservation and visibility of the cultural resources contained within the region (Figure 8.5).

Quantitative examination demonstrates only a weak relationship between modern agricultural intensity and the number of LTF sites identified per square mile, when examined by county (Figure 8.6). Thirteen counties located within the study area were excluded from computation due to the absence of sites with evidence for LTFs identified within their boundaries (Figure 8.6). The mean percentage of farmland within these counties was 86.6%. The results indicate that only Eddy County (NM) demonstrates a strong correlation between low agricultural intensity and a high number of LTFs identified per square mile (Figure 8.6). Eddy County is unusual, however, in that the federal government manages a high proportion of the land within its jurisdiction. The county is atypical, therefore, of land ownership in the wider region and not a useful indicator of the effects of agricultural activity on cultural resources. A simple Pearson correlation coefficient test reveals no significant relationship (P=0.840) between the percentage of land farmed by county and the numbers of LTF sites identified per square mile.

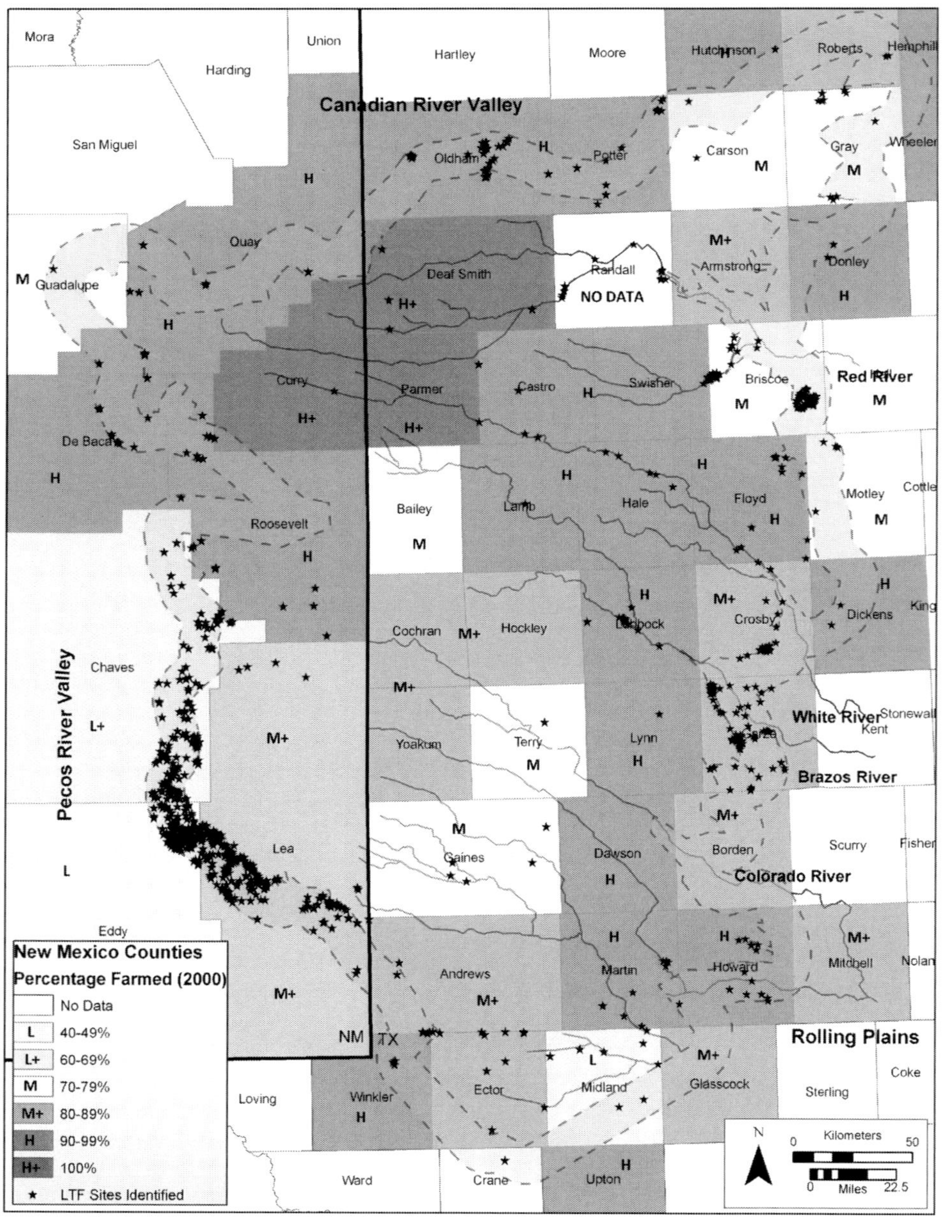

Figure 8.5. Spatial distribution of sites overlaid by modern farming intensity by county (data U.S. Census Bureau, Census 2000).

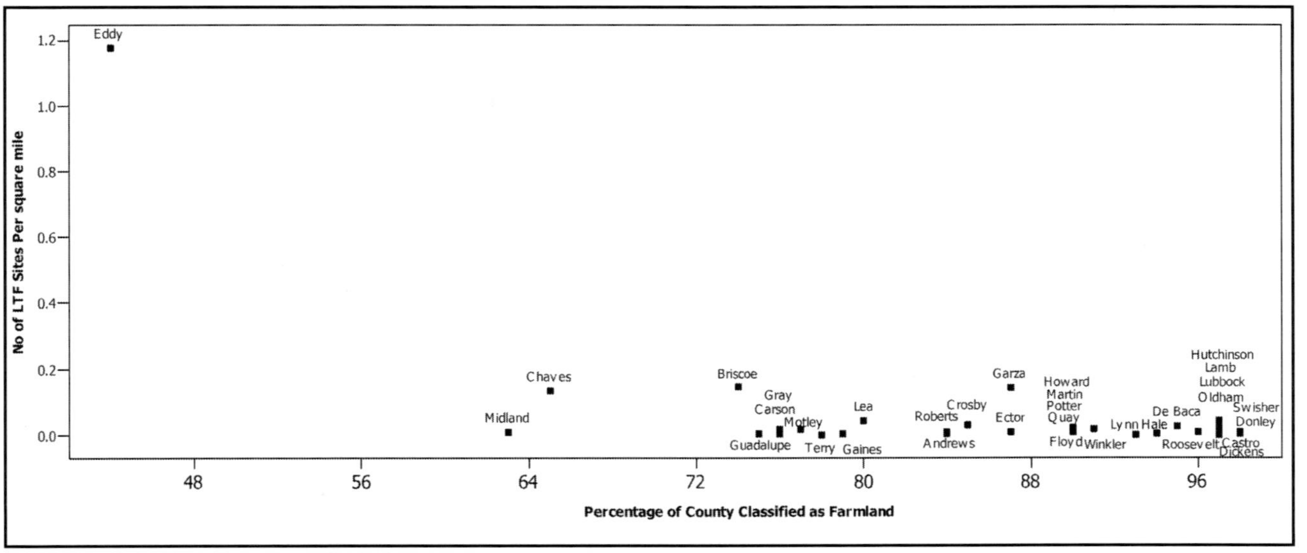

Figure 8.6. Bivariate scatterplot showing percentage of County classified as farmland versus the number of LTF sites identified per square mile, excluding Armstrong, Bailey, Borden, Chochran, Dawson, Glasscock, Hall, Hemphill, Hockley, Mitchell, Upton, Wheeler, and Yokam Counties due to the lack of identified sites (agricultural data from the U.S. Census Bureau, Census 2000).

Conclusions drawn from these data should be interpreted with the caveat that private land ownership of vast tracts of land across the Southern High Plains for the purposes of agriculture has undoubtedly depressed the numbers of hunter-gatherer campsites reported and therefore identified by the current project. It is most likely that archaeological sites located in the vicinity of the seasonal Playa lakes within the interior of the region have been most adversely affected by this process.

Federal Land

Sites identified in the archaeologically dense zone at the south-western flank of the investigation area appear much more likely to be identified on the Bureau of Land Management (BLM)-owned section of the land (Figure 8.7). The distribution of sites raises important questions as to the factors that have resulted in the increased visibility of hunter-gatherer fire technology in this area of New Mexico. Specifically, were fire-related resources such as wood or distinct socio-political groups spatially confined to this section of landscape? Answers to these questions are unfortunately not straightforward and are clearly the result of a complex set of circumstances that have resulted in the increased visibility of sites in some areas (i.e. the BLM land) and not others.

Firstly, from a geographic perspective the geomorphic setting of these sites is along an erosional band, known formally as the Mescalero Plain (Sebastian and Larralde 1989). This plain makes up much of the eastern portion of the Pecos River Valley and the surficial deposits are primarily defined as Plains-Mesa Sand Scrubland (Figure 8.8). The modern surface is extremely dynamic and is characterized by rapidly shifting deposition and erosion processes (Sebastian and Larralde 1989:7). As already noted in chapter 4, the dominant southwest wind pattern results in the eastward transport and deposition of the loose sediments along the Pecos river valley on to the adjacent High Plains. The shifting and generally erosional setting of the Mescalero Plain appears, therefore, to have been conducive to the exposure of hunter-gatherer sites with evidence for domestic fire technologies at the modern ground surface. Although largely the result of geomorphic processes, the resultant sites still obviously require archaeologists to record them. This may explain the discrepancy between the density of sites located on the BLM section of land and the relative paucity in areas of farmland in what are essentially the same geomorphic settings (Figures 8.7, 8.8).

The second factor underlying this distribution then, is clearly political. The BLM lands are federally owned and managed by the government of the United States and are therefore subject to national laws regarding the management of the cultural resources within its boundaries (particularly the Archaeological Resources Protection Act of 1979). A political climate sympathetic to conserving cultural resources on federal land has resulted in the need for planning documents for these vast regions. The practical method of achieving this goal has historically been systematic archaeological survey. Multiple cultural resource surveys have therefore been conducted on the BLM land within the research area (see Katz and Katz

2000 for a recent review of surveys). The intensity of the survey effort can be examined by referencing the site-survey forms. For instance, site LA43319 in Lea County has been surveyed seven times in the period 1988 to 2003. Similarly, site LA83680 has been surveyed eight times in the period 1990 to 2002. The emphasis of these surveys is on identifying (and re-identifying) sites and there is little need to excavate, as the land is generally not threatened by development.

Large Development Projects

In areas outside of the BLM-administered landscape, large-scale developer-led projects have clearly influenced the spatial distribution of LTF sites (Figure 8.9). These largely survey-based projects include, from north to south: the Diamond Shamrock pipeline project in Potter County, the Tule Canyon/MacKenzie Reservoir survey, Swisher and Briscoe Counties (Katz and Katz 1976); the Canyon Lakes redevelopment, Lubbock County (Johnson and Stafford 1976); the L7 ranch survey, Crosby County (Nickels 2002); and the mid-continent pipeline project, Andrews and Midland Counties (Brethauer 1977). Additionally, the development of two state parks has also resulted in significant site densities being reported, both at Palo Duro Canyon State Park (Boyd 1995) and Caprock Canyons State Park (Mercado-Allinger 1982).

All the identified projects were undertaken at the periphery of the Llano Estacado in the rough breaks and canyonlands. The exceptions are the Canyon Lakes project in Lubbock County, which was a consequence of a unique need for urban renewal (i.e. the devastation wrought to a large section of the City of Lubbock by an F5 tornado); and the Mid-Continent Pipeline, which runs through the south of the region, servicing the Permian Basin oil industry (Brethauer 1977).

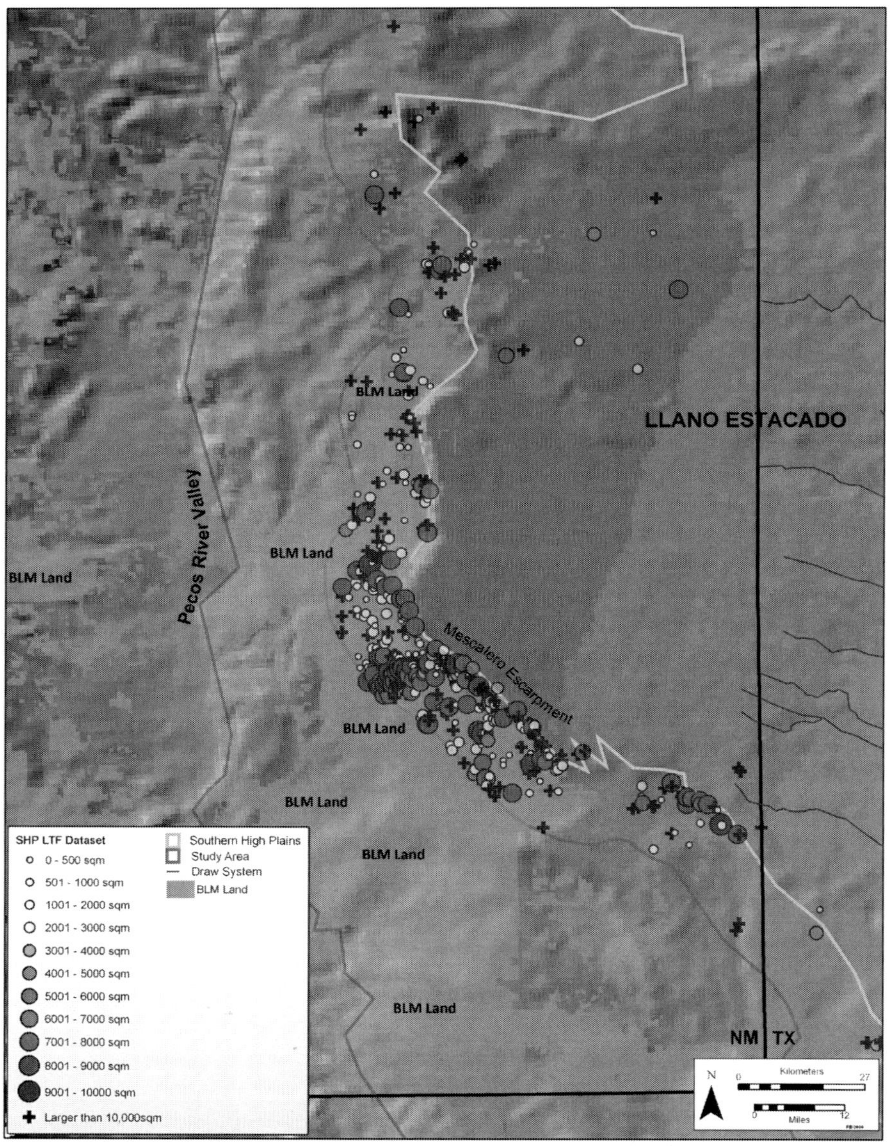

Figure 8.7. Distribution and size of sites identified along the Mescalero Escarpment and adjacent dunes relative to Bureau of Land Management (BLM) land.

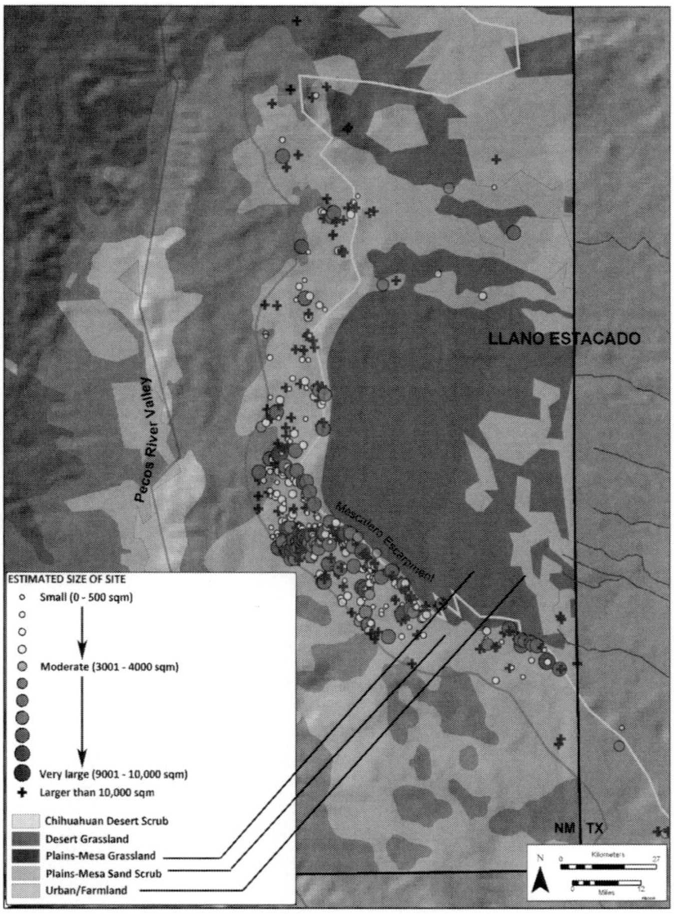

Figure 8.8. Distribution and size of sites identified along the Mescalero Escarpment and adjacent dunes relative to modern land cover (vegetation data from the Natural Resources Conservation Service, www.nrcs.usda.gov).

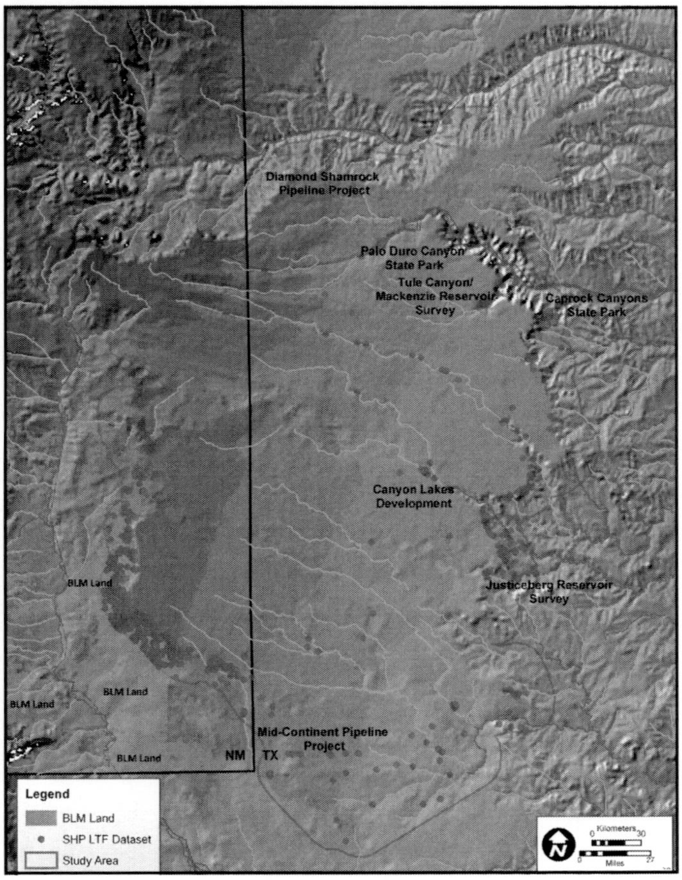

Figure 8.9. LTF sites identified as large-scale survey projects in the Texas portion of the study area.

Archaeological Community

Despite a strong trend toward commercialisation and institutionalisation, archaeological sites on the Southern High Plains can be and still are identified and recorded by interested avocational archaeologists. These individuals are distributed fairly randomly in the wider landscape, oftentimes examining sites located nearby to the city and county in which they live. The distribution of people with an interest in recording archaeological sites is loosely controlled by a number of factors, including the presence of a research institution such as a museum or university in the local community, or the establishment of an archaeological group or society that is interested in recording archaeological sites. The location, then, of research centres, groups, and individuals has the potential to affect the distribution of sites recognised in the regional dataset in unpredictable ways.

Examination of the current dataset indicates that the archaeological community has played a significant role in the formulation of the spatial distribution, an excellent example being the 53 sites recorded by Emmett Shedd of Aspermont, Texas, between the late 1950's and late 1970's. Very little information is available as to the types of localised thermal features encountered at the sites he recorded. However, it is important to note that these sites comprise over 5% of the total number identified in the regional dataset and the majority of sites along the south-eastern caprock escarpment (mainly Garza County, see Figure 8.3 for visual confirmation of this contribution).

Geomorphology

The geomorphology of the Southern High Plains is well known (see Chapter 4) and it has already been demonstrated that the local conditions in the dune fields along the Mescalero escarpment clearly affect the visibility of archaeological sites in this area (e.g. Sebastian and Larralde 1980; Hall 2002). Recent geoarchaeological investigations have clearly demonstrated that virtually all archaeological sites identified along the Mescalero dunes are associated with a single depositional context of eolian sands, identified as Unit-2 (Hall 2002:21). Optically Stimulated Luminescence (OSL) dating on this deposit has indicated that it effectively comprises a palimpsest of cultural material that spans the last 5,000 years of human occupation.

Geomorphic processes in the remainder of the study area have most likely also radically affected the visibility of archaeological material and sites, the most geomorphically dynamic setting being the eastern escarpment where the slow westward creep continues (Gustavson and Simpkins 1989:18).

High-quality well-buried stratified sites are most likely, to be preserved and therefore encountered in the draw systems within the interior of the region (e.g. Holliday 1988a, 1988b, 1990, 1995, 1997). Unfortunately, the deeply buried nature of these sites, (terminal Pleistocene deposits are often more than 5 meters below the modern ground surface), means that access to the oldest deposits is logistically difficult (Stafford 1981, Backhouse and Johnson 2009). Sites that have intensively excavated at these depths, such as at Lubbock Lake or Blackwater Draw Locality #1, have been done so due to fortuitous external circumstances (dredging of the valley at Lubbock Lake and commercial quarrying at Blackwater Draw). Furthermore, the nature of the archaeological record (hunter-gatherer material culture) and the site settings are not appropriate for geophysical detection methods. Targeting of these deeply buried deposits therefore requires considerable research effort in order to achieve the invasive sampling required with the resolution necessary to identify the often-ephemeral traces for hunter-gatherer activities (Backhouse and Johnson 2009). Despite high potential, research in the Draws has been limited due to the practical difficulties involved in its development. For the foregoing reasons, the frequency of sites identified within the draw systems are most likely radically under-represents the true number of available sites in this geomorphic setting.

Spatiotemporal

The temporal association of the sites identified in this dataset is not well known (Figures 8.10) and nearly 60% of the sites cannot be placed in any form of temporal framework. For the remainder of sites, artefact typological information is the most common form of temporal control accounting for some 36% of the total dataset (Figure 8.10).

Unfortunately, less than 1% of the reported site-survey forms contain any absolute dating information (such as the radiocarbon method). The apparent lack of absolute dating is most likely to be the result of the lag time between the submission of archaeological site-survey forms (typically fairly rapidly after the completion of a project) and the time required to find funding, submit, and receive dates (typically around between 3 and 6 months after submission). These dates, therefore, are simply not available at the time of completing the site-survey form. The situation is unfortunate and there is currently little incentive in either Texas or New Mexico to amend site-survey forms with this critical archaeometric information at a later date. At present, the only way to receive such dates is to trawl through the mountain of grey literature (where available) and extract pertinent dates as appropriate. This exercise was undertaken for the site components examined in detail for technological analysis. However, until such records are made available electronically, as part of more sophisticated state-maintained databases, then the limited temporal information available in the present dataset must be utilised to examine the record from a coarse-grained perspective.

Site data, comparable to the current project, was generated by Sebastian and Larralde (1989) in an effort to characterise the cultural resources located within the large area managed by the BLM in the south-west portion of the study area. The temporal distribution of sites identified in both the BLM survey and the LTF database is remarkably similar and direct comparison between these datasets highlights several important factors. Firstly, this similarity indicates that the wider regional dataset is significantly controlled by the density of sites identified along the eastern flank of the Pecos river valley. Secondly, the temporal composition of the dataset has not changed significantly in the intervening twenty years of investigation. Fine-grained comparison indicates that the current LTF dataset conforms more closely to Sebastian and Larralde's (1989) south-eastern New Mexico Pecos Valley dataset than it does to the south-eastern New Mexico Plains dataset (Table 8.1). The influence of the Pecos river valley site distribution is, therefore, clearly highly significant when interpreting the results of the wider LTF site distribution.

Table 8.1. Frequency of sites identified in the present study versus those surveyed in south-eastern New Mexico, presented by cultural component.

Temporal Component	LTF Dataset	South-eastern New Mexico Plains Dataset†	South-eastern New Mexico Pecos Valley Dataset†
Paleoindian	0.29	12.7	1.2
Archaic	5.8	13.7	8
Ceramic/Jornada Mogollon	28.8	32.1	32
Protohistoric	1.3	0.5	1.3

† data from Sebastian and Larralde (1989; Table 8.3)

Spatiotemporal examination of the entire regional dataset (Figure 8.11) reveals no clear patterning in the location of particular temporal components throughout the landscape. The distribution is not easily interrogated because of the locally tight spatial distribution of sites identified (e.g. the density of sites identified along the Mescalero Escarpment). In order to examine the temporal distribution in more detail, it is, therefore, necessary to examine each temporal component separately.

Throughout the following, I have deliberately not concentrated on key sites, such as Lubbock Lake (41LU1). This decision is primarily because the well-recorded components at these sites dominate the regional record for any given temporal period. I return to them, however, when considering specific examples of LTF technology in the subsequent technologic section.

Paleoindian (11,500 – 8,000 RCYBP)

Four possible Paleoindian campsites are identified in the regional dataset (Figure 8.12). Three of these are located in the New Mexico portion of the study area. With so few sites identified in this time period it is impossible to infer any meaningful regional patterning. The paucity of data for Paleoindian campsites suggests that: there was a low Paleoindian population density; the use of low-visibility or archaeologically invisible domestic fire technologies was taking place; the eradication of the archaeological evidence for LTFs by site-formation processes has taken place; the inability of archaeologists to identify Paleoindian LTFs has depressed the number identified, limited access to Paleoindian-aged deposits has lowered the number of sites identified, or a combination of one or more of these factors is acting on the archaeological evidence.

Of the four sites, LA3324 (Blackwater Draw Locality 1) and LA6209 (the Milnesand site) are the most widely known and published (see Sellards, 1955; Warnica and Williamson 1968; Johnson *et al.*, 1986; Buchanan *et al.* 1996; Hester 1972; Holliday 1997; Hill 2002). At both sites, LTFs are seemingly documented in stratigraphic and contextual association with Paleoindian material culture or extinct bison.

The evidence for LTFs within the Paleoindian levels at Blackwater Draw Locality 1 is minimal. A single brief description (Hester 1972:178) suggests LTFs at this site were associated with Folsom-period campsite debris.

If confirmed, this would be the earliest evidence for domestic fire technology available for the Llano Estacado. Additionally, the Milnesand site (Sellards 1955) and nearby Ted Williamson site (Warnica and Williamson 1968) also appear to offer limited evidence for examining Paleoindian-period domestic fire technology (see Technological: Paleoindian).

Unfortunately, very little information is available for the Lenorah site (41MT1). The presence of hearths is recorded on the site-survey form (Collins 1966) but their relationship to the ca. 35 Plainview projectile points that were recovered is unclear. Similarly at LA43391 in Lea County, hearths are recorded with fire-cracked rock and diagnostic Paleoindian projectile points. All artefacts were recorded on the modern ground surface and the relationship between the projectile points and the features is unknown.

Archaic (8,000 – 2,000 RCYBP)

The number of identified sites rises significantly in the Archaic period (Figure 8.10). One density focussing along the Mescalero dunes to the southwest of the region is noted; outside of this area, sites are spread out fairly evenly around the periphery of the region (Figure 8.13). Very few of the sites are located away in the interior of the region, with only two sites over 30 miles from the nearest caprock escarpment.

Assuming evidence for aboriginal fire technology is a useful proxy indicator for human population in a landscape (see chapter 5), then coarse grain observation based on the spatiotemporal distribution of sites identified as 'Archaic' suggests that there is increasing evidence for human population, as compared to the preceding Paleoindian site density (Figure 8.12). Qualitative examination of the site-survey forms that comprise the archaic dataset reveals no reported radiocarbon dates. The fifty-nine sites therefore represent typological determinations based on various reporting criteria (Table 8.2).

It is useful to note that the association of diagnostic projectile point types is the most commonly stated method for assigning a temporal period to these sites (51.7%). Survey is the dominant method of encounter and only rarely are buried features reported. The high number (43.1%) of sites assigned to this temporal period without any explicit criterion is obviously worrying. Unfortunately, it is almost impossible to objectively assess the reasons these sites were assigned to the Archaic period. It is likely the case that many sites were assigned to this category on the basis that no Paleoindian projectile points, no ceramics, and no

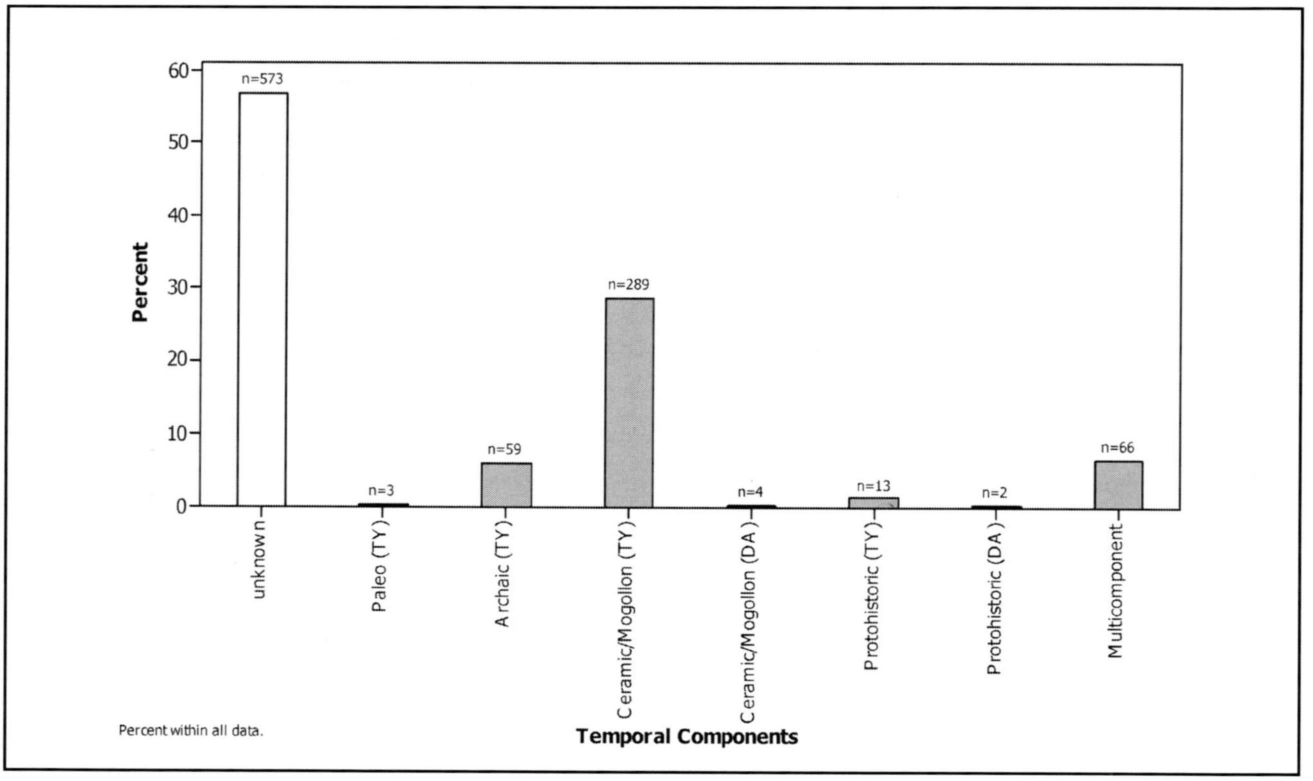

Figure 8.10. Temporal distribution of prehistoric sites with LTF components within the study area.

historic material were observed (see also Sebastian and Larralde 1989 for a similar discussion of the problem concentrating on the south-eastern portion of New Mexico; and also Shelley 1994 for a regional perspective).

The likelihood that this scenario has affected the spatiotemporal association of sites identified as Archaic is somewhat confirmed by the clustering of sites in arbitrary geopolitical areas that are very unlikely to be proportionally representative of Archaic population dynamics. The high number of Archaic sites identified in Briscoe County, within what is now Caprock Canyons State Park (Figure 8.13) is probably a result of a particular temporal identification criteria policy, rather than of direct evidence for a population concentration during the Archaic.

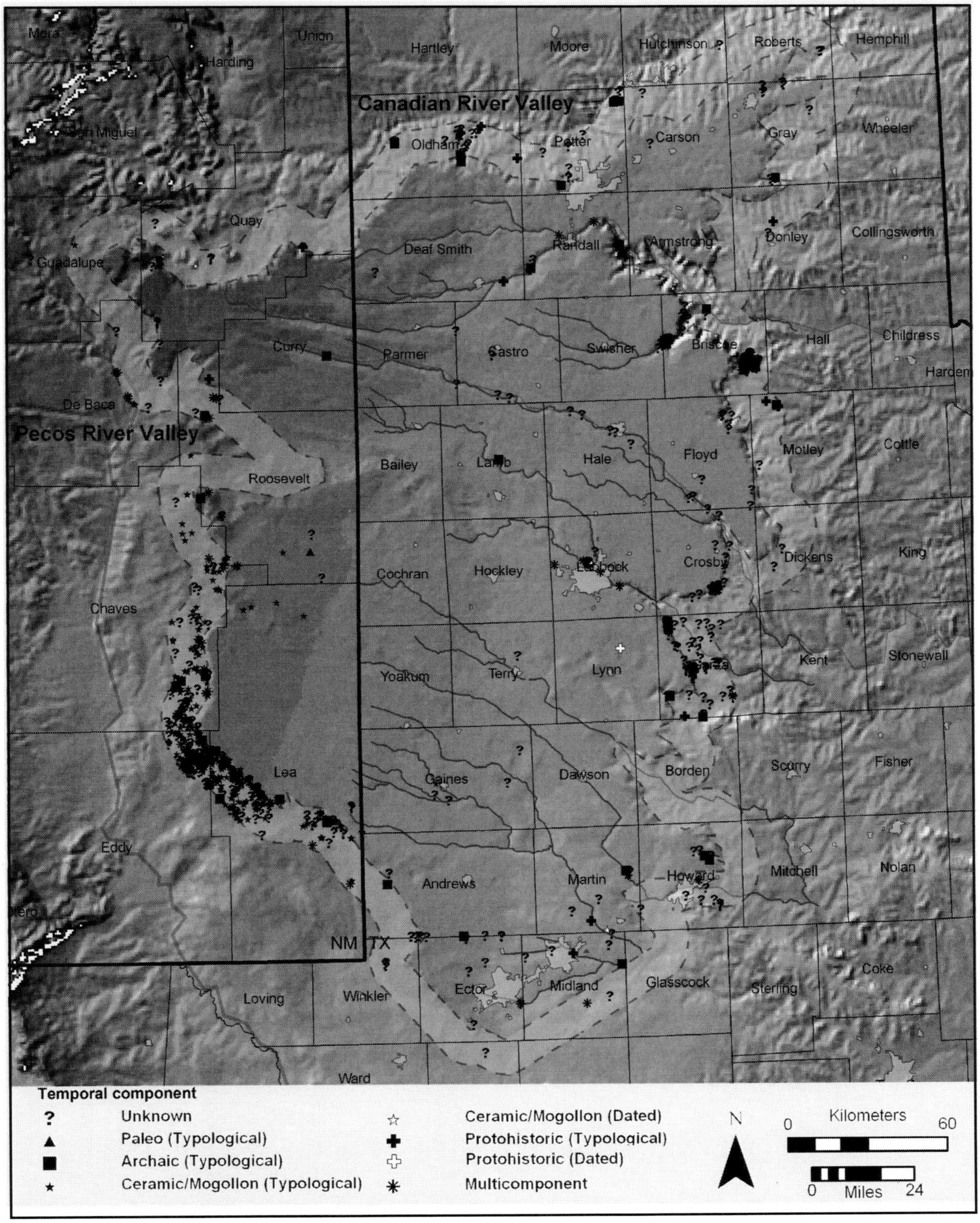

Figure 8.11. Spatiotemporal distribution of archaeological sites with LTF components within the study area.

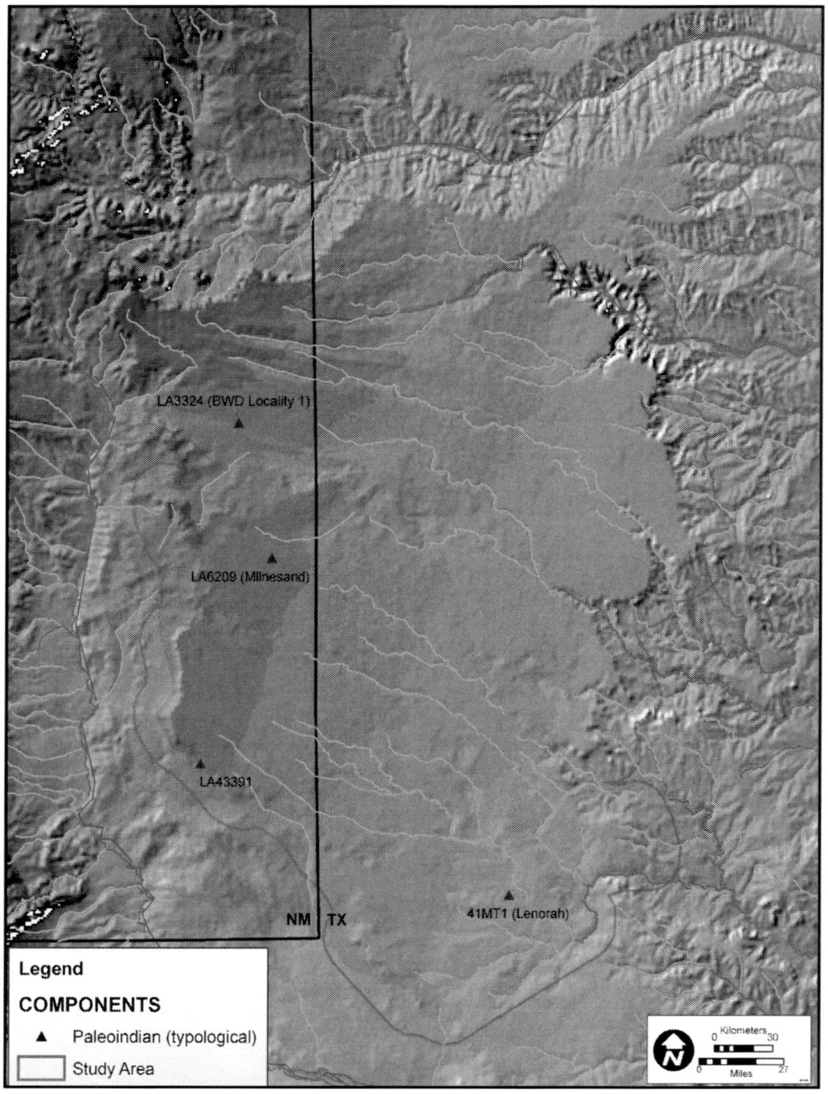

Figure 8.12. Distribution of Paleoindian archaeological sites with LTFs within the study area (BWD Locality 1 added).

Table 8.2. Summary of Archaic period sites identified in the regional LTF dataset.

Basis for Temporal Association	Percent of Sites (%)	Mode of Investigation (number of sites)	Range of Years Investigated	Range of Sites*
Stratigraphic Location	3.4	Survey (1) Testing (1)	1977-1992	41LA1, 41MT21
Diagnostic projectile point Types and Stratigraphic location	1.7	Survey	1998	41PT185
Diagnostic projectile Point Types	51.7	Survey (25) Testing (5)	1968 - 2002	41AD25, 41AD48, 41BI59, 41BI291, 41BI339, 41BI430, 41BI630, LA109928, LA66373, LA66909, 41CB47, 41HW36, 41HW38, 41BI294, 41GR659, 41MY18, 41PT124, 41RD34, LA105829, LA108916, LA43393, LA50483, 41GR640, 41OL29, 41OL39, 41OL178, 41GR659, 41GR623, 41RD3, 41RD50
Unknown/Unspecified	43.1	Survey (23) Testing (1) Excavation (1)	1957 - 2001	41GR120, 41BI10, 41BI143, 41BI153, 41BI175, 41BI22, 41BI298, 41BI312, 41BI319, 41BI323, 41BI332, 41BI365, 41BI390, 41BI395, 41BI397, 41BI400, 41BI466, 41GY41, 41MD32, LA107646, LA48272, LA54114, LA75225, LA79987, LA87022

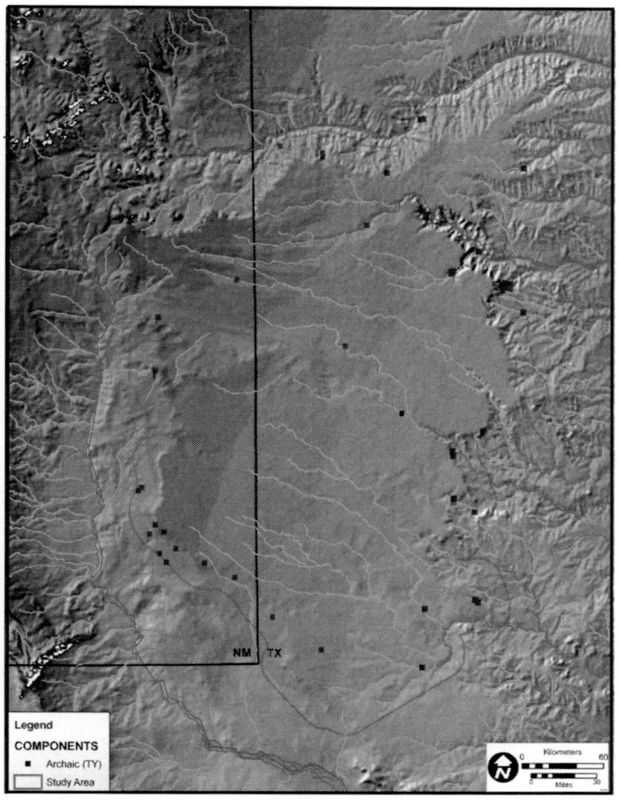

Figure 8.13. Distribution of Archaic archaeological sites with LTFs within the study area.

Ceramic Period (ca. 2,000- 500 RCYBP)

The Ceramic/Mogollon period is the best-represented temporal period identified in the LTF dataset (Figure 8.10). Furthermore, the distribution of identified sites appears to be strongly clustered, with a focus in the southwestern portion of the study area (Figure 8.14). The number of sites identified in this time period is several orders of magnitude more frequent than for earlier periods and appears at face value to be suggestive of a dramatic increase in population.

The assignment of sites as Ceramic is obviously largely dependant on the identification of discarded pottery. A crude assessment of the size of isolated pottery fragments and diagnostic lithic materials (such as projectile points) suggests that they are not overly dissimilar. All other things being equal, the opportunity for archaeologists to encounter either artefact type might reasonably be expected to be similar. Given the inherent spatial bias likely to be operating on the current dataset, then we might reasonably expect material culture and, therefore, sites representing all the cultural time periods to be equally well represented in this section of the study area. The Ceramic/Mogollon LTF site dataset shows this to be spectacularly not the case. The corollary of this argument is that the Ceramic period does appear to represent the first intensive evidence for domestic fire technologies in a spatially concentrated portion of the Southern High Plains landscape.

The pattern of sites identified in the LTF dataset is in line with more focused studies of the Jornada-Mogollon culture group (e.g. Whalen 1994; Miller and Kenmotsu 2004). During this period, archaeologically observable cultural activity appears to have shifted toward a more sedentary lifestyle, with populations focused in semi-permanent residences for the purposes of agriculture and craft specialization. The collapse of this way of life, around 500 years ago, was also dramatic and appears well evidenced in the distribution of sites plotted for Ceramic/Jornada Mogollon and the later Protohistoric periods (Figures 8.14, 8.15). Miller and Kenmotsu (2004) have utilized an extensive radiocarbon dataset (n > 1000 samples) to provide a more fine-grained chronology for the thermal features in the eastern Trans-Pecos area. There data indicates a peak in LTF construction and therefore aggregated populations by ~1300 B.P. followed by a marked decrease after ~700 B.P (Miller and Kenmotsu 2004:251). The elevation of the features identified in the southwest of the current study area appears consistent with the interpretation that Ceramic period populations mapped onto areas where edible desert plant species were most abundant (i.e. areas over 3,500 feet). It is likely, therefore, that the majority of the features identified in this area represent the remains of facilities for plant processing.

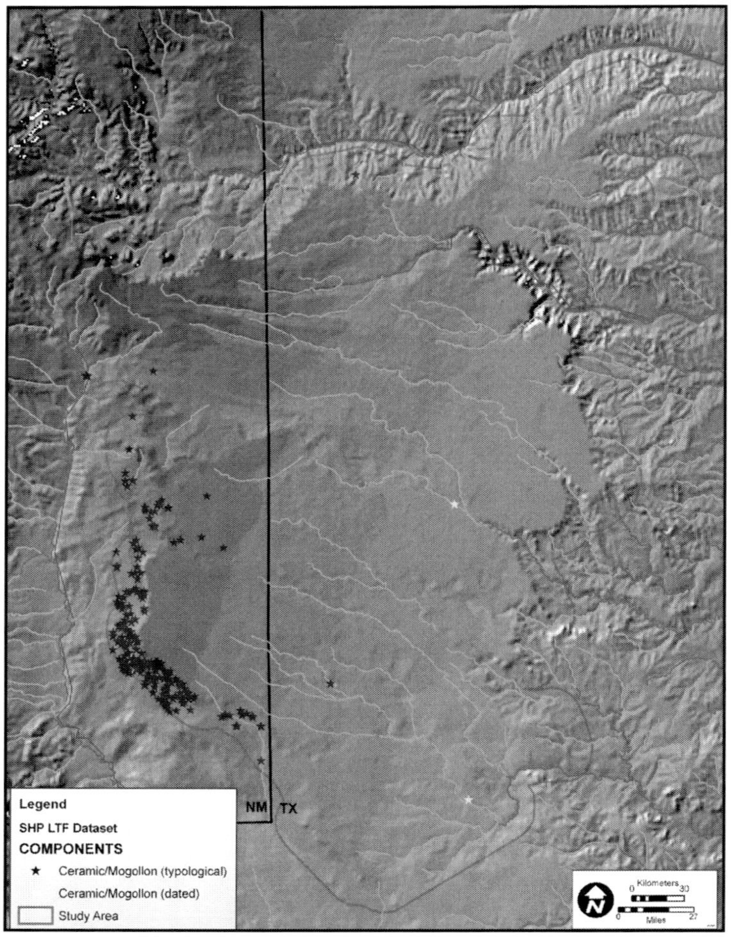

Figure 8.14. Distribution of Ceramic/Mogollon archaeological sites with LTF components within the study area.

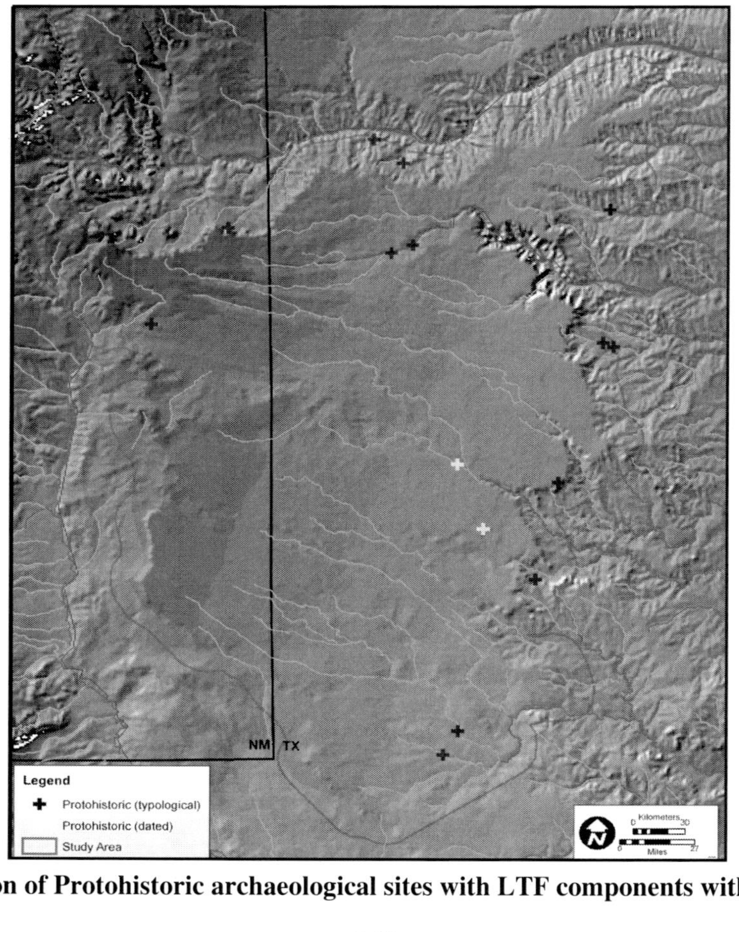

Figure 8.15. Distribution of Protohistoric archaeological sites with LTF components within the study area.

Protohistoric Period (ca. 500 - 300 RCYBP)

The distribution of sites during the Protohistoric period appears to represent a massive decrease in the number of archaeological sites with evidence for domestic fire technologies (Figure 8.10). The apparent decrease should, however, be tempered by the relatively short period of time represented by the Protohistoric (200 years). Nevertheless, the spatial distribution of Protohistoric sites indicates a radically different distribution to that observed during the preceding Ceramic period (compare Figures 8.14 and 8.15). On the basis of only this dataset then, it appears that a return to a more Archaic type of land use has occurred (compare Figure 8.15 with 8.13).

During the Protohistoric, it seems then that camps were typically located at the periphery of the region with a focus on the eastern draw systems and rough canyon-lands that demark the eastern boundary of the study area. The previously identified and seemingly populous portion of south-western New Mexico appears to have undergone a radical depopulation at the end of the Ceramic period ca. 1500AD. Many theories have been put forward to explain the apparent collapse of the Jornada Mogollon and these will not be commented on further here (for a recent summary see Miller and Kenmotsu 2004). However, the regional scale of the current dataset does permit the course-grained observation that populations may have once again become more mobile (and therefore less archaeologically visible), thus resulting in the ephemeral traces of activity at disparate locations within the wider landscape. This scenario appears to fit well with the Protohistoric site distribution as presented here (Figure 8.15).

Technological

Analysis of the technological component of the LTF dataset is to some extent controlled by the frequency and mode of archaeological research undertaken throughout the region. Examination of the maximum mode of investigation, therefore, reveals a similar overall approach to site investigation across both states (Figure 8.16). Both state datasets are dominated by survey-based investigations that comprise the vast majority (85%) of the overall regional dataset. More intensive modes of investigation appear to be common in the Texas portion of the study area, although the overall quality of investigations is clearly more variable in this area with a small number of sites unable to yield even basic information as to the mode of investigation undertaken.

Examination of the regional dataset clearly reveals that the detailed systematic recording and excavation of LTFs at archaeological sites on the Southern High Plains is extremely rare. This realisation severely limits the interpretative possibilities available when examining these features from a fine-grained technological perspective.

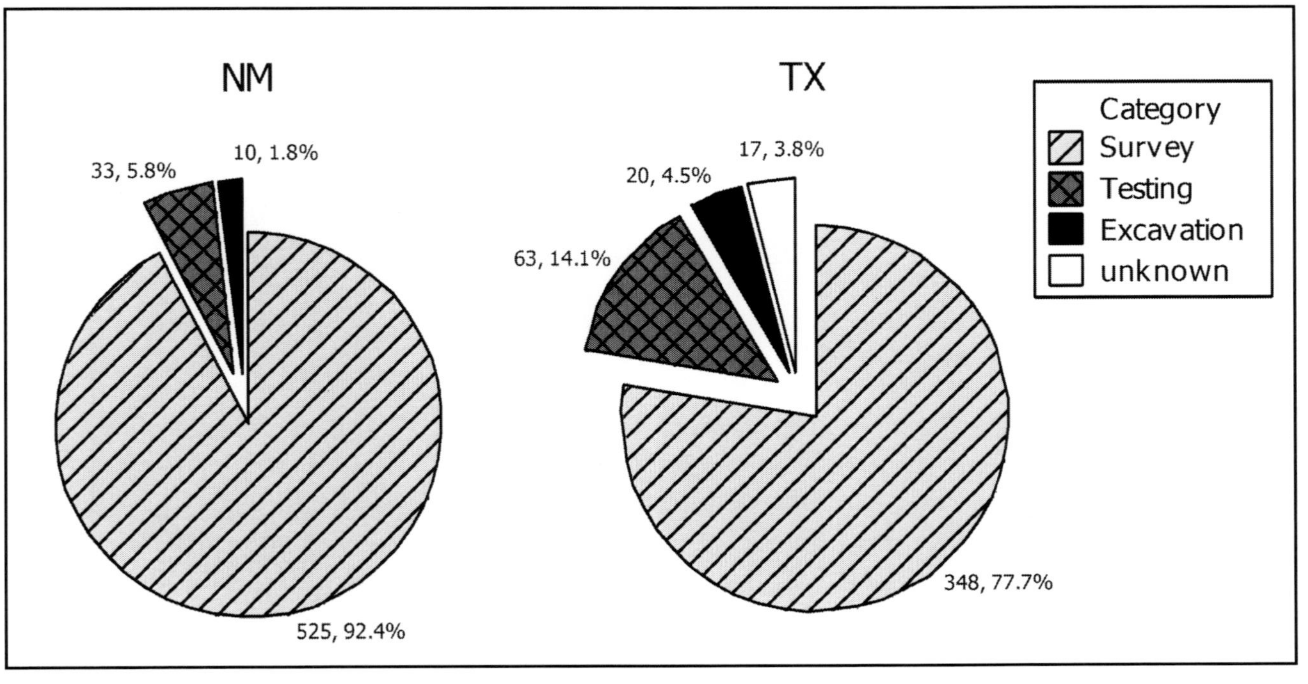

Figure 8.16. Mode of archaeological investigation undertaken at sites with LTFs by state. Note: where more than one mode of investigation occurs then the most intensive is selected regardless of date.

Number of Features Per Site

The number of LTFs identified per site indicates that low numbers of features dominate the regional distribution (Figure 8.17). A single LTF appears to be typical, although sites with between two and five features are also well represented. Unfortunately, at almost 30% of sites no determination as to the number of features originally present or preserved in the archaeological record was recorded. This is perhaps not surprising given that the record is dominated by hot-rock clusters. Only a handful of sites (n=8) identified over 20 features, fewer still (n=5) noted over 30 features, and 40 features were recorded at just four sites in the database (Figure 8.17). The spatial distribution of sites with different frequencies of identified LTFs does not appear to exhibit any obvious patterning (Figure 8.18) other than that already identified for the regional dataset (Figure 8.11). The frequency distribution map does graphically illustrate the greater ambiguity of the data in the Texas portion of the study area (visible by the number of question marks plotted in the Texas portion of the area).

Variation in the type of physical evidence for different LTF technologies at a site level appears to be fairly low across the entire dataset (Figure 8.19). The distribution of technological components would appear to suggest that a significant level of technological homogeneity exists both over time and also across space. More likely, however, is the realisation that site-formation processes, archaeological visibility, and recording practices, or a combination of these factors, is skewing the dataset toward the identification of technologically homogenous sites. This apparent technological determinism is tempered by the observation that hot-rock clusters consistently dominate the regional dataset (Figure 7.1).

Primary LTF Component

Geographical examination of the classification of the primary LTF component (Figure 8.20) demonstrates several trends in the distribution of sites. Features identified along the western escarpment and dunes of the south-west of the region are dominated by hot-rock clusters. Significantly, but at a lower order of density, the sites on the Texas side of the state line in both Andrews and Winkler counties demonstrate the same characteristics (i.e. hot-rock clusters and dispersed hearths). This pattern is encouraging and suggests that the classificatory procedure (i.e. the hierarchical key) is operating fairly evenly between the two state-level datasets.

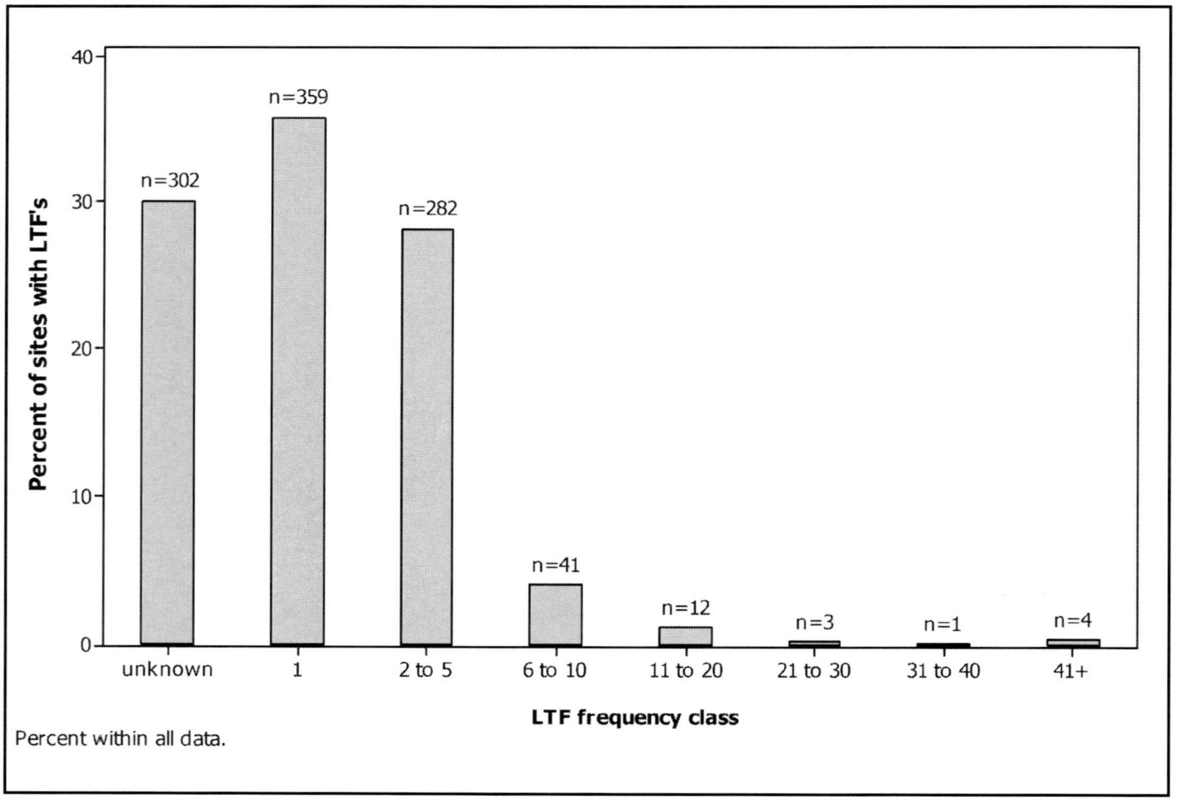

Figure 8.17. Frequency of LTFs at prehistoric archaeological sites within the study area.

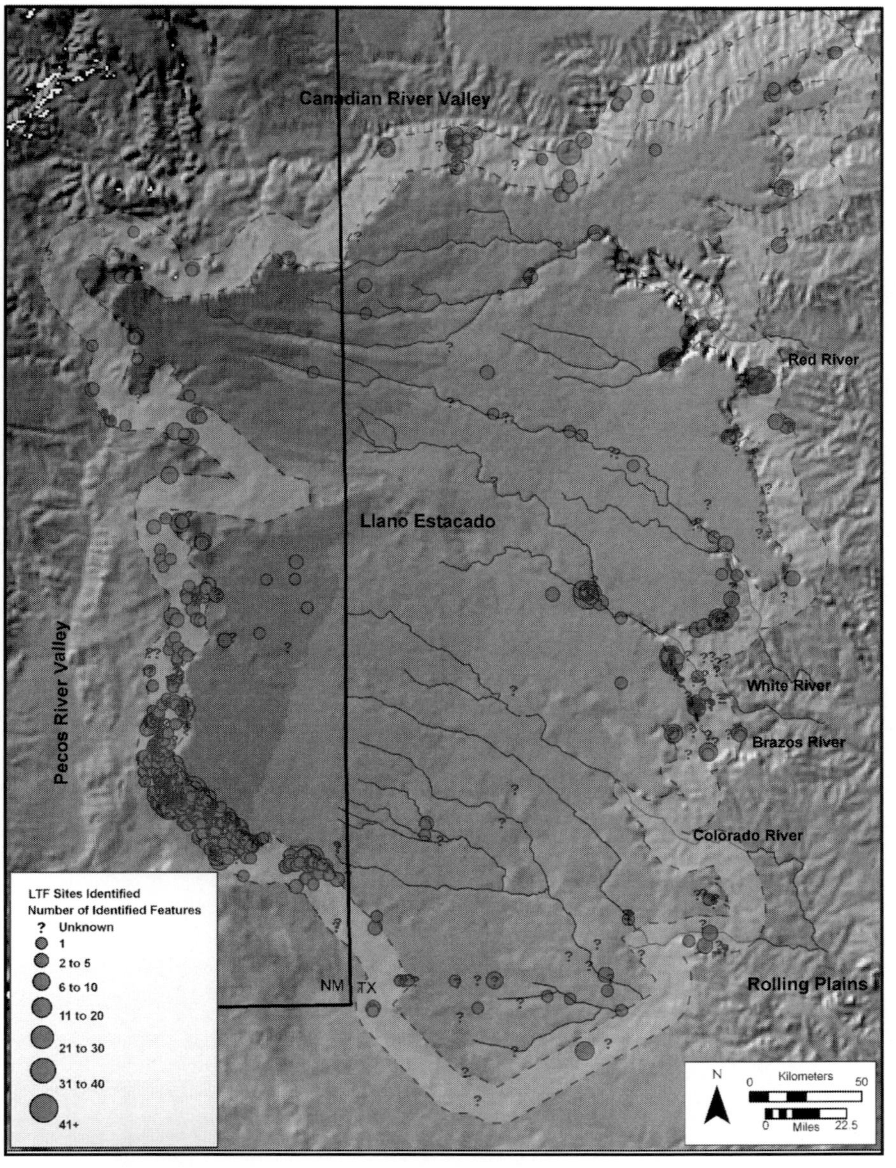

Figure 8.18. Frequency of LTFs by geographic location at prehistoric archaeological sites within the study area.

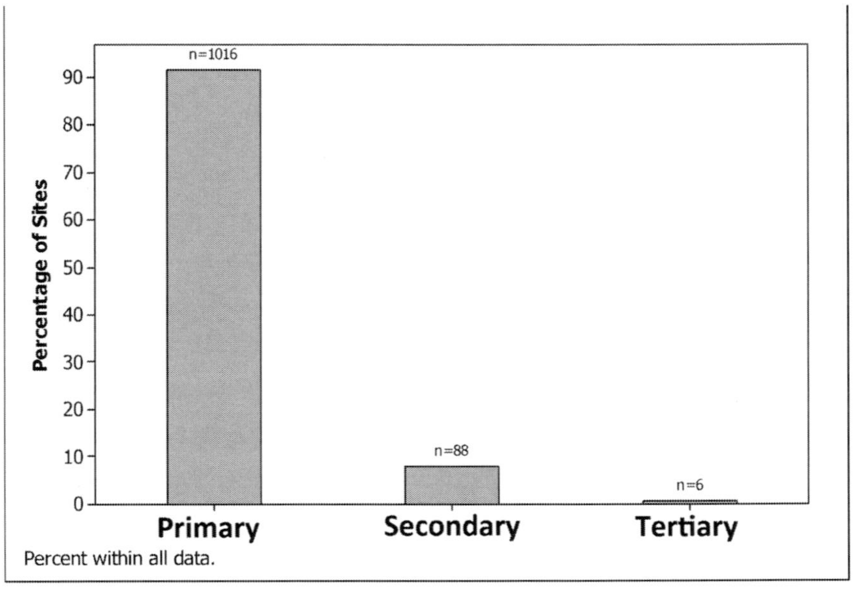

Figure 8.19. Frequency of different LTF components at prehistoric archaeological sites within the study area. Note: the lack of variation in identified components indicating that in most cases only one type of LTF was discernable when applying the hierarchical key.

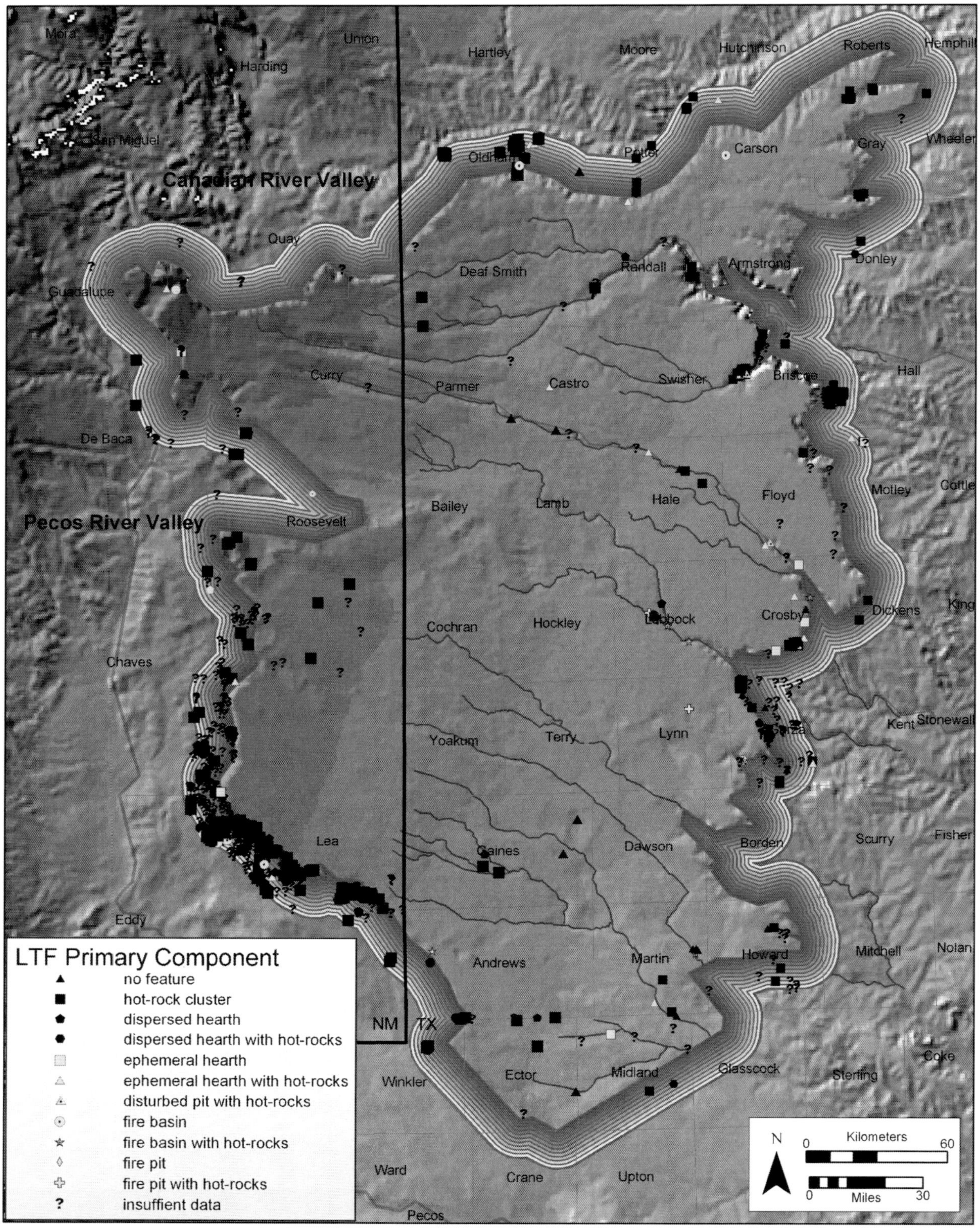

Figure 8.20. Classification of the primary LTF components at prehistoric archaeological sites within the study area.

Examined at a regional scale, the dataset also appears to show little evidence for skew between the two investigated states (Figure 8.20). Proportionally, more variation is perhaps evident in the Texas database and this is likely to be the result of the previously detailed (see Chapter 3) free-form recording procedures. Closer examination of the primary LTF component by state (Figure 8.21) underscores that significant variation exists between the two datasets when considered at the level of individual classifications. Specifically a 'no-feature' classification is most likely to occur in the Texas dataset, whereas an 'insufficient-data' classification are more likely to occur in the New Mexico dataset. The large percentage difference in these two categories is again almost certainly the result of the specific recording procedures. These two anomalous categories can in some ways be conflated as they represent opposite ends of the LTF classificatory scheme and both effectively indicate that in the cases of these sites, very little can be inferred as to hunter-gatherer fire technology.

Perhaps most significant is the realisation that a number of categories are only recognised in one or the other of the state datasets (Figure 8.21). Two categories, 'hearths with hot-rocks' and 'disturbed pits' are only represented in sites located in New Mexico. In contrast 'disturbed pits with hot rocks', 'fire basin with hot-rocks' 'fire pits', and 'fire pit with hot-rocks' are only identified at sites located in Texas. Crucially, the distribution appears to be skewed by decision process 1 (excavated into a former ground surface, see Chapter 3) of the hierarchical key classificatory structure. It would appear that features with some subsurface structure are much more evident or strongly reported on Texas site-survey forms than their New Mexican counterparts. This results in a skew to the right of the hierarchical key structure (Figure 3.7) in the New Mexican dataset, whereas there is a skew to the left in the Texan dataset. Despite this slant, the occurrence of overlap in classifications in nine (60%) of the fifteen categories suggests that the classificatory structure is generally applicable regardless of reporting context. The framework can therefore be regarded as being largely successful in fulfilling its design parameters (see Chapter 5).

The simple interpretation of the observed skew in the results is that it is more likely to be that LTFs are exposed on the modern ground surface in the New Mexico portion of the study area, whereas features in Texas tend to be encountered in excavation or otherwise exposed in section. The preceding statement is supported by returning to the modes of research undertaken in the two areas (Figure 8.16). These data indicate that subsurface explorations (testing and excavation) are more than three times more likely in the Texas section of the region. Further qualitative support is provided by referencing the distribution of regionally important hunter-gatherer sites located within the study region (e.g. Figure 4.9), which also demonstrate a clearly Texas-centric distribution, suggesting that focussed excavation (and therefore the potential to find intact buried features) has occurred more frequently in Texas than in New Mexico. Lastly, it is important to again note that a significant number of the New Mexican sites are located in Bureau of Land Management (BLM)-maintained land that has been subjected to significant erosion processes that have resulted in the mixing of cultural material of different time periods at the modern ground surface (Figure 8.18).

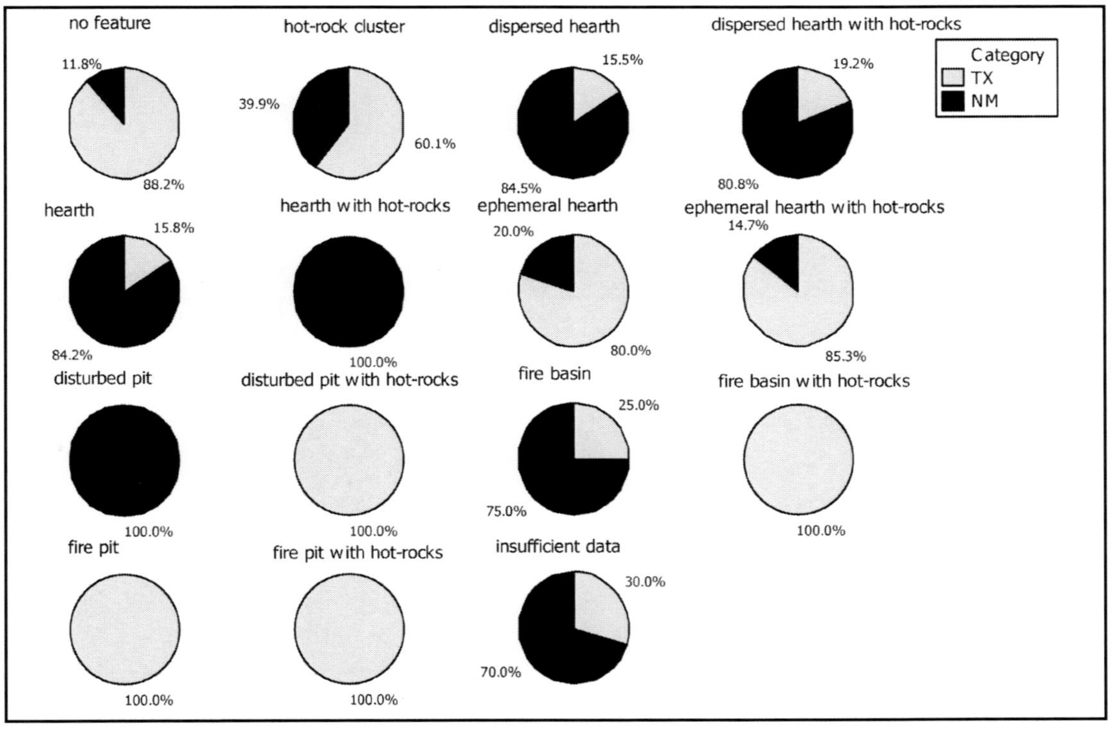

Figure 8.21. Distribution of the primary LTF components identified at sites and presented by State (note: including BWD, Locality 1).

Secondary and Tertiary LTF Components

The largely homogenous nature of the evidence for hunter-gatherer domestic fire technology is reinforced by the realisation that around 90% of the sites identified are comprised of only one technological component (Figure 8.19). The limited evidence for different fire technologies at individual sites is evidenced by secondary (less than 10% of all sites) and tertiary components (less than 5% of all sites).

The classificatory distribution of secondary LTF technologies largely mirrors that observed for the primary components (compare Figures 7.4 and 8.22). The secondary components, therefore, consist largely of dispersed hearths and hot-rock clusters. The qualitative range in types represented does indicate that some intra-site variation in LTF technology is sometimes recognised and reported by site investigators. The minimal presence of tertiary components (Figure 8.23) reinforces the potential to observe and report a range of LTFs at the site scale.

Paleoindian LTF Technology

Only minimal evidence for Paleoindian LTF technology was identified within the study area (Figure 8.10) and three of the four site records provided inadequate information for the purposes of assigning an LTF type using the hierarchical key classificatory structure (Figure 8.24).

Blackwater Draw Locality 1 has a complex excavation history (see Holliday 1997 for a summary). LTFs appear to have been recognised during the Warnica and Shay excavations in 1954; details, however, are extremely limited. James Hester provides the only published description:

> "Circular shallow firepits are reported from the Brown Sand Wedge along the eastern pond margin. These pits, 36 inches in diameter and 8 inches deep, may also be associated with the Folsom occupation" (Hester 1972:178).

The features were reportedly filled with oxidised sand and impressions of the tools used to excavate them were apparently still visible in the walls during excavation (Hester 1972:46).

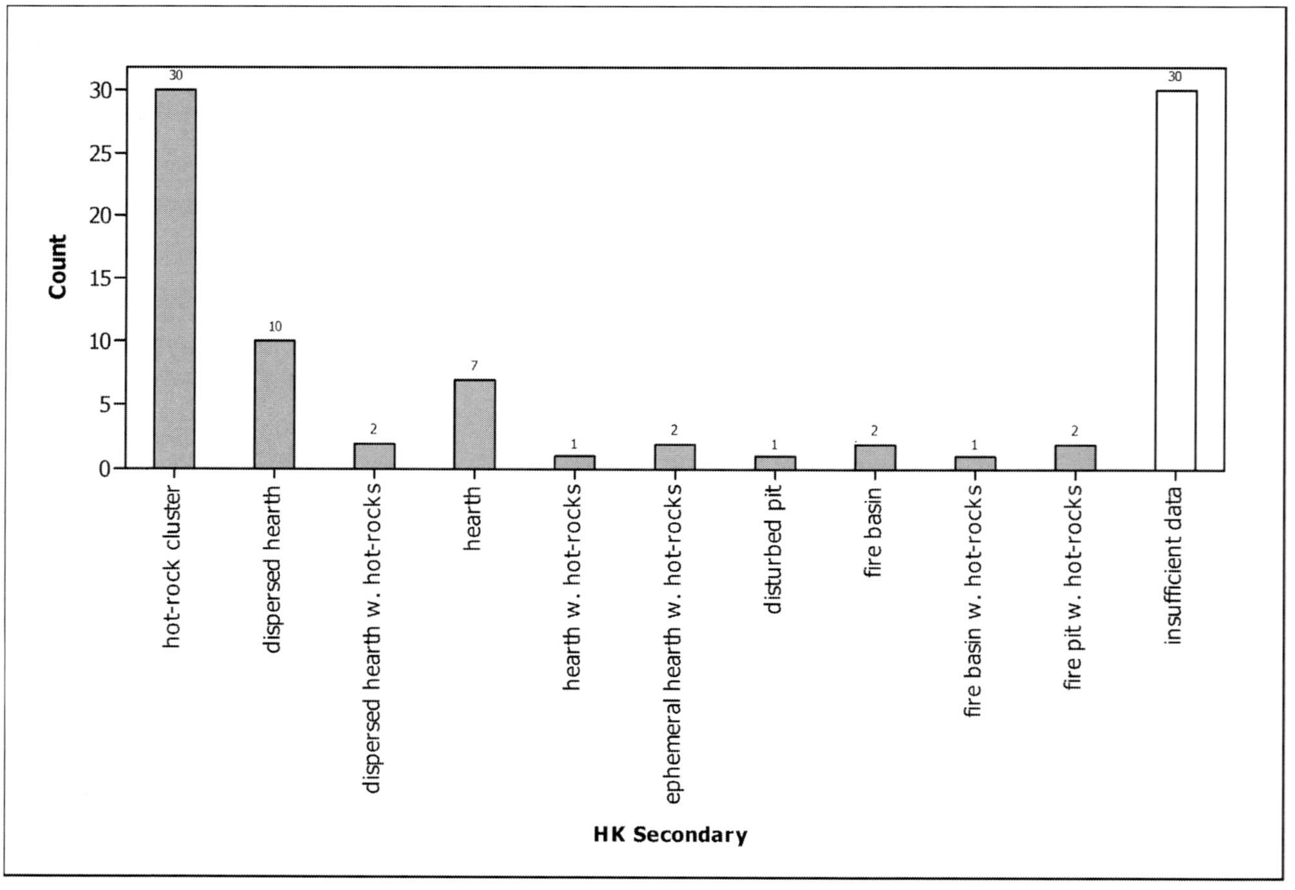

Figure 8.22. Classification of the secondary LTF components at prehistoric archaeological sites within the study area.

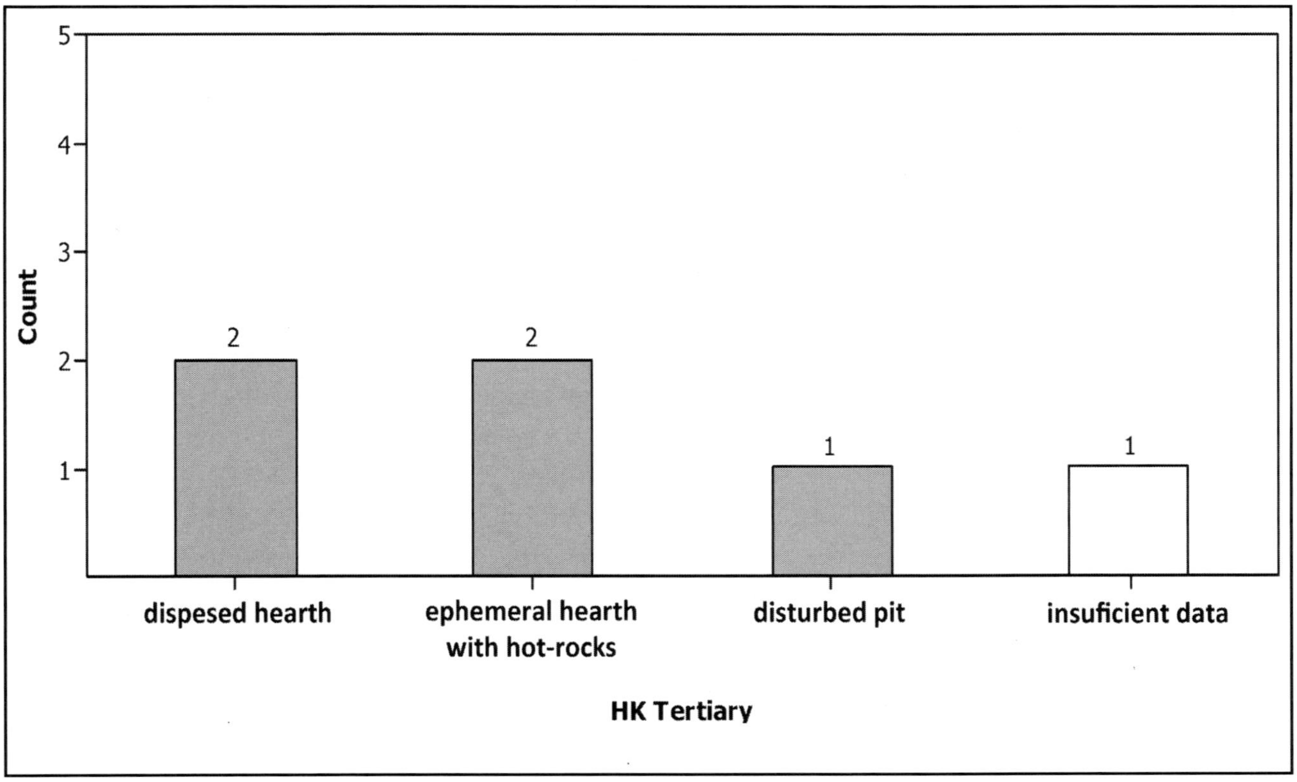

Figure 8.23. Classification of the tertiary LTF components at prehistoric archaeological sites within the study area.

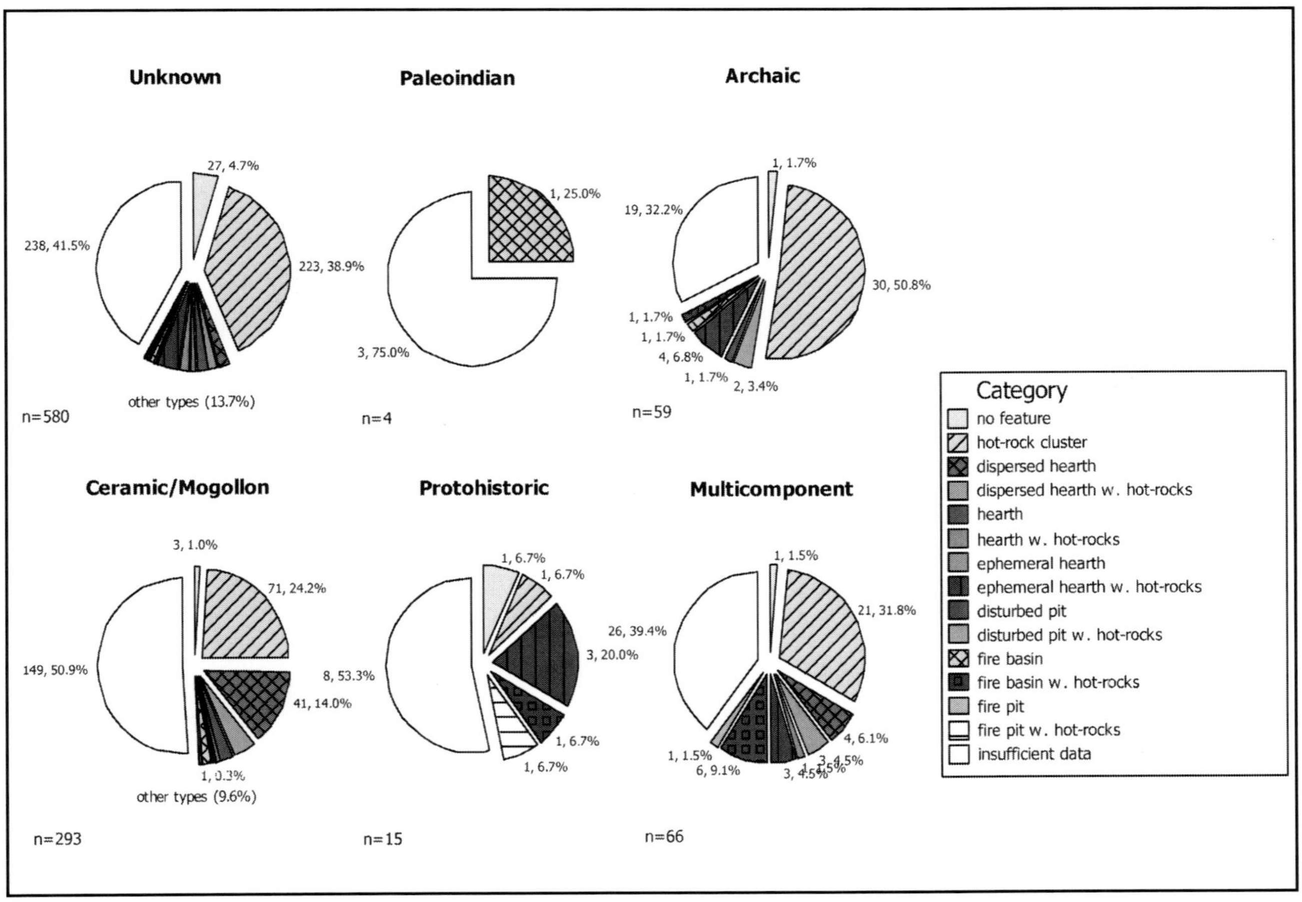

Figure 8.24. Temporal distribution of primary LTF components at prehistoric sites within the study area (note: including BWD, Locality 1).

Dating the deposition of the "Brown Sand Wedge" unit has been characterised as 'problematic' by Holliday (1997:62), but there is little reason to doubt that these features do represent Paleoindian LTFs. Hester's description suggests that the features individually could be classified as fire basin-type features. It is perhaps important to note that neither hearthstones nor burned bone were associated with these facilities. The total number of LTFs identified at Blackwater Draw Locality 1 is unfortunately unknown.

The Milnesand site (LA6209) is the most extensively investigated and concisely published of the three sites (Sellards 1955; Warnica and Williamson 1968; Johnson *et al.* 1986; Buchanan *et al.* 1996; Holliday 1997; Hill 2002). The earliest evidence for hunter-gatherer domestic fire technology on the Southern High Plains is, therefore, currently reliant on the observations and records from this important site.

Elias Sellards' initial report in *American Antiquity* (Sellards 1955) indicates that charred bison bone was recovered from a hearth "at the margin of the blowout and may be a little higher in section and hence somewhat later in time than the bone bed" (Sellards 1955:338). The results of radiocarbon determinations on this charred bone were not published in this report. James Warnica and Ted Williamson revisited the site in 1965 and provide a more detailed description of the location as a background to the substantial lithic artefact collection made by the second author in the years since Sellards' visit. Their article, also published in *American Antiquity* (Warnica and Williamson 1968) provides more detail as to the LTFs observed at the site.

Warnica and Williamson identified a hearth containing bone and charcoal in the vicinity of Sellards' original excavations (Warnica and Williamson 1968:16). Subsequent excavation of the hearth indicated some structural integrity and an elliptical outline measuring between 40-45cm in diameter (Warnica and Williamson 1968:16). The excavations appear to have identified two use episodes with hearth fill. Comparison of the excavated profile to Sellards' original indicated to Warnica and Williamson that some 20.3cm of sediments had been lost through erosion in the 12 years since Sellards' work (Warnica and Williamson 1968:17). If this is the case, then it would indicate that original dimensions of the feature may have been much larger than those recorded in 1968. Nevertheless, the revisit of the Milnesand site provides sufficient detail to classify the observed feature as a 'fire basin type' using the hierarchical key structure.

Photographic evidence of the fire basin described by Warnica and Williamson in 1968 was recently identified in the photographic archives of the Museum of Texas Tech University in an envelope of photographs submitted by Mr. Warnica (Figures 8.25 - 8.27). These photographs appear to corroborate the early descriptions of the LTF present at the site. An additional photograph in the envelope (Figure 8.28) appears to be from the poorly documented Williamson site located nearby the Milnesand site (Holliday 1997). The photograph clearly illustrates another LTF with a similar morphology to the one observed at Milnesand; however in this instance, no bone is associated with the feature.

Figure 8.25. Photograph of the fire basin at the Milnesand site. Text caption hand written by James Warnica states 'Ted Williamson, Sr. Standing at hearth at the Milnesand site Feb. 21, 1965. Photo was Taken from Sellards 1954 Test Pit, Looking to the N.E.' (photograph: James Warnica, courtesy of the Anthropology Division Museum of Texas Tech University).

The evidence from Blackwater Draw, Milnesand, and Williamson sites would appear to indicate evidence for the use of small basin-type features fuelled by wood or bone (or both) by the terminal Pleistocene. Unfortunately, at all these sites the association of the LTF with diagnostic artefacts is problematic. The Blackwater Draw Locality 1 features were excavated over 50 years ago and are very badly documented.

At Milnesand, the LTF, Paleoindian-type projectile points and *Bison antiquus* remains are not necessarily all associated, as the remains of later period occupations have been observed in the area (Johnson, Personal Communication, 2006). Radiocarbon assay on a sample obtained by Sellards yielded an age of 5730±100 years B.P. a result believed to be the product of contamination (Hill 2002). Mathew Hill Jr.'s (2002) examination of the limited bone collection agrees with a Paleoindian (ca. 10,000 B.P.) association for the site on the basis of morphological similarities of the LTF features described in 1955 and 1968 with a site located off of the Southern High Plains (Hill 2002:334). Hill goes on to suggest that a number of cultural processes could account for the burned bison bone recovered from the feature: used as a fuel, cooking, or accidentally burned after being thrown into the fire (Hill 2002:334). Despite this, Sellards himself had noted that the hearth he recorded was stratigraphically higher and "probably later in time than the bone bed" (Sellards 1955:338).

Figure 8.26. Photograph of the fire basin at the Milnesand site. Text caption hand written by James Warnica states 'Bottom of hearth exposed at the Milnesand site. Finger of Ted Williamson, Sr. points to charred bone. Feb. 21, 1965' (Photograph: James Warnica, courtesy of the Anthropology Division Museum of Texas Tech University).

Conclusions from the Paleoindian LTF dataset indicate that very little is currently known about Paleoindian domestic fire technology within the study area. Neither Blackwater Draw Locality 1 nor the Milnesand site exhibit evidence that is sufficiently robust enough with which to definitely associate the identified LTFs with the Paleoindian populations that undoubtedly once occupied both these sites. The importance of modern well-documented excavation and of obtaining sufficient radiocarbon determinations is underlined by these examples. Unfortunately, renewed investigations by the Museum of Texas Tech University (Johnson *et al.* 1986; Buchanan *et al.* 1996; Holliday 1997) appeared to indicate that little or no cultural material remains at the Milnesand site.

Figure 8.27. Photograph of the fire basin at the Milnesand site. Text caption hand written by James Warnica states 'Bottom of hearth exposed at the Milnesand site. Finger of Ted Williamson, Sr. points to charred bone. Feb. 21, 1965' (Photograph: James Warnica, courtesy of the Anthropology Division Museum of Texas Tech University).

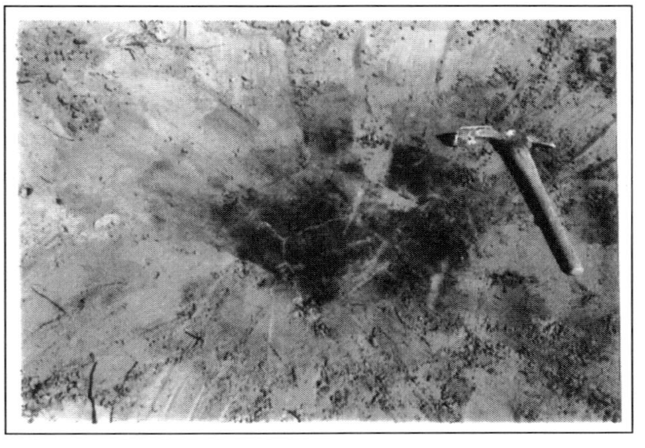

Figure 8.28. Photograph of the fire basin, most probably at the Williamson site. Text caption hand written by James Warnica states 'View of hearth at a Folsom site on the former game refuge S.E. of Milnesand Feb. 23, 1965. The sand was burned but it was not collected. There were no burned bones in the hearth' (Photograph: James Warnica, courtesy of the Anthropology Division Museum of Texas Tech University).

Archaic LTF Technology

Limited evidence for Archaic LTF technology was identified within the study area (Figure 8.10). Although no absolute dates are available in the regional databases, the site data were sufficient to characterise 66.1% of the primary LTF components observed (Figure 8.24). The Archaic period marks the first definite appearance of hot-rock technology in the regional dataset, with hot-rock clusters comprising over half of the known features for this time period. Other identifiable LTF types comprise only one or two examples each. At almost a third of the sites identified, insufficient data exist with which to ascertain even the basic classification for the purposes of the hierarchical key (Figure 8.24).

Unfortunately, descriptions of Archaic LTFs are not common and excavation data are rarer still. Examples of the more completely recorded features include: 41MY18, where two small rock-lined hearths were being exposed by erosion in 2002 (Cruse 2002); 41CB47, where four burned-caliche hearths, which varied from 1 to 2m were scattered across eroded surfaces (Nickels 2001); LA107646, where 16 fire-cracked rock concentrations were observed (Centennial Archaeology 1994); 41AD48, where at least two identified features were observed with charcoal staining and burned caliche (Tucker and Winters 1990); 41OL178, where one feature was identified and consisted of a dispersed hearth about 1 meter in size, comprised of burned caliche (Kotter and Cruse 1983); and 41GR120, where six LTFs were partially exposed at the modern ground surface and were reportedly all covered with sandstone slabs (Howard 1985).

These limited descriptions suggest that Archaic LTF technology often is associated with the use of hearthstones and the excavation of basin or pit features to contain them. No examples of Early Archaic (8500 to 6400RCYBP) LTFs are recorded in the regional dataset and none are known to exist for this time period. A Middle Archaic (6400 to 4500RCYBP) LTF (FA16-1) was excavated from Area 16 at Lubbock Lake (41LU1). The feature, measuring 1m in diameter and 40cm in depth, was filled with charcoal-stained sediment (Johnson 1987a; Johnson and Holliday 1995; 2004). A layer of burned caliche and a sandstone metate were recorded covering this feature. No animal bones were recovered from the feature fill and the function has been interpreted as a vegetal oven, providing evidence of adaptation to the harsh Altithermal conditions that affected the region during the middle Holocene (Johnson and Holliday 1995, 2004). In the well-stratified late quaternary valley fill that defines the regions' draw systems (see chapter 4) Late Archaic (4500 to 2000 RCYBP) LTFs are not easily separated from later Ceramic period occupations due to the relative landscape stability through this time period (e.g. Johnson and Holliday 1995, 2004). Very little is known, therefore, about individual LTF features built during this time period.

An excellent example of a Late Archaic LTF was, however, excavated during recent LLLRRP investigations at PLK-Locality 19 (Backhouse and Johnson 2007b:31-32). The site is located within a strip of dense Juniper that abuts the caprock escarpment along much of the eastern section of the Llano Estacado (Figure 8.4, 8.29). Survey at the site undertaken during the winter months of 2006 identified several, apparently intact, hearthstone features eroding from the banks of a large arroyo. A single feature (FA19-1) was selected for complete excavation (Figure 8.30a). The feature was chosen because charcoal was observed eroding from the exposed sediments and it was decided that the feature would, therefore, be an excellent candidate for absolute dating.

Excavations revealed a shallow basin that was containing more than 600 caliche hearthstones. The scatter of hearthstones observed down-slope from the feature suggests that it was originally much deeper and had subsequently been eroded to its excavated morphology. Nevertheless, the recovery of large amounts of intact charcoal from the base of the feature suggests that the fuel had been subjected to an oxidizing environment, which would be expected in an open-type feature. The fuel appears to have been quickly covered, as is common at the base of earth-oven features. No bones were recovered from within the feature and a broken burned sandstone block strongly suggests that the feature was utilized for baking plant materials. Samples were collected for macrobotanical and phytolith analysis. A comprehensive programme of radiocarbon dating of the charcoal recovered from this feature suggests a date of ca. 4,000 RCYBP (Eileen Johnson, Personal Communication, 2007).

No artefacts were recovered in association with the excavated feature at PLK-Locality 19 and it is therefore very likely that this feature was not located centrally within a campsite. The quantity of hearthstones used in the construction of the feature suggests that considerable effort was expended in its construction. The results of environmental analysis of this feature may be extremely instructive as to the plant materials being processed by people moving through this important ecological setting in the Late Archaic.

Figure 8.29. Fuel rich Juniper landscape found off the caprock along the eastern edge of the Llano Estacado. (P.N. Backhouse)

Figure 8.30. Excavation of a Late Archaic LTF (fire basin with hot-rocks) at PLK-Locality 19: a) field crew recording and mapping the LTF; b) half-section profile revealing the blackened sediments and largely intact structure at the base of the feature. (P.N. Backhouse)

Hot-rock technology appears to be consistently associated with LTFs observed in Archaic contexts and hearthstones were identified at 85.75% of Archaic sites. Examination of the frequency of sites exhibiting hot-rock technology by temporal period indicates an increase in this technology during the Archaic and subsequent Ceramic periods. Although more Ceramic period sites with hearthstones have been identified, proportionally, hot-rock technology is more often represented at Archaic period sites (hearthstones were identified at 67.58% of Ceramic period sites). The reasons for the reduction in the visibility of hot-rock technology during the Ceramic period are likely to be complex and may be the result of the increased site density and therefore greater diversification or specialization in the type of sites represented in the archaeological record. No evidence for the utilisation of specific vegetal or faunal resources in LTFs during this period is known (in the form of contextually associated burned and calcined bone).

Hot-rock technology appears to be consistently associated with LTFs observed in Archaic contexts and hearthstones were identified at 85.75% of Archaic sites. Examination of the frequency of sites exhibiting hot-rock technology by temporal period indicates an increase in this technology during the Archaic and subsequent Ceramic periods. Although more Ceramic period sites with hearthstones have been identified, proportionally, hot-rock technology is more often represented at Archaic period sites (hearthstones were identified at 67.58% of Ceramic period sites). The reasons for the reduction in the visibility of hot-rock technology during the Ceramic period are likely to be complex and may be the result of the increased site density and therefore greater diversification or specialization in the type of sites represented in the archaeological record. No evidence for the utilisation of specific vegetal or faunal resources in LTFs during this period is known (in the form of contextually associated burned and calcined bone).

Ceramic and Jornada Mogollon LTF Technology

LTFs constructed between 2000 and 500 RCYBP comprise the vast majority of the regional dataset (Figure 8.10). Archaeological evidence for these features and the domestic fire technologies that they represent should, therefore, reasonably be more advanced than for both preceding and succeeding time periods. Unfortunately, inspection of the regional dataset indicates very few details are available with which to ascertain the physical characteristics of the LTF features built during this time period (Figure 8.24).

The majority of the Ceramic/Mogollon temporal dataset is comprised of sites with insufficient data (51.5%), hot-rock clusters (24.2%), and dispersed hearths (13.5%). The cumulative percentage of these informationally limited categories accounts for almost 90% of the temporal dataset. This realisation severely limits the potential to investigate domestic fire technology during this time period. For instance, only three features have reported diameters and less than five have reported depths. The paucity of physical information is matched by the number of radiocarbon dates associated with the sites records (n=3, it should be noted that it is likely that more sites have associated radiocarbon dates but these are not available in the site-survey records).

The huge number of sites identified during this time period along the margin of the Pecos river valley are generally poorly documented. However, their topographic location is consistent with the botanical interpretation that they represent the remains of a resource intensification that focussed on the acquisition of desert plants (Katz and Katz 1985:42). Because of the coarse-grained technological resolution of the regional dataset, the interpretation of LTF technology for the Ceramic period is unfortunately reliant on a number of recently excavated examples largely from the east of the study area.

PLK-Locality 4. An excellent example of an early Ceramic LTF was recorded during field survey at PLK-Locality 4, located on the rolling plains just off of the southeastern Caprock escarpment in Garza County (Backhouse and Johnson 2007b). The feature (FA4-5) was identified along a recently cut shallow erosional channel (Figure 8.31). The internal structure of the feature appears, therefore, to have been deflated onto the modern ground surface as a dense hearthstone scatter. Very few artefacts were in contextual association with the LTF and it is likely that the depositional setting may have resulted in smaller artefacts being displaced downstream during heavy rain events. Nevertheless, a programme of detailed recording was undertaken to determine if the function and age of the feature could be determined. A clear void is visible at the centre of the feature, indicating that it most likely was once a hearthstone-lined pit or basin (Figure 8.31).

Figure 8.31. Early Ceramic LTF (hearth with hot-rocks) identified along an eroded intermittent drainage channel at PLK-Locality 4. (P.N. Backhouse)

A range of different hearthstones material types was used in the construction of the feature and it appears that no one material type was favoured over the others. Perhaps most surprising , given the context of recovery, was the identification of charcoal specimens around and under the hearthstones towards the centre of the structure. Radiocarbon dating on a sample recovered from this context indicated that the feature was constructed ca. 2,000 RCYBP (Eileen Johnson, Personal Communication, 2007). Furthermore, the identification of blocks of charcoal at the base of the feature supports the interpretation that it was used as an oven. No animal bones were recovered; however, the recovery of a single burned Mesquite bean (P*rospis glandulosa*) provides tantalizing evidence that the feature was used to process Mesquite (Backhouse and Johnson 2007b:30). If this interpretation is correct, then it suggests that Ceramic-period hunter-gatherer groups may have been mapping onto this important seasonal resource during the late summer. Technologically, the feature is relatively labour intensive and appears to confirm that vegetal matter was a constituent component of early Ceramic-period diets.

Lubbock Lake (41LU1, Lubbock County, Texas)

Regionally, the highest frequency of archaeologically identified and archaeometrically dated LTFs have been investigated within the Lubbock Lake Landmark, a National Historic and State Archeological Landmark (41LU1). This site is located in the upper Brazos river drainage along the south-eastern portion of the study area and has been intensively studied and reported during the same programme of research that has spanned over thirty years (see chapter 4). Over this programme of research, Ceramic period LTFs have been identified at six separate sites and areas within this 300-acre property (Johnson 1987a) and because of the long-term research programme, it is possible to examine the technological components of these features on a methodologically similar basis (Table 8.3).

Table 8.3. Summary of the physical characteristics of the Ceramic period LTFs investigated at the Lubbock Lake Landmark and nearby sites.

Site/Area	Feature	Radiocarbon Dated RCYBP	Hot-Rock Technology	Burned Bone	Max Diameter (cm)	Max Depth (cm)	Reference
41LU1	FA1-18	660±40 (SMU-2445)	yes	~	55	10	Johnson 1993:74, 185
41LU1	FA1-19	885±70 (SMU-2352)	yes	~	40	6.5	Johnson 1993:74, 187
41LU1	FA65-2	1590±40 (SMU2639)	yes	~	30	22	Johnson 1995:115
41LU1, Area 39	FA39-5	690±30 (SMU2449) 650±40 (SMU2450)	yes	~	50	5	Johnson 1993:64, 66, 70
41LU1, Area 10	FA10-1A	975±124 (DRI3060) 733±47 (DRI3120)	yes	yes	115	19.5	Buchanan 2002:143
41LU1, Area 10	FA10-1B	947±103 (DRI3598) 875±40 (A11166) 795±55 (ETH19566) 775±55 (ETH19567) 769±101 (DRI3596)	yes	yes	138	~	Buchanan 2002:143; Backhouse 2002:186
41LU1, Area 10	FA10-1C	1030±65 (DRI3203) 791±47 (DRI3123)	yes	yes	129	18.75	Buchanan 2002:143
41LU1, Area 10	FA10-1D	876±79 (DR3062) 862±104 (DRI3274)	yes	yes	27		Buchanan 2002:143
41LU1, Area 10	FA10-1E	960±55 (ETH17509)	yes	yes	92	10	Buchanan 2002:143
41LU1, Area 10	FA10-1F	990±50 (ETH17510)	yes	yes	~	~	Buchanan 2002:143
41LU1, Area 10	FA10-1H	~	yes	yes	103	~	Backhouse 2002; Selwood and Holliday 2002
41LU1, Area 10	FA10-1I	965±80 (A11167)	yes	yes	118	10	Backhouse 2002:186; Selwood and Holliday 2002
41LU1, Area 10	FA10-1J	~	yes	yes	130	<6	Backhouse 2002; Selwood and Holliday 2002
41LU1, Area 10	FA10-5	905+105/-100 (A11170) 861±160 (DRI3595) 801±73 (DRI3061)	yes	yes	47	25	Buchanan 2002:143; Backhouse 2002:186; Selwood and Holliday 2002
41LU1, Area 10	FA10-6	630±60 (A11164)	no	yes	66	15	Backhouse 2002:186
41LU65	FA65-1	730±40 (SMU 2638)	yes	yes	25	16	Johnson 1995:109, 115
41LU74	FA74-1	1040±50 (SMU2658)	yes	yes	50	16	Johnson 1995: 121
41LU75	Pit A	~	no	yes	76.2	45.7	Brown 1999:41
41LU75	Pit B	793+70 (SMU2710)	no	yes	102	38.1	Brown 1999:41

41LU1 Area 10 represents the most intensively studied, most LTF numerous, and best-dated hunter-gatherer campsite located within the Southern High Plains study area (Johnson 2002). The opportunity, therefore, afforded by examining the archaeological evidence of aboriginal domestic fire technology at this site is regionally without parallel. Despite nearly a decade of investigation at this location (1993-2000, 2006), excavations are not yet incomplete. Nevertheless, nearly 30,000 artefacts have been mapped and recovered *in situ* or from matrix screening of the excavated sediments (Buchanan 2002). The methodological treatment of hearthstones as individual artefacts during this excavation has resulted in a wealth of data not normally available for hunter-gatherer campsites. The preceding statements underscore that sufficient research has been undertaken to comment on the range of domestic fire technologies represented.

Area 10 is located on the surface of the Llano Estacado and occupies a portion of the uplands overlooking Yellowhouse Draw. Perhaps most important is the realisation that this specific portion of the draw has been a source of water at least since the late Pleistocene (Johnson 1987a). The camp was therefore most probably positioned to access both the water sources and the various fauna that also were reliant on its continued availability for their existence. The majority of the evidence indicates that the camp was occupied periodically throughout the late Ceramic period (ca. 1000 – 500 RCYBP) and the dynamic upland environment has most likely resulted in the superposition of various camping events over one another (Buchanan 2002). Nevertheless, the structure of at least ten LTFs has been determined through careful excavation and a systematic programme of radiocarbon-dating assays (Table 8.3; Figure 8.32).

A clear technological difference is apparent between the majority of the LTF features located to the east of the area (FA1-1) and a single, apparently isolated, example to the west of the site (FA4-4). The features within FA1-1 are ca. 1m in diameter (mean = 96.5cm, SD = 38cm) and 15cm deep (mean = 14.89cm, SD = 6.65cm), with feature fill typically consisting of dense frequencies of burned-caliche hearthstones (n>2000), burned bone, charcoal, and low frequencies of lithic debitage (Backhouse 2002). The vast majority (ca. 99%) of the rocks used for hearthstones appear to have been gathered nearby (<500m) most likely to be from an exposure of the shallowly buried Ogallala Fm. A very limited number of sandstones have been identified within the features. If confirmed, these may represent the broken remains of vegetal- processing equipment.

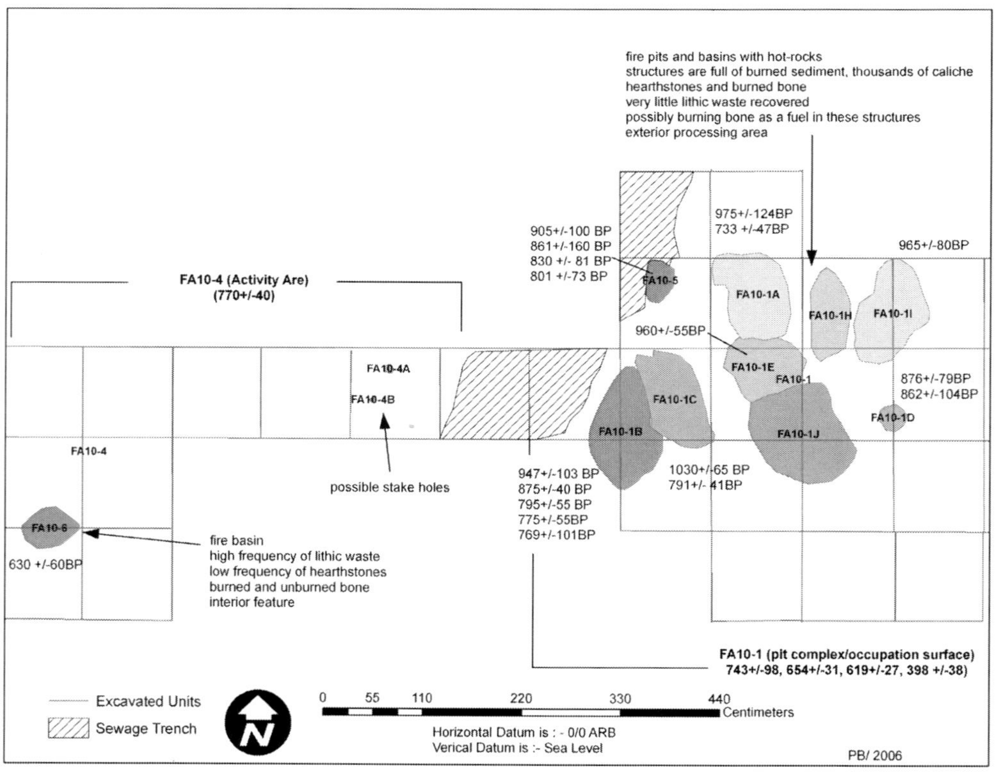

Figure 8.32. 41LU1, Area 10 plan drawing showing the relationship of the overlapping LTFS (fire basins with hot-rocks) to the east of the area and the more domestic focus to the west of the site; note radiocarbon assays are listed for individual features from oldest to youngest (data Backhouse 2002:table 10.5; Buchanan 2002:table 7.1).

The bone recovered from within the fill has been identified almost entirely as modern bison (*Bison bison*), although other smaller mammals were also noted (Buchanan 2002:145; see Table 8.4). The highly fragmented nature of the bones has lead to the speculation that specialized processing of the bones for marrow or grease was occurring (Hamilton 1998). Alternatively, the bone may have been utilised as a fuel source (cf. Théry-Parisot 2002). Faunal elements identifiable within the features are dominated by bison rib pieces, although other elements are present and include teeth and diaphyseal segments (Buchanan 2002).

Regardless of the specific function represented, a great deal of energetic investment has been exerted in the construction of each LTF and these features clearly represent complex expressions of hunter-gatherer domestic fire technology. The hearthstone and bone-choked nature of the pits suggests that they have either been backfilled following utilization, or that the more likely purpose of these features was primarily to cook on top of them, or to heat the hearthstones for external utilization (i.e. indirect boiling technologies). The range in radiocarbon dates (Figures 2.9, 8.32) suggests that this area of intensive human activity may well have been revisited multiple times throughout the late Ceramic period.

The archaeological record in the western portion of the area (FA1-4) is entirely different to that of the eastern. A single fire basin LTF has been identified in this portion of the area. The feature is shallow and the fill comprises both burned and unburned material. Over 1,000 tiny lithic flakes and debris were recovered from the feature fill as well as numerous burned and unburned bone scraps (Backhouse 2002:189). The feature is interpreted as being located within a domestic area and was most likely to be within the interior of a structure. Support for this interpretation is provided by the high degree of tiny lithic debitage recovered in the feature fill, the result of cleaning a domestic living space (cf. Binford 1978, 1980, 1987). Based on the high fraction of tiny carbonized fragments observed in the excavated sediments, wood appears to have been the primary fuel source. The slightly later date obtained on the fire basin LTF in the west of the area (Table 8.3) may indicate that the two occupations were not contemporaneous.

Table 8.4. Summary of faunal resources identified in association with LTFs on the Southern High Plains by major time period.

Taxon	Paleoindian	Archaic	Ceramic/ Mogollon	Protohistoric	References
Amphibians					
Order: Testudines			Yes		Selwood and Holliday 2002:175
Terrepene Carolina puttnami (extinct box turtle)	Possible				Johnson 1977:74
Reptiles					
Crotalus atrox (western diamondback rattlesnake)				Yes	Johnson 1987b:137
Birds					
Class: Aves			Yes		Backhouse 2002:192
Mammals					
Order: Rodentia			Yes		Backhouse 2002:192
cf. *Syvilagus* spp. (cottontail)				Yes	Johnson 1987b:137
Lepus californicus (blacktail jackrabbit)			Yes	Yes	Selwood and Holliday 2002:175, Johnson 1987b:137
Canis Latrans (coyote)				Yes	Johnson 1987b:137
Canis lupus (gray wolf)				Yes	Johnson 1987b:136
Taxidea taxus (badger)				Yes	Johnson 1987b:137
Antilocapra americana (pronghorn antelope)				Yes	Johnson 1987b:137
Bison antiquus (extinct bison)	Possible				Warnica and Williamson 1968:16; Hill 2002:334
Bison bison (modern bison)			Yes	Yes	Backhouse 2002:192; Brown 1999:41; Johnson 1987b:136

An analogous site structure was recorded at 41LU75 where two intersecting LTFs were excavated and radiocarbon dated to the later Ceramic period (Table 8.3). From the high quantity of broken and burned bone recorded in their vicinity (Brown 1999:39), it seems that bison bones were being heavily utilised within the features. No carbonised wood fragments were recovered within the feature fill and the excavator speculates, therefore, that buffalo chips were the primary fuel. The results from sites such as 41LU75 and Area 10 underline the need for intensive and patient methodologies if significant insights into hunter-gatherer domestic fire technology are to be obtained.

Protohistoric LTF Technology

Despite the strongly documented historical presence of Native American communities living on and around the Southern High Plains during the Protohistoric Period (see Chapter 6), very little archaeological evidence for their presence has been systematically examined. This statement also holds true for domestic fire technology and over 62% of the LTFs identified during this period were not recorded sufficiently enough to determine even a basic type using the classificatory hierarchal key (Figure 8.24). Quality site-excavation data are again best represented from sites along the southeastern portion of the study area and at the Lubbock Lake site in particular (Table 8.5).

Important technological evidence is present in these investigations. Charcoal recovered from a 'fire basin with hot-rocks' feature excavated at Tahoka Lake (41LY52) was identified as Honey Mesquite (*Prospis glandulosa*) (Backhouse 2003:28). Mesquite is common within the region today; however, it was not as prevalent prior to Anglo settlement (see Chapter 4) and its use as a fuel for Native American groups has not been previously demonstrated archaeologically. Mesquite is not the only available wood source and dried bison faeces (buffalo chips) have often been implicated as the primary fuel for Plains groups (cf. Holland 1984). The feature at Tahoka is also unusual as the construction involved the placement of large rimming hearthstones (caliche) around the circumference of the basin (Backhouse 2003). This construction method has also been noted at other later prehistoric sites in the region (e.g. Brown 1986).

Table 8.5. Summary of the physical characteristics of the Protohistoric period LTFs investigated at the Lubbock Lake Landmark.

Site/Area	Feature	Radiocarbon Dated	Hot-Rock Technology	Burned Bone	Diameter (cm)	Depth (cm)	Reference
41LU1, Area 8	FA8-6	530 BP 475 BP	Yes (largely structural)	Yes (limited)			Johnson and Hartwell 1989:192
41LU1,	FA19-2	380 BP	Yes (largely structural)	~			Johnson and Hartwell 1989:192
41LU1, Area 14	FA14-1	315+50BP (SI2701)	Yes (limited)	Yes	~	~	Johnson *et al.* 1977:86
41LU1, Area 15	FA15-1	285+60BP (SI2703)	Yes	Yes	70	35	Johnson *et al.* 1977:86
41LU129	FA1-1	350±40 (A12934) 225±90 (A12934.1)	Yes	No	70	15.25	Backhouse *et al.* 2005a
41LY52	FA14-1	470±45 (A11987.1) 460±55 (A11985.1) 380±50 (A11983) 270±35 (A11987) 255±80 (A11981) 240±95 (A11982)	Yes (largely structural)	No	91	21	Backhouse 2003
41BI83	1	360±80 (BGS284)	Yes	Yes	80	8	Katz and Katz 1976:136

It is interesting to note that hearthstones do not appear to have been utilised as heating elements in several of the excavated features (41LY52-FA14-1, 41LU1- FA8-6, FA14-1, FA19-2). In the case of 41LU1-FA8-6, FA19-2, and 41LY52-FA14-1, the caliche appear to have been used to provide a structural lining rather than as a heating component (Johnson 1987b, Johnson and Hartwell 1989, Backhouse 2003). The lack of hearthstones in these features hints then at a possible increase in direct container boiling technologies; or alternatively that these features represent more carefully delimited facilities constructed within habitation structures (cf. Binford 1983). Analysis of the faunal elements surrounding the LTFs at the Lubbock Lake site indicated to Johnson (1987b:136) that "secondary processing of select limbs was occurring". Furthermore, she hypothesises that in addition to bison, smaller mammals appear to have been transported to camps for specialised processing (Table 8.3). Johnson speculates that this may have included cracking open bones in the production of grease, or brain and marrow extraction (Johnson 1987:136). Processing is most likely to have taken place in paunch/hide containers or ceramic vessels, both of which artefact classes are unlikely to be directly archaeologically visible.

Conclusions

The results of the technological analysis of LTFs for the four major prehistoric cultural time periods on the Southern High Plains is largely reliant on a very small number of high-quality well-dated examples that mostly are located to the east of the study area. In contrast, very few high-resolution studies have been completed in the southwest of the study area where evidence for LTFs is abundant. Given the low research status, results of the classificatory structure proved a useful mechanism for identifying variation in the regional dataset (Table 8.6). Furthermore, this variation appears to be behaviourally significant indicating broad shifts in hunter-gatherer subsistence practices both through time and across space.

The basic classifications provided by the hierarchical key do not, however, address the full potential of the investigation of this important feature class. Four key areas of localised thermal feature technology (Function, Structure, Hot-rock technology, and Fuel; Figure 8.33) are identified from the archaeological record and within these areas, significant further research is still required. Focussed excavation programmes throughout the study area are needed in order to add qualitative data and, therefore, develop a truly regional model for hunter-gatherer domestic fire technologies. The examples derived from the long-term multidisciplinary LLLRRP clearly demonstrate the potential for such programmes to be successful.

Table 8.6. Summary of the archaeological evidence for hunter-gatherer domestic fire technology by major cultural time period.

Characteristic	Paleoindian	Archaic*	Ceramic/Mogollon*	Protohistoric*
Hot-rock technology	unknown	frequent	frequent	frequent
Hot-rock clusters	unknown	frequent	infrequent	rare
Dispersed hearths	unknown	-	infrequent	-
Dispersed hearths with hot-rocks	unknown	rare	rare	rare
Hearth	unknown	-	rare	-
Hearth with hot-rocks	unknown	rare	rare	-
Ephemeral hearth	unknown	rare	rare	-
Ephemeral hearth with hot-rocks	unknown	rare	rare	infrequent
Disturbed pit	unknown	-	rare	-
Disturbed pit with hot-rocks	unknown	rare	rare	rare
Fire basin	rare	rare	rare	-
Fire basin with hot-rocks	unknown	rare	rare	rare
Fire pit	unknown	-	-	-
Fire pit with hot-rocks	unknown	-	-	rare

*where <10% of features identified = rare; 11 – 49% = infrequent; and >50% of sites identified = frequent

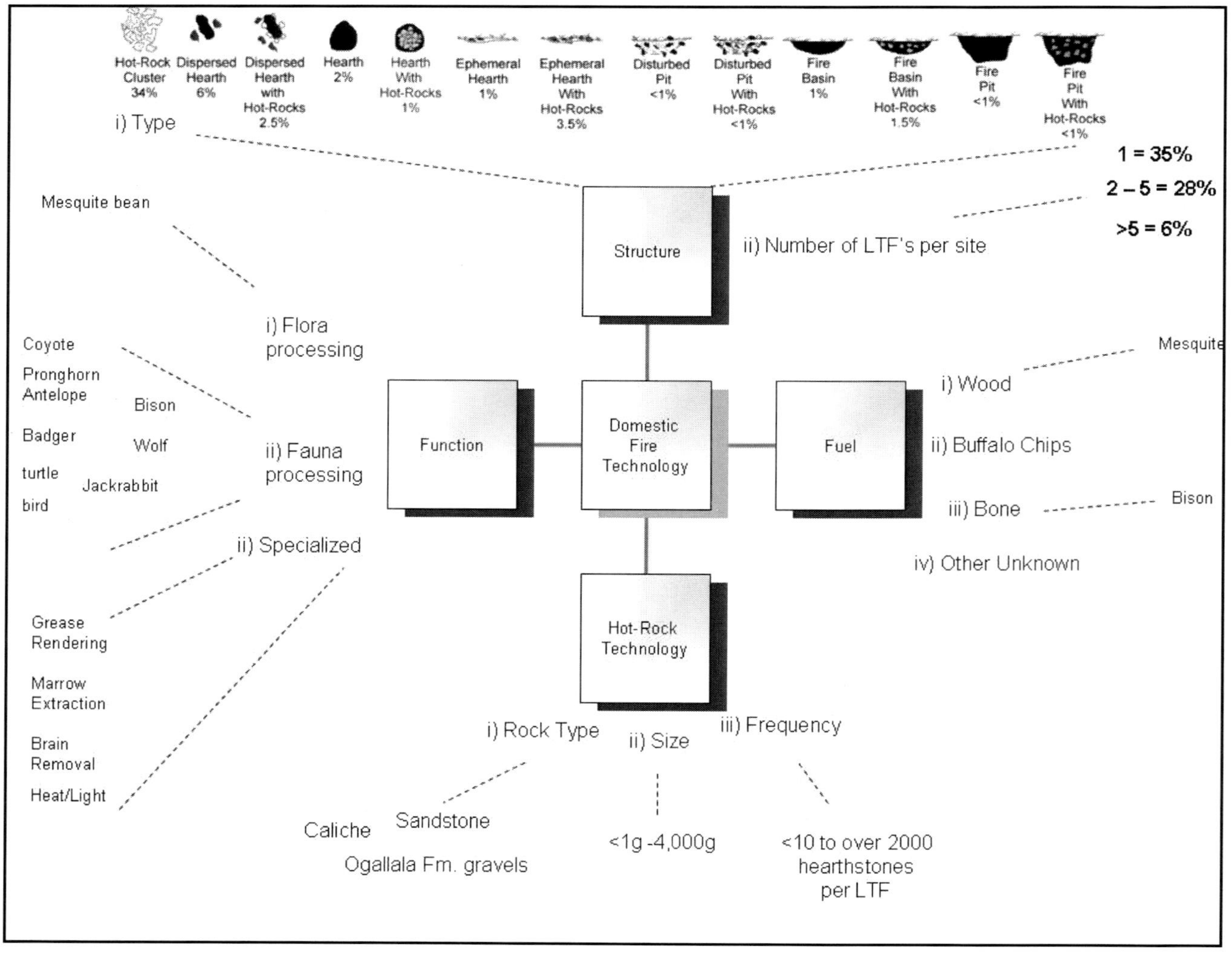

Figure 8.33. Summary of the archaeological evidence for localized thermal features identified within the Southern High Plains study area.

Synopsis of the Archaeological Results and Conclusion

The Southern High Plains study area exhibits a rich record for hunter-gatherer occupation and the use of LTFs as proxy markers for campsites appears to be a useful mechanism for modelling the presence of human populations on the landscape over deep time scales. State site files comprising nearly one hundred years of archaeological research in Texas and New Mexico were useful for modelling the distribution of hunter-gatherer campsites over time and across space. Unfortunately, the level of detail contained within these documents is often extremely limited and in many cases is not supported by any supplemental documentation. Nevertheless, by analysing the results of the application of the taxonomic key and referencing detailed case studies, then sufficient research has taken place to develop a course-grained model of hunter-gatherer domestic fire technology and to identify areas in which further research is critically needed.

The archaeological evidence for domestic fire technology is fundamentally controlled by a great many site-formation processes that have profoundly altered the physical structures that archaeologists must engage with. The realisation that LTFs are not equally preserved either by their age of construction, the construction technique used, or in a particular geographic area of the region is extremely important. The structural preservation of an LTF is a relatively rare occurrence on the Southern High Plains and typically, if a researcher encounters such a feature it is because it is being exposed either at the modern ground surface or in a cut bank. In either case, the opportunity to properly investigate such features during surveys is rarely seized upon.

CHAPTER 9 - DISCUSSION AND POTENTIAL FOR APPLICATION TO OTHER REGIONS

Abstract

The results of the four analytical strands of research that comprise the research project are drawn together and presented as a synthesis of hunter-gatherer fire technology on the North American Southern High Plains. Ethnographic data suggest that the physical structure of LTFs is often behaviourally significant and insists on the need to consider such features as the archaeologically most visible component of a wider, more complex framework of human fire technology. LTF features are found to be the most commonly occurring feature type at archaeological sites recorded in the study area. Furthermore, the application of the taxonomic key confirms that the archaeological signature of these features varies significantly across both time and space. Analysis of the regional dataset in light of the archaeological and experimental evidence indicates that the physical expression of LTFs appears to be a result of the complex interaction of processes of which prehistoric technological strategies are only a minor component part. Nevertheless, the results demonstrate that prehistoric populations consistently utilised the periphery of the region to locate their campsites. Throughout the study area, the physical structure of LTFs generally becomes more elaborate over time and strongly suggests a widening diet breadth with increasing reliance on plants and greater population tethering from the Archaic period onwards. The huge number of LTFs identified in the southwest of the area during the Ceramic period appear to be primarily the result of the specialised processing of desert succulents that occur only in that area. The ability to place the interpretation of LTFs in a regional context, therefore, represents a significant new avenue for hunter-gatherer research on the Southern High Plains. Application of the project to other geographic regions is assessed by a pilot study, which examines the evidence for hunter-gatherer domestic fire technology in the latest post-glacial record of the southern portion of the British Isles. The results suggest that in this region, also, significant potential exists for exploring variation in domestic fire technologies.

Introduction

The preceding chapters have identified, defined, and developed a research model for exploring localised thermal features (LTFs) as culturally relevant technological phenomena. All four analytical strands of research applied in the investigation of these features (development of the hierarchical key, experimental, ethnographic, and archaeological) are referenced in the discussion of the results of the project. The results illustrate clearly that by tacking between archaeological, experimental, and ethnographic contexts, the discussion of LTFs can progress significantly beyond that of basic description.

Drawing together the strands of evidence for domestic fire technology on the Southern High Plains within a typical Culture History framework, I critically discuss the implications of the presented project for both understanding the hunter-gatherer inhabitants of the region during the late Quaternary and also the way that we must engage with the material record. I then explore the potential of the interpretative model as a tool for the examination of the archaeological record of an entirely different geographic region. Lastly, I discuss the implications of the presented research for the placement of domestic fire technology within hunter-gatherer studies.

Drawing Together the Evidence for Domestic Fire Technology on the Southern High Plains.

The interpretative resolution achieved by drawing together data from nearly one hundred years of research, across two states, over a represented cultural time span of some 12,000 years, is necessarily coarse-grained. The project represents the first attempt at seriously considering LTFs as having the potential to be technologically meaningful cultural markers when considered at both macro and regional levels. The technological complexity observed in the ethnographic record (Chapter 6) reinforces the strength of this proposition. Research potential is, however, tempered by the realisation that features are themselves differentially preserved and that site-formation processes have themselves actively structured the physical evidence (as addressed by experimental research in Chapter 7).

The regional archaeological resolution, as set out in this project, is a function of the number of relevant sites identified within the study area, the nature and intensity of the fire technology represented, the way in which it was recorded for the purposes of the states' record systems, and the effect of site-formation processes on the preservation and visibility of the features themselves. Within this broad framework, higher-resolution sites, and experimental and ethnographic approaches, all facilitate the development of fine-grained snapshots of technological strategies at particular times and in particular places. Unfortunately, very few high-resolution excavations have taken place on and around the Southern High Plains in the last twenty years. Because of this, the results of the present study are heavily reliant qualitative datasets generated by the detailed investigations of the LLLRRP, mainly in the southeastern portion of the area (see Chapter 8).

Nevertheless, the regional record indicates that fire technologies are physically apparent in the archaeological record from the Paleoindian through Protohistoric cultural periods. Using these broad time-spans as conceptual brackets to constrain research, models of hunter-gatherer domestic fire technology can be proposed (Figures 8.33)

Implications for Understanding Prehistoric Populations and Technology

Archaeological research on the Southern High Plains is driven by interpretation derived from a handful of 'key' sites located on top of the vast High Plains escarpment (see Chapter 4; Figure 4.9). In contrast, research programmes seeking to assess the cumulative regional record of sites recorded on survey forms held on file with the federal government are rare. Furthermore, no previous research has attempted to contextualise these sites in relation to those identified in areas surrounding the escarpment (see Boyd 1997 for a limited application to the late prehistoric period along the eastern escarpment). The results of using a contextual approach to define the locations of hunter-gatherer campsites throughout the region were illuminating, indicating that more than 90% of the identified sites are located at the periphery of the region (Figure 8.2).

The spatiotemporal distribution of hunter-gatherer campsites is in large measure necessarily a function of the availability of critical resources necessary to the survival of hunter-gatherer groups. Primary amongst these is the availability of water and the presence or absence of fuel, without which there can be no fire and by extension no LTFs for archaeologists to observe in the archaeological record. The parsimonious explanation, therefore, for the locations of hunter-gatherer campsites observed in the current study, is that they are a response to the distribution of fuel and other critical resources necessary for hunter-gatherer lifeways.

There is no doubt that the most predictable water sources were located at the periphery of the region where stream-heads emerge from the vast Ogallala Fm aquifer that underlies the geology of the region (Gustavson and Simpkins 1989). In addition, high-quality lithic resources necessary for the production of stone tools are also only found in these borderlands (Holliday 1997) and rocks suitable to be used as hearthstones would have been difficult to obtain throughout much of the interior. Examination of the modern biodiversity underscores that the margins are the most biologically productive in terms of leafed-flora abundance (Figure 9.1). The range of modern tree flora confirms that these areas would have been predictable sources for the collection of wood (Figure 9.2). The longevity of Mesquite in the local flora continues to be a critical research question, given that the modern distribution appears to have a clear spatial relationship to the archaeological record (Backhouse et al. 2001). Nevertheless, it would appear that the gross distribution of hunter-gatherer campsites and the presence of wood fuel resources (as well as other critical resources) are positively correlated.

While the preceding ecologically conditioned model is intuitively satisfying, it does not take into consideration that critical resources were available (but perhaps less predictable) in the interior of the region. More than 25,000 Playa lakes held water (Sabin and Holliday 1995) and the large herds of slow-moving bison were clearly an important source of food (Johnson 1987a). Excrement from the latter could also have been utilised as a fuel source in the form of chips (dried dung) (Holland 1984) and dried bone (Backhouse 2002). The use of chips as a fuel has some ethnographic analogue in the form of the nineteenth-century *Cibolero* camps that also used chips in their campfires because of the lack of trees. Critical resources were also predictably located along the draw systems that incise the region, along which natural springs were sometimes located. Sites with springs were often heavily utilised by prehistoric populations (e.g. Lubbock Lake, Mustang Springs).

Prehistoric mobility strategies were clearly complex.

While not the focus of the current study, the identification of sites with LTFs as campsites provides a useful model by which mobility strategies can be assessed both over time and across space. The availability of different fuel resources would clearly have played an important role in how hunter-gatherer groups moved around the landscape. At this point, it is not possible to say definitively that hunter-gatherer groups preferentially utilised the various wood resources at the edge of the region. The apparent absence of evidence within the interior may be a result of less substantial features being required because of the type of food being processed (i.e. primarily meat). These features would have a lower chance of survival over time and would therefore be less likely to be identified archaeologically. The interior of the region also generally lacks hearthstone resources and suffers from low archaeological resolution due to the private ownership of much of the land (Figure 8.5). The relationship between fuel type and campsite location does however appear to be significant and requires further investigation.

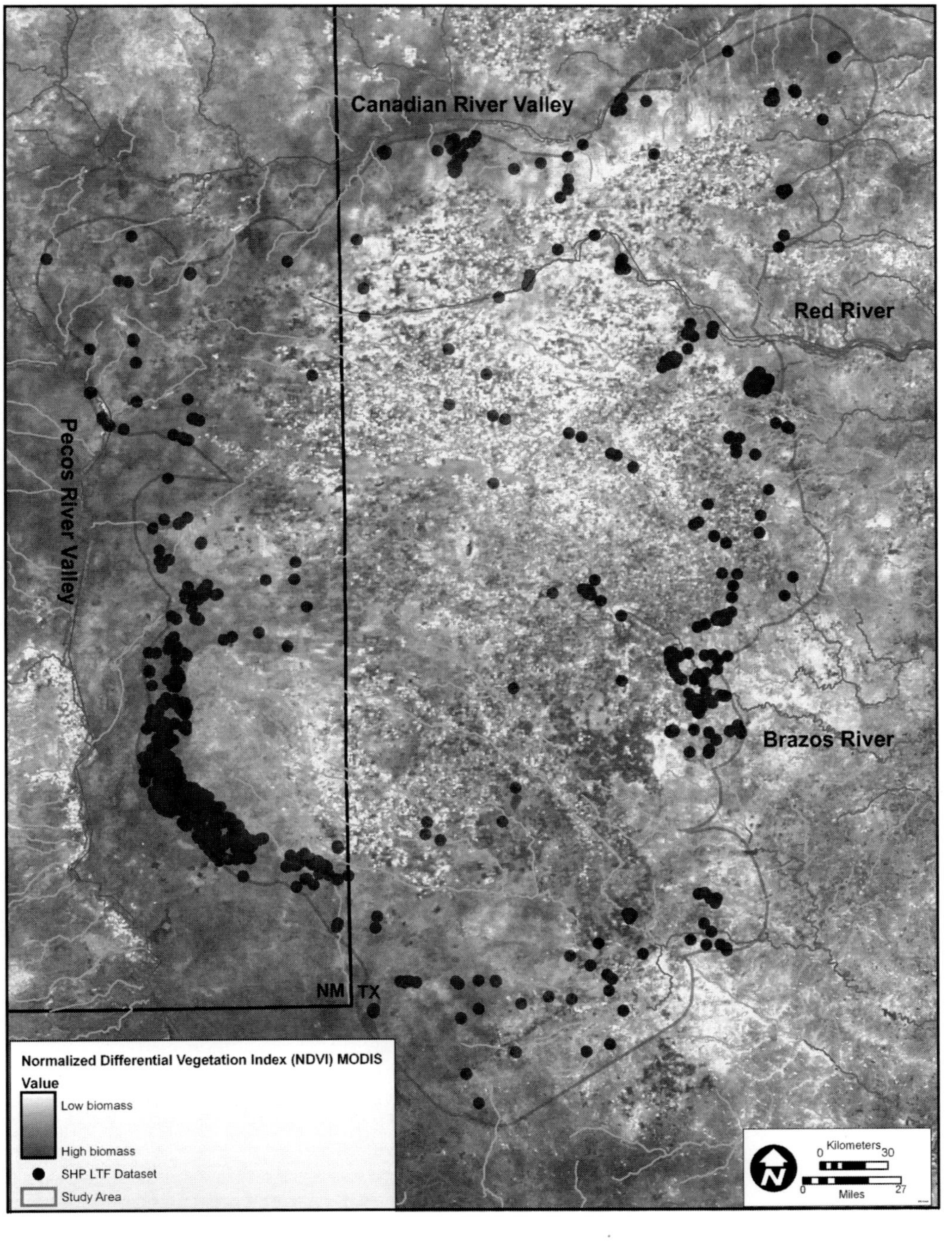

Figure 9.1. Hunter-gatherer campsite distribution and modern biomass index for the Southern High Plains Study area; Note that the darker shaded areas indicate areas of higher biomass(data: Normalized Differential Vegetation Index (NDVI) acquired from NASA's Moderate-resolution Imaging Spectroradiometer (MODIS), averaged data acquired between 1990 and 2005 and available from the USGS Seamless Data Server (www.seamless.usgs.gov)).

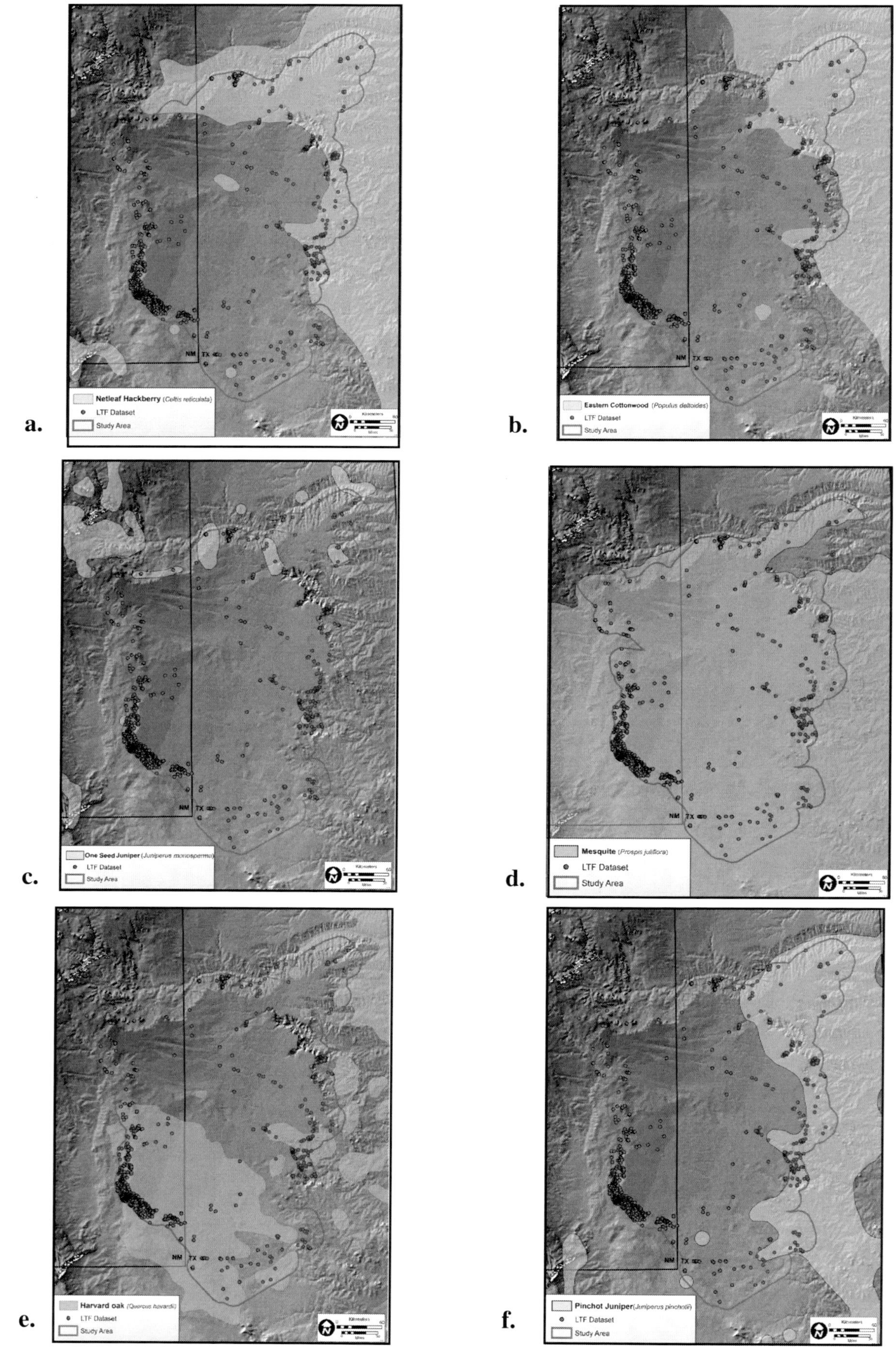

Figure 9.2. Hunter-gatherer campsite distribution and modern ranges of major tree species that may be suitable as fuel: a) Netleaf Hackberry (*Celtis reticulata*); b) Eastern Cottonwood (*Populus deltoids*); c) One Seed Juniper (*Juniperus monosperma*); d) Mesquite (*Prospis prospubescens*); e) Harvard Oak (*Quercus havardii*); f) Pinchot Juniper (*Juniperus pinchotti*) (data: modified from the "Atlas of United States Trees" by Elbert L. Little, Jr. (and other publications) freely available from http://esp.cr.usgs.gov/data/atlas/little/).

The spatial distribution of hunter-gatherer campsites as identified by this study suggests that human populations mapped onto the abundant and predictable resources at the edge of the region. High frequencies of sites identified along the incised draw systems suggest that these were used as resource-rich arteries along which groups could travel in order to obtain the herd animals that occupied the interior of the region. This broad subsistence model is constrained by site preservation and visibility biases that have conspired to favour the identification of sites at the periphery of the region relative to the interior. Unfortunately, the scale of mechanised farming within the interior of the region (Figure 8.5) most likely has left very few sites with which to address whether the abundance in campsites at the periphery of the region is truly a function of hunter-gatherer subsistence strategies.

The results of the presented research suggest a spatiotemporal regional model for resources use intensification expressed in terms of a widening diet breadth, most likely a result of increasing population and greater competition for resources, as a function of time (Figure 9.3). This interpretation both supports and adds quantitative support to current models, developed at a continental scale by Thoms (2003, 2008, 2009).

Paleoindian

Viewed from a traditional Culture History perspective, the archaeological evidence (or lack thereof) suggests the presence of limited human populations throughout the Paleoindian period (Figures 8.12, 9.3). The archaeological footprint of these population(s) appears very slight, even when considered in the broader terms of hunter-gatherer research.

Dating of the construction of the few possibly Paleoindian LTFs is problematic. In all cases, the features have not been dated by absolute dating assay or have produced results that are too young for a Paleoindian association (e.g. Hill 2002). Contextually at least, none of the sites contain any evidence for pre-Clovis (older than 12,000 RCYBP) domestic fire technology (as no Pre-Clovis sites have been identified in the study area). Furthermore, the limited evidence suggests the LTFs represent cultural activity only in the later Paleoindian (Folsom and later) periods. The following discussion of Paleoindian fire technology is presented with the caveat that the Southern High Plains dataset is minimal and urgently requires the efforts of field archaeologists to obtain samples for reliable dating of any suspected features as a matter of some priority.

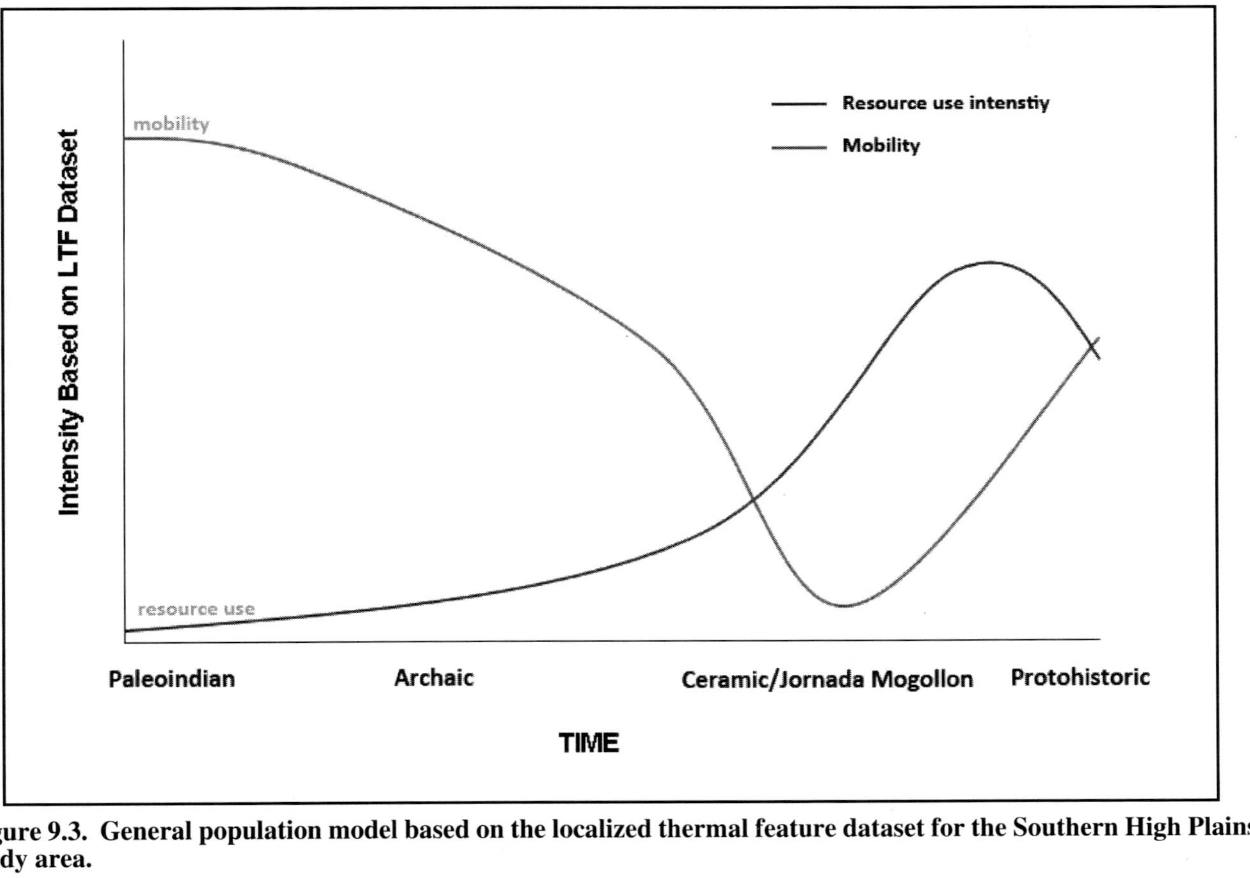

Figure 9.3. General population model based on the localized thermal feature dataset for the Southern High Plains study area.

The archaeological record suggests that domestic fire technology during the Paleoindian period may have been largely non-intensive and therefore have left only ephemeral physical evidence for archaeologists to find (*cf.* Sergant *et al.* 2006). Hearthstones, the key thermal feature marker on the Southern High Plains, do not appear to be a constituent component of Paleoindian fire technology. This realisation reinforces the interpretation that the Paleoindian populations most likely utilised radically different fire technologies from the populations recorded by Europeans in the historic period ,which themselves comprise the basis for ethnographic observations.

All other things being equal, then a relationship between time and preservation might be expected to explain the apparent lack of sites with evidence for domestic fire technology i.e. Paleoindian features are generally less likely to be intact than Protohistoric features. While this assumption is most likely to be largely true, the scale of feature construction also is a significant factor. The limited data for this period suggest that features were built directly on the ground surface or in shallow basins, as appears to be the case at Blackwater Draw Locality 1, Milnesand, and the Ted Williamson sites (Figures 8.25 - 8.28). These types of features, were examined experimentally and are demonstrated to be highly susceptible to erosion processes. Furthermore, experimentation has indicated that the organic by-products of burning (carbonised wood, fragments of burned bone and seed) can be quickly flushed from the feature matrix, by heavy rainfall or high-wind events (Backhouse and Johnson 2007a). Preservation bias, a function of the type of technology originally represented, the geomorphologic setting of the site, and the time depth involved, is therefore undoubtedly a significant factor in archaeologists' likelihood to encounter Paleoindian-aged LTFs.

Viewed at a wider scale, there is very little evidence for Paleoindian-age thermal facilities on the North American Great Plains (Thoms 2009). Jason LaBelle has undertaken the most comprehensive review of features, examining 118 LTFs from the Plains and bordering regions (LaBelle 2005). As in the present study, only limited numbers of features were recorded sufficiently to determine individual morphologies (LaBelle 2005: 230). Just off the Plains (and within LaBelle's dataset), the Barton Gulch site in southwestern Montana has yielded a unique glimpse at (late) Paleoindian domestic fire technology.

A total of seventy-five LTFs were recorded at this site, of which eighteen were classed as 'hearths' and 57 as 'roasting pits' (Armstrong 1993:8). The investigators distinguished between these two types of features on the basis of slight differences in morphology, *in situ* evidence for firing, and the presence of plant macrofossils, which appear to have been recovered more commonly in the roasting-pit features (see Armstrong 1993:Table 1). Hot-rock technology does not appear to be present at this site. The minimal morphological variation observed among features ensures that the application of the taxonomic key to this dataset results in all the features being classified as fire basins (Type 11). This result demonstrates that the taxonomic key is not itself interpretative and detailed site and feature scale observation must also be factored into the investigation of LTFs if fine-grained behavioural interpretation is to be achieved. The investigators suggest that the hearths were used to generate coals that were then added to the surrounding roasting pits to form the layers between which plant food was cooked. There is no analogy, however, for this type of technology in the Southern High Plains dataset.

The range of macrobotanical remains recovered from Barton Gulch suggests that Slimleaf goosefoot (*Chenopodium leptophyllum*) and prickly pear cactus (*Opuntia polyacantha*) were the primary foods being cooked in the features (Armstrong 1993:13). The inhabitants appear to haved used grasses and rushes within the features, perhaps to wrap and therefore insulate the actual foodstuffs being processed. Additionally, there was almost no evidence for the utilisation of faunal resources (other than small mammals e.g. rabbit) (Armstrong 1993).

The Barton Gulch feature data contribute substantially to the 36% of Paleoindian-aged features that could be morphologically determined by LaBelle's (2005) analysis. The results of this analysis are in broad agreement with the present dataset and indicate that the morphology of most features are relatively standardised with shallow depths and diameters typically less than 75cm (LaBelle 2005:230). LaBelle concludes quite correctly that these features, due to their small size, apparently low-firing intensity, and lack of hot-rocks are unlikely to represent intensive (large-scale) plant-processing or water-boiling facilities. Instead, it appears they were built to process small but varied amounts of plant material on a relatively expedient basis (LaBelle 2005).

Fire is obviously not necessary to process all foodstuffs and Wrangham and Conklin-Brittain (2003:36) report that "up to 56% of plant roots eaten by African foragers were sometimes consumed in their raw state". However, if meat contributed a large part of the late Paleoindian diet, as has often been suggested (see Grayson and Meltzer 2003 for a recent discussion), then domestic fire technology may have been necessary to tenderise the food, therefore allowing maximum intake. These authors furthermore suggest that heating meat to temperatures above 80°C results in less chewing and, therefore, greater consumption efficiency (Wrangham and Conklin-Brittain 2003:40).

Following this line of reasoning, experimental data collected for the present study clearly indicate that Mesquite is a high-energy fuel resource easily capable of obtaining temperatures of 80°C or more (Backhouse *et al.* 2005b; Backhouse and Johnson 2007c). Additional experimental work is required to assess the heating output of various dried animal excrement (see Holland 1989 for an experiment using bison manure as a fuel source) and also the contribution that bone may have made as a fuel. Experiments undertaken by Théry-Parisot (2002) indicate that the addition of bone (a potential fuel that is possibly represented at Milnesand; Warnica and Williamson 1968:16) to wood-fuelled LTFs can result in significant increases in combustion time. These experiments suggest that the combustion of bone results in convection- and radiation-based heat-transfer, which lends itself to direct cooking technologies (such as grilling) as well as the production of heat and light (Théry-Parisot 2002:1419).

The processing of meat, by direct cooking, would probably leave little archaeological evidence and any discarded bone would in most circumstances not be likely to survive in the archaeological record. Experimental research has indicated that the small basin-type structures, that appear to be typical of the Paleoindian period, might be further explored by examining the relationship of the surrounding artefact scatter to the feature. The plotted distribution may provide useful information as to the role of the LTF in spatially organizing activities that occur around it. This is particularly true if these features were primarily constructed for direct cooking while providing necessary warmth and light for the comfort of the humans at the camp (Backhouse 2004). This humanistic approach requires higher-resolution datasets detailing the geometry of individual features as well as the fuel used in Paleoindian LTFs, followed up with groundtruthing in actual archaeological situations. Unfortunately, sites with potential for the high-resolution investigation of Paleoindian LTFs have yet to be systematically investigated by modern excavation techniques within the study area.

It is of interest to note that the types of early Holocene features identified in North America are not themselves typical of northern hemisphere late Pleistocene hunter-gatherer domestic fire technologies. Disparate Old World datasets indicate that hot-rock technology was well established by the Upper Palaeolithic, at least in portions of Europe and the Middle East (Petraglia 2002; Thoms 2009). Hearthstone technology in these contexts appears to be independent of subsistence mode, and various technologies are represented at temporary camps as well as in more sedentary contexts. The absence of evidence for hot-rock technology in North America is, therefore, potentially significant. It suggests that either domestic fire technologies were less intensive or varied than in later time periods, or that taphonomic factors are distorting our ability to identify the earliest evidence.

Clearly, much is left to be learnt about Paleoindian domestic fire technology, and while plant processing does appear to be present (e.g. Armstrong 1993), it has not yet been identified on the Southern High Plains. At present, the minimal archaeological record appears to support the traditional interpretation of small highly mobile groups moving frequently across the landscape and subsisting mainly at the hunting end of the hunter-forager continuum (e.g. Hoffman 1989; Johnson and Holliday 2004).

Archaic

The Archaic represents a period of ca. 6,000 years and traditionally is characterised as a period of population growth, increased diet breadth (including both plant and meat resources), and greater regional tethering of the population (e.g. Hoffman 1989). In contrast to adjacent Plains regions (e.g. Frison 1983; Black and Ellis 1997), very little is known about hunter-gatherer domestic fire technology for much of the Archaic period within the study area. This is particularly true when considered in context with the length of time represented by the period. What evidence there is appears to point to the intensification and increasing complexity of domestic fire technologies. The plotted distribution of Archaic period campsites (Figure 8.13) indicates that populations were perhaps making more intensive use of the Pecos and Canadian River valleys to the west and north of the region as well as the rugged canyonlands to the east. The early Archaic (8,000-6,500 RCYBP) record contains no absolutely dated LTFs and is therefore a critical area for future research.

The association between oven features and an increased reliance on vegetal food resources (Johnson and Holliday 1986, 1995, 2004) has had a profound impact on our understanding of hunter-gatherer lifeways during the Middle Archaic (6,500-4,000 RCYBP) (*cf.* Shelley 1994). Synthesis of hunter-gatherer adaptations to the perceived harsh conditions of the Altithermal neatly fit this model, implying that populations were effectively constrained at key resource areas for longer periods of time than during the preceding, apparently more mobile, Paleoindian period. Middle Archaic populations might logically have expanded their diet breadth during these times of stress (e.g. Winterhalder 2001). The result of such a strategy is presumed to be a greater reliance on lower-ranking food resources, such as plant foods, which require longer processing times (Hoffman 1989; Wandsnider 1997; Wrangham *et al.* 1999).

Unfortunately, while the LTF dataset does appear to support greater population intensification in areas where edible plant resources are most abundant, the archaeological evidence is extremely limited and relies heavily on the contextual associations such as the fragmentary mano (grinding tool) recovered in association with the well-excavated fire basin with hot-rocks at the Lubbock Lake site (41LU1; Johnson 1987). More features need to be identified, macrobotanical samples should be undertaken as a matter of priority, and absolute temporal controls are necessary to avoid Type-I errors (too small of a sample size) when interpretation at the regional level is attempted.

The suspected development of hot-rock technology during the Archaic does appear to be consistently identified at a regional scale. A number of plausible scenarios may be framed to explain this apparent technological innovation, such as: a response to lower fuel availability; rising populations; an increase in plant material in the diet; the need to cook desert plants longer to make them digestible; intensification in the processing of faunal elements (e.g. grease, marrow, brain); decreased mobility and therefore increased resource intensification; the influence of a new cultural group into the area; indigenous technological development; or any combination of the above. The most logical scenario suggests that rising population levels coupled with decreased population mobility, decreasing fuel availability, less-predictable faunal resources, and encroaching desert plant species resulted in increased resource intensification, and therefore the necessary stimulus for the increased reliance on hot-rock technology (cf. Thoms 2003).

The physical structures of LTFs identified at Archaic sites appear in general to be poorly preserved and, therefore, they are not well understood. Classification by means of the hierarchical key has indicated that hot-rock clusters comprise more than half of the data-set (Figure 8.24). Other feature types are also recognised and these too tend to contain hearthstones. This classification does not itself equate with the interpretation that vegetal-processing ovens dominated domestic fire technology in the Archaic period. The large number of hot-rock clusters and ephemeral hearths with hot-rocks (6.8%) might suggest that taphonomic processes have acted more aggressively to destroy the physical form of a large number of Archaic features (see also Frison 1983 for a perspective from the northern section of the high plains). The realisation that the Archaic is the least-researched period (in terms of excavation, Figure 9.4), has only added to the problem of refining our understanding of the types of structures represented.

Ethnographic data have underscored that an increase in hot-rock technology need not necessarily directly imply a greater reliance on plant resources. Furthermore, these accounts suggest that a diverse range of functions were associated with the use of hearthstones, not all of them directly related to subsistence technologies. Nevertheless, the covariance of Archaic sites with evidence for hearthstone technology in geographic areas of high biotic diversity (Figure 9.1) suggests that subsistence activities may in large part account for the observed distribution.

Archaic domestic fire technology appears to exhibit greater complexity as compared to the preceding Paleoindian period. The technological development of hot-rock technology during this time appears to be significant and is most likely to be an indicator of a wider shift in diet. Unfortunately, the geomorphic setting of many Archaic sites has largely resulted in only the structural base or hearthstones from individual features being preserved. It is essential, therefore, that whenever Archaic period LTFs are encountered that detailed archaeometric, environmental, and faunal analysis are conducted to determine the function of these features.

Ceramic

The trend toward decreased mobility appears to have fluoresced along the Pecos river valley of southeastern New Mexico during the Ceramic ca 2,000 – 500 RCYBP (Figure 8.14). At this time, populations appear to have intensively utilised the resources in the southwestern portion of the area and a wide diet breadth should be expected to be represented in the archaeological record. verall, although Ceramic period LTFs are frequently observed, little is known of the specific dietary habits of the people that built them. The use of hearthstones as a component part of domestic fire technology is again evident throughout the region.

As demonstrated from the archaeological record (Chapter 8), the broad spatial pattern of identified Ceramic-period sites with evidence for LTFs appears to conform reasonably well with previously projected (but largely untested) models for the spatial extent of the Jornada Mogollon archaeological tradition (Figure 9.5). The presented dataset, however, lacks the temporal and cultural resolution necessary to determine more accurately the lifeways of the peoples occupying this area. It is very likely that peoples along the Pecos River valley were increasingly influenced, both by the more sedentary craft-orientated communities to the west and the high plains hunter-gatherers to the east. Unfortunately, synthetic examinations of what must have been a culturally dynamic geographic area are badly in need of updating (see Katz and Katz 2000 for a recent discussion of the issues). In this regard, botanical data clearly indicate that desert succulent species suitable for processing in large oven features should be expected in locations above 3,500 feet (Katz and Katz 1985:42). Ceramic-period sites located over 3,500 feet above sea level comprise the majority of the New Mexican portion of the investigated dataset. In this area, only three sites fall below the 3,500 feet threshold and an average elevation of 3,875 feet. Although more detailed analyses are clearly required for specific features in this area, a strong relationship does appear to be indicated by the botanical range of succulents and the locations in which sites with Ceramic-period LTFs are located.

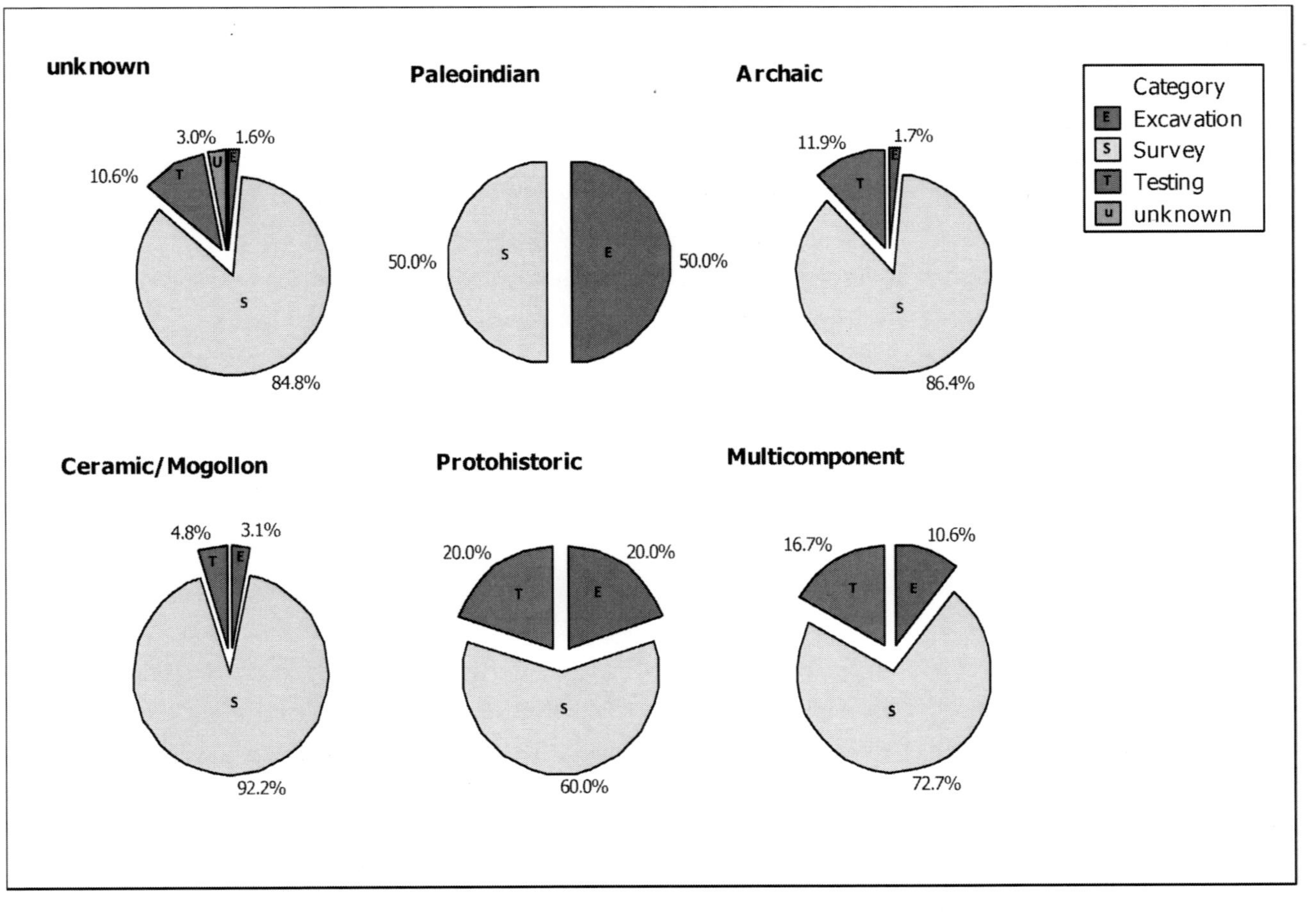

Figure 9.4. Modes of engaging the archaeological record on the Southern High Plains by major temporal period.

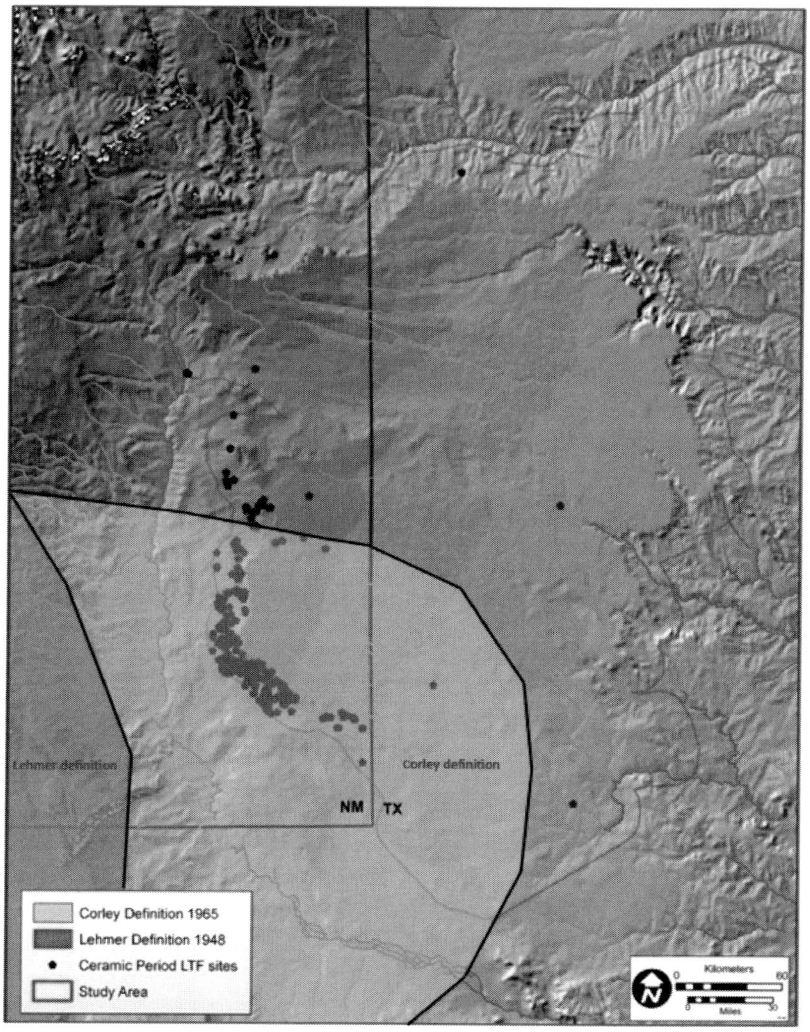

Figure 9.5. Spatial distribution of Ceramic period sites with evidence for LTFs and relationship to previously developed models for the spatial extent of the Jornada Mogollon culture (data Corley 1965).

In contrast to the Mescalero dunes area, very little evidence for hunter-gatherer populations is available from the High Plains portion of the region. Where identified, however, the evidence supports the intensive processing of food resources, perhaps as a result of increased group size or a greater need to store game products for transport and trade.

Classification of the regional dataset indicates a broader range of features are represented than the preceding period and there appears to be less reliance on hot-rock technology (Figure 8.24). Over half of the dataset could not be classified, suggesting much more emphasis should be placed on recording the physical structures observed at individual sites during surveying activities. Increased observations of non-hearthstone categories such as dispersed hearths (14%) suggest that preservation of Ceramic-period features may be enhanced in relation to the preceding Archaic. This is likely to be a simple function of the time elapsed between abandonment and observation of the features by archaeologists.

Specialized large-scale LTFs known locally as ring middens (Figure 9.6) are well documented along the Pecos River Valley in the southwestern portion of the study area (Katz and Katz 1985; Sebastian and Larralde 1989). Unfortunately, probably due to low levels of consistency in their recording, these features are not clearly identified by the hierarchical key classification. Nevertheless, these distinctive structures do appear to be an important component of Ceramic-period domestic fire technology and are likely to be represented within the current dataset. Within the study area, only the survey form filed for LA25668 in Lea County, New Mexico explicitly indicates the presence of a ring midden. Though the 10-meter by five-meter feature recorded at LA130927, in Eddy County,
In contrast to the Mescalero dunes area, very little evidence

for hunter-gatherer populations is available from the High Plains portion of the region. Where identified, however, the evidence supports the intensive processing of food resources, perhaps as a result of increased group size or a greater need to store game products for transport and trade.

Figure 9.6. Example a of a 'Ring Midden' feature recorded along the western margin of the Pecos river. Note feature were located directly to the west of the present study area. (courtesy of George MacDonell, Bureau of Land Management, Carlsbad Office).

Compilation of radiocarbon dates assayed on ring middens by Katz and Katz (2000) suggests that this type of feature was increasingly utilised throughout the period (Figure 9.7). It is of note that the distribution indicates a gradual increase from the late Archaic with a peak in production around 1000 RCYBP.

The function of ring midden-type features is not straightforward. The assumption that they represent the remains of earth oven-type structures for the processing of desert succulents (xerophyes), such as lechuguilla, sotol, and prickly pear, is based on well-preserved macro-botanical evidence in the lower Pecos area (e.g. Dering 1999). These xeric species are representative of the Chihuahuan biotic province and the closest modern extent of this biotic province to the study area is the Delaware basin in southern Eddy County (ca. 40 miles south of the present study area). The prehistoric distribution and relationship to the area of these species is largely unknown. It is very likely, however, that the range of these species expanded and contracted within the broader paleoclimatic patterns already identified (Chapter 4). Katz and Katz (2000) compile data from several sites investigated by VanDevender to provide useful supporting evidence that all three species were present along the eastern portion of the Guadalupe Mountains in western Eddy County throughout the Holocene. In any case, these species appear to have been available to populations living along the Pecos River Valley at least by the Late Archaic and probably throughout much of the Holocene.

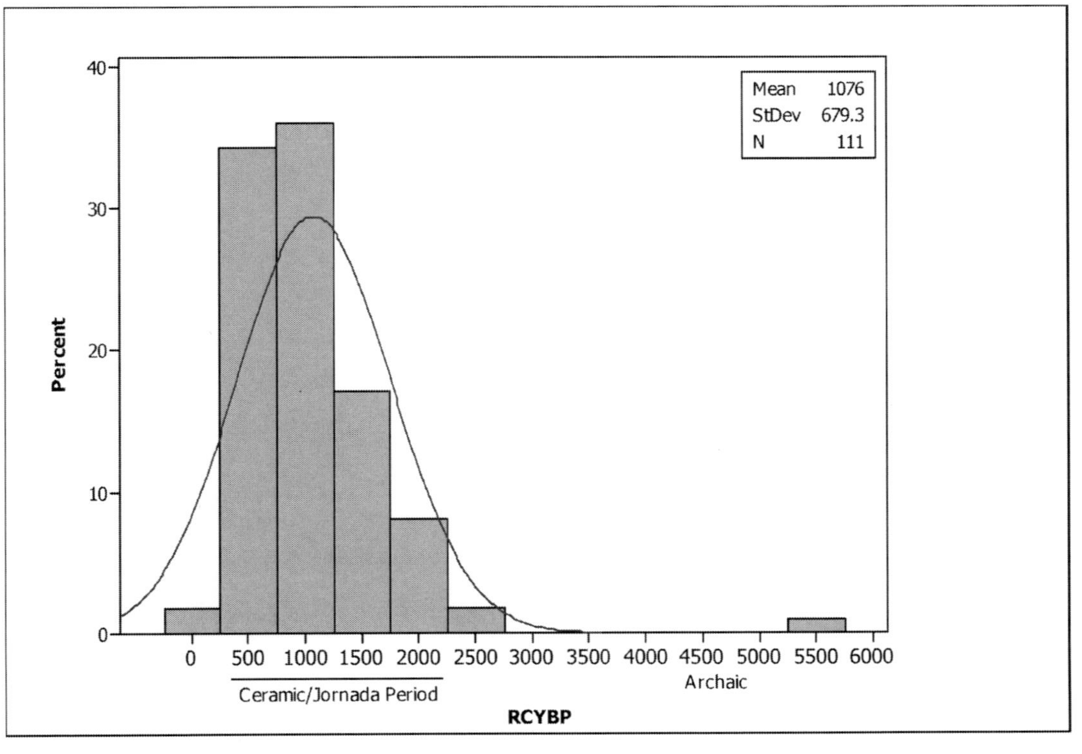

Figure 9.7. Frequency of radiocarbon determinations obtained on Ring Midden features identified in southeastern New Mexico (data: Katz and Katz 2000, table 35).

Ethnographic research emphasises the substantial investment in labour and resources (in terms of fuel and the slow-growing plant communities that were being processed) required in the construction of large earth ovens (see Chapter 6). Additionally, xerophytic plants (plants which require very little water for survival) require heating for long periods between (typically between 36 and 48 hours) to remove toxins and, therefore, become consumable by prehistoric populations (Dering 1999: 661). If the ring midden features observed in the Pecos River Valley portion of the study area do consistently represent large-scale xerophyte processing facilities, then the implication is that the constant utilization of this technology would limit population mobility at certain times of the year when these plants are harvested, notably such as winter-spring (lechuguilla). Energetic return values from cooking these resources are, however, relatively low and experiments conducted by Dering predict that a single earth oven would only supply sufficient food to sustain a small family group for up to two days (Dering, 1999:665).

The large cost in terms of labour and fuel and resultant visibility of the archaeological signature for these features arguably biases the archaeological record for these structures and may not, therefore, be representative of the actual mobility patterns of hunter-gatherer groups in the Pecos valley area. Quantitative morphometric, data calculated from the features encountered during the present research (Chapter 8) appear to corroborate this finding (Table 9.1.). The size of LTFs, sampled in this study by excavation, increases throughout the Holocene until the Ceramic period, after which they appear to decrease in size (Table 9.1).

Table 9.1. Summary of the size of features recorded by major temporal period.

Period	Mean LTF Surface Area (cm^2)	Number of Sites Represented
Paleoindian	4010	2
Archaic	4717	12
Ceramic/Mogollon	8401	25
Protohistoric	8891	7

Overall, the evidence derived from the archaeological observation of fire technology during the Ceramic period suggests an increased population, decreased population mobility, increased diet breadth, and increased logistical foraging (*cf.* Bird and O'Connell 2006). Although the archaeological record is clearly biased toward the preservation of the evidence for resource-intensive activities that appear to mainly have occurred in the Pecos River Valley, it most likely greatly underestimates evidence for the hunting portion of the subsistence base that likely took place on the Llano Estacado itself. The increasing evidence for the construction of earth ovens throughout the period suggests either:

- That the population size was gradually increasing;
- That diet breadth was necessarily increasing as a function of diminishing numbers of high-ranked resources;
- Access to high-ranking resources on the High Plains was being restricted by other human population(s);
- Culinary preferences increasingly favoured the consumption of desert plant species as a function of time.

The excavation, analysis, and dating of more sites along the Pecos River Valley is critical in order to better understand the nature of the cultural adaptation in this area.

Outside of the Pecos River Valley, the evidence for Ceramic-period hunter-gatherer campsites is extremely limited and the extent to which the well-excavated campsite at Lubbock Lake (area 10; Figure 8.32) is representative of the wider pattern of domestic fire technology at this time is unfortunately largely unknown. Evidence from this excavation does appear to indicate that the intensive utilisation of resources was not only limited to plant materials but also extended to animal resources as well.

Protohistoric

The collapse of the Ceramic-period economies is clearly evident in the Protohistoric dataset with populations apparently abandoning the previously densely occupied south-eastern portion of the region and returning to a pattern of increased mobility (Figure 8.15). Katz and Katz identify a peak in ring midden construction ca. 500 RCYBP as evidence for the first arrival of Athapaskans in the area (Katz and Katz 2000:57). The archaeological record for domestic fire technology during this time does not, however, corroborate the presence of these new populations. The low number of identified Protohistoric sites is most likely to be the result of research bias (e.g. Surovell and Brantigham 2007) as much as the displacement of populations in the Pecos valley area.

The lack of evidence means that generalisations about regional patterns in domestic fire technology must be developed from the limited archaeological record and applied at larger scales. This situation is far from satisfactory and additional Protohistoric sites clearly need to be identified. Nevertheless, the decreased reliance on hot-rock technology for heating purposes during this time period would appear to indicate that groups were more mobile and less reliant on direct container boiling technologies as is also suggested by the ethnographic evidence (Chapter 6). Hearthstones appear to have taken on a structural importance and were used to line the interior or define the perimeter of excavated features (see excavated examples in Chapter 8). The use of hearthstones in these structures may represent a continuation of a technologic element after its functional importance has been superseded by the introduction of ceramic cooking vessels.

Ethnographic evidence contributes especial importance to the interpretation of the Protohistoric time period and some of the domestic fire technologies recorded by early travellers, traders, military expeditions, and missionaries should logically be expressed in the Protohistoric archaeological record of the region. It has already been noted that the large oven structures built by the Mescalero Apache and recorded by Edward Curtis in the early nineteenth century (see Chapter 6) are not present in the Protohistoric dataset. Additionally, the camps of the nomadic buffalo-hunting peoples encountered living on the High Plains proper do not appear in the dataset other than at the Lubbock Lake site. Again, much additional fieldwork is required to determine the nature of the Protohistoric dataset.

Gender

A brief comment on the visibility of gender roles in fire technology is appropriate, as the results are entirely situated in the ethnographic record and therefore relate only to human use of fire technology in general terms. Ethnographic data reviewed for this project consistently indicate that gender roles are actively played out throughout the life-cycles of domestic fire technologies (e.g. the examples noted by De Vaca as he passed through southern Texas in the sixteenth century, Chapter 5). These observations suggest that women are more likely to be responsible for the operation of LTFs and in the preparation and cooking of food items. The assignment of simple gender roles is, however, typical of early anthropological recording and do little to situate fire features as loci for human action. Important questions remain regarding the likely shifting gender roles during the initial design, construction, and resource procurement phases of feature life-cycles. Nevertheless, fire technology appears to be a highly structured arena of gender roles and clearly delineated cultural patterns are strongly implicated by the ethnographic record. This is a positive outcome and a potentially useful avenue for future research as there appears to be little reason to suggest that gender roles remained static either over time or across space. The available data, therefore, provide very few clues that are suitable for historical or agency-based interpretation that attempt to identify the biological structure of hunter-gatherer groups living in the remote past.

Conclusions

The investigation of prehistoric domestic fire technology on the Southern High Plains of North America has proved an excellent tool for exploring variation in the fire features constructed by the hunter-gatherer populations that occupied this region (Figure 9.8). The process of interrogating the regional dataset has drawn together disparate evidence for domestic fire technologies into a broad cultural ecological framework. This framework indicates that evidence for LTFs is intimately bound up in a largely homogenous or idealised concept of hunter-gatherer campsites. This concept is challenged by the presented strands of evidence, which suggest that fire technology has consistently been a critical, technologically heterogeneous, and often archaeologically observable component of human activity within the study area. By considering the ubiquitous LTFs recorded on the Southern High Plains in more active terms than as passive site furniture, research directions can be pursued that are simply not attainable from the examination of stones and bones alone. Furthermore, the conceptual linkage of the positive evidence for LTFs with the placement of hunter-gatherer campsites has resulted in the development of a novel demographic tool that illustrates the potential for developing feature-orientated research strategies.

Implications for Understanding Site-Formation Processes and Regional Sources of Bias

The growing realisation during the development of this region has been that the regional dataset is as much a product of political (landownership), research (type, institution, and method), time dependant, and geomorphic factors as it is the result of past cultural technological activities. These diverse factors clearly have an incredibly complex effect on the structure of the regional record and any explanation of hunter-gatherer fire technology should be wary of the role of that these may have had on the patterning observed data.

Any attempt to interpret the regional record necessarily requires an active assessment of how these sources of bias have structured the sites, features, and objects observed during fieldwork. Experimental research (Backhouse and Johnson 2007a) has indicated that individual processes may be isolated and usefully simulated in order to understand how they may have acted on the archaeological residues left over from past human occupation.

It should be noted that the regional dataset does not represent a complete inventory of all LTFs located within the study area and this was not the aim of the project. It is also noted that features are a component of the larger classificatory concept of archaeological sites. The assumption *a priori* that the identification of features and sites are positively correlated is implicit in the presented project. While this relationship would be difficult to demonstrate, it is logical that where artefacts and features are identified, they are recorded individually or in groups as sites. As far as the author is aware, features are not considered isolates (as sometimes artefacts are) and, therefore, their identification should always be synonymous with the designation of that location as a site. This logic indicates that querying the regional database for sites with evidence for localised evidence of fire technology (LTFs) is reasonable and should not be adversely biased by the realisation that site and not feature recording was the primary goal of the original recorder.

The project has undoubtedly, though hopefully rarely, omitted sites at which good evidence for LTFs was recorded in the field but where these observations did not translate to the submitted site-survey records (as is often the case in follow-up research conducted after the submission of the original site-survey form), or application of the hierarchical key has resulted in the omission or misclassification of sites and features.

Application of standardised identification and classification methodologies suggest the dataset is a representative sample of the types of documentary evidence for these features that was available for research at the time of querying. No doubt, more sites will be identified as a function of time, increasingly refined database querying tools will be developed to isolate pertinent data more efficiently, and other factors will alter the shape of the current data. It is interesting to note that the misspelling of a feature during data entry could result in its misidentification by electronic key word searches. Regardless of the expected ongoing recovery, access, and availability efforts, it is essential that we design research tools that maximise the return from the dataset already generated. Although these tools must necessarily use simple and nominal data structures, their application is essential to developing a regional approach.

Lastly, it should be made explicitly clear that the investigation of hunter-gatherer fire technology is a novel methodology designed explicitly to provide a complementary approach to more traditional modes of research, which tend to focus on objects and sites (see Chapter 5). The interpretation of the Southern High Plains record presented here is based solely, for better or for worse, on the available evidence for LTFs identified in the archaeological record. No attempt to present a neutral model for site patterning on the Southern High Plains was attempted although this would be a useful exercise. The presented dataset does not represent a holistic Culture History model of the archaeological evidence for prehistoric populations on the Southern High Plains (see Hoffman 1989; Hughes 1989; Sebastian and Larralde, 1989; Johnson and Holliday 1995, 2004 for regional discussions) and that was not its purpose.

Potential for the Investigation of Other Regions

The investigation of LTFs on the Southern High Plains of North America has arguably demonstrated the potential of the interpretative project to provide an inovative way of exploring a largely ignored facet of hunter-gatherer technology in a holistic manner. The applicability and usefulness of this approach to the investigation of hunter-gatherer fire technology in other regions is however untested.

Hunter-gatherer populations are evident in the archaeological record of nearly all geographic areas of the earth. Therefore, can the method for investigating domestic fire technology outlined in this study, be usefully applied to other regions? If so, how do the results differ from those observed on the Southern High Plains, and what does this mean for any attempt to resituate fire technology as a behaviourally relevant human phenomenon (*cf.* Gheorghiu 2002)?

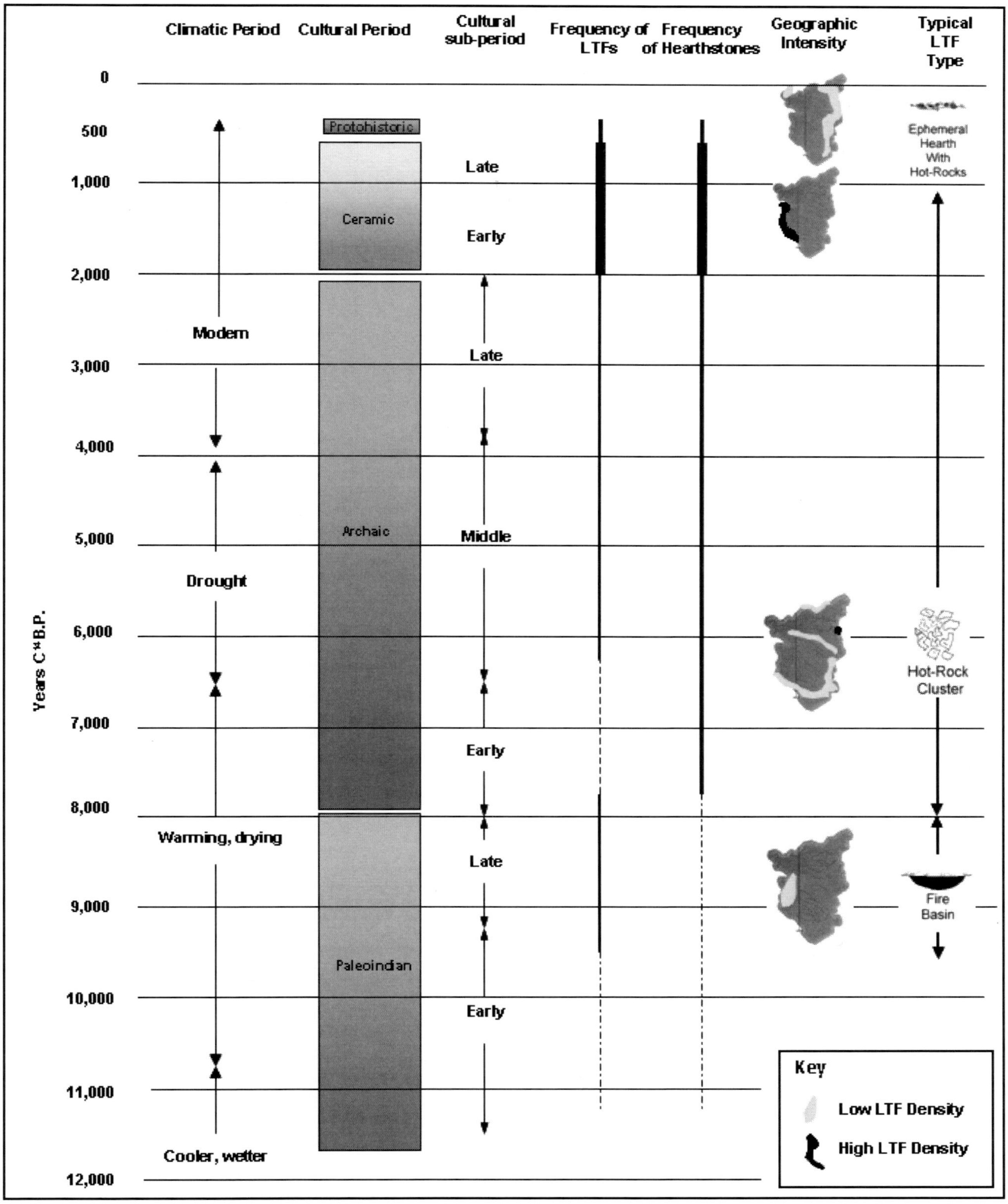

Figure 9.8. Summary of the regional record for hunter-gatherer fire technology on the Southern High Plains and surrounding area (climate and Cultural Periods after Johnson and Holliday 1995, 2004).

Put another way, a regional approach is, in my opinion, key to understanding hunter-gatherer adaptations (see also Smith and Winterhalder 1992; Kelly 1995; Winterhalder 2001) and research directions therefore must be regionally situated. If we cannot integrate the data we have already collected at this scale, then we are forced to engage, and therefore interpret, the archaeological record based only on the latest most scientifically or theoretically modern excavation techniques. The interpretative project presented for the Southern High Plains attempts to work from the existing dataset (no matter how limited the component parts might be) by utilising the regional perspective as a jumping-off point for focussed research. In order to achieve this goal, both method and theory must be able to operate on all datasets and not just the high-quality ones. For these reasons the hundred-plus years of archaeological research in the southern portion of the British Isles is an ideal laboratory in which to assess the potential of hunter-gatherer fire technology studies to be culturally informative beyond the Southern High Plains.

Southern British Isles Dataset

The test area encompasses a range of physical landscapes, which to a large extent are conditioned by the underlying geology and have been significantly altered by a long history of human intervention and manipulation (e.g. Lowenthal and Prince 1964; Muir 1998). The north of the area stretches from the Cotswolds and the Severn Estuary in the west to the Fens of Norfolk in the east. Major drainages include the Exe, the Tamar, and the Parett rivers in the west; the Test, Avon, Arun, and Thames in the central southern portion; and the Great Stour, Stour, and the Nene in the eastern portion. The south and southeast of the study region is most intensively farmed for crop production whilst the southwest is predominately pasture. Modern climate systems can be generally classified as cool temperate oceanic type (D1) with rain possible year round and occasional extremes of heat or cold (Sweeney and O'Hare 1992). Seasonal variation in temperature is pronounced with frost common during the winter months (Lowenthal and Prince 1964). This climate pattern supports both deciduous and coniferous trees.

Paleoenvironment

Palaeoenvironmental reconstruction for the post-glacial period in the British Isles in general and Southern England in particular has received much recent attention (e.g. Atkinson *et al.* 1987; Bennett 1989; Momber 2000; Spikins 1999). The application of multidisciplinary research designs has allowed significant refinement to long-held assumptions of environmental change (Spikins 1999:80). The enormity of the transformative changes in terms of climate, ecology, and geomorphology following the last glacial maximum should not be underestimated. Various proxy datasets appear to indicate a rapid warming in temperatures following the Younger Dryas cold phase (ca. 10,000 RCYBP), which continued throughout the Mesolithic period; additionally, smaller scale variability was also are a component of this weather system (Meese *et al.* 1994).

Models of forestation during the post-glacial indicate rapid spread of species introduced from the south and east. A general pattern of the colonisation by birch and pine woodland is likely to have given way to a mixed woodland of pine, oak, hazel, elm, and lime. By 7000 RCYBP, alder becomes the dominant tree species (Rackham 1976; Smith 1992; Spikins 1999). The rapid rise in temperatures and melting of the ice cap resulted in sea level and the drowning of vast landscapes (Momber 2000). The south and east of the study area would have undergone the most extensive transformation during this process with the final detachment of Britain from mainland Europe occurring ca. 8500 RCYBP (Lambeck 1995).

Site Distribution

Evidence for the most recent human colonisation of Britain is currently accepted to have occurred by around 12,600 RCYBP. The archaeological signature of these late Upper Palaeolithic populations, although distinct, is largely restricted to cave sites in the Cresswell Crags region and to the Devonian formations of the southwest (Smith 1993; Darvill 1996). Whilst exceptional but generally isolated anomalies to this scheme exist (Barton 1992), it is not until amelioration of the climate at ca.10,000 RCYBP that conditions conducive to the preservation of archaeological materials, and therefore sites, begin to increase. Wymer's (1977) gazetteer of Mesolithic sites in England and Wales remains a dated but much-used resource in Mesolithic research; however, the lack of contextual i.e. feature information is problematic in assessing research potential for this project.

Appropriate sites with hunter-gatherer components and evidence for LTFs were identified by a scrutiny of the available literature; correspondence with County-level Sites and Monuments Records Offices (SMRS); and access to two publicly available national-level electronic databases. Period and keyword searches were performed on the Archaeology Data Service (ADS) database hosted by the University of York and available online at [http://ads.ahds.ac.uk] and the Pastscape database hosted by English Heritage and available online at [http://www.pastscape.org.uk]. Searches were performed between January 2003 and October 2005 and resulted in the identification of 88 records (Table 9.2). Several geographically clustered sites (i.e. the Oakhanger group or Wawcott sites) are grouped together for the purpose of this survey and the true number of sites is actually, therefore just short of 100.

Table 9.2. Summary results of literature search of post-glacial hunter-gatherer sites with positive evidence of LTFs in the Southern Britain Investigation area.

County	Site*	Reference(s)
Berkshire	Greenham Dairy Farm/ Faraday Road	Sheridan et al. 1963; Carter 1976; Reyneir 1993; Ellis et al. 2003
	Moorfarm, Holypot	ADS (EHNMR-628126)
	North bank of river Kennet	ADS (EHNMR-653167)
	Thatcham, sites 1-5	Healy et al. 1992; Carter 2001
	Wawcott Sites	Froom 1972 ; 1976
Buckinghamshire	Eton College	ADS (EHNMR-655974)
Cambridgeshire	East Waste Landfill Site (MIL EW 94)	ADS (EHNMR-1074116)
	Peacocks Farm	Smith et al. 1989
Cornwall	Hudder Field	NMR (NAT-INV-426256)
	Memmoan Field, Callean Farm	NMR (NAT-INV-426248)
	Windmill Farm	NMR (625839)
Devon	Stocklands	Devon SMR (rec no. 34354)
	Three Holes Cave No. 7.	Rosenfeld 1964
	Westward Ho!	Churchill 1965
Dorset	Culverwell Mesolithic Site	Palmer 1971; 1976; 1999
	Culverwell Old Lower Lighthouse, Site 1	Palmer 1971; 1976; 1999
	Hampreston	Powell 1999
	Hengitsbury Head, Powell Mesolithic Site	Barton 1992
	Irwene Minster	Summers 1941
	Mother Siller's Channel	Palmer 1999; ADS (EHNMR-650737)
	Sweethill	Palmer 1999; Bellamy 2000
	Ulwell	Calkin 1952
	Whitcombe Hill	Palmer and Dimbleby 1979; Palmer 1999
	Winfrith Heath	Palmer and Dimbleby 1979
East Sussex	Eridge Rocks	ADS (EHNMR-1347438)
	Heathrow, Terminal 5	Framework Archaeology, forthcoming
	Hermitage Rocks	ADS (EHNMR-626346)
	Selmeston Sand Pit	ADS (EHNMR-626396)
	Streat Fishing Lake	ADS (EHNMR-1124411)
Greater London	A13 Thames Gateway DBFO: Movers Lane	ADS (EHNMR-1362596)
Greater London	BAQ90 (GAZ 208)	Sidell et al. 2002
	Bexley Rugby Football Club (TNT98)	Pre construct Archaeology, forthcoming
	Jackson's Common	ADS (EHNMR-647931)
	Marlborough Grove (GAZ 204)	Sidell et al. 2002
	Three Ways Wharf	Lewis 1991
	West Heath	Collins and Lorimer 1989
Gwent	Goldcliff	Bell 1994; Burrow 2003
Hampshire	Bowmans Farm	Green 1991
	Broom Hill	O'Malley and Jacobi 1978
	Chilworth #1	ADS (NMR_NAT-INV-230137)
	Churchplace Inclosure	ADS (NMR_NAT-INV-226296)
	Clanfield Reservoir	ADS (EHNMR-627581)
	Fort Wallington	Palmer 1971; Hughes and ApSimon 1978
	Longmoor 1	ADS (EHNMR-627743)
	Shortheath	Rankine 1953
	The Warren, Oakhanger Sites I, II, III, V, VII, VIII	Rankine 1952; Rankine et al. 1960
	Wakeford's Copse	ADS (EHNMR-651929)
	Wallington Way	ADS (EHNMR-1038360)
Hereford and Worcester	Madawg	Burrow 2003
Hertfordshire	Glaxos, Ware	ADS (EHNMR-656209)
	Northaw	ADS (EHNMR-638544)
	Sandy Lodge	ADS (EHNMR-1084924)
	The Grove Estate	AOC Archaeology, forthcoming
Isle of Wight	Bouldnor Cliff	Momber 2000
	Brook Bay	IOWSMR(rec. no.4)
	Chilton Chine West	IOWSMR(rec. no.265)
	Werrar	IOWSMR(rec. no.952)
	Whale Chine	IOWSMR(rec. no.201)
Kent	Abbey Fields	ADS (EHNMR-1358707)
	Addington #1	ADS (NMR_NAT-INV-412518)
	Beechbrook Wood (ARC BBW 98)	Museum of London Archaeology Service 1998
	Sandway Road	Anonymous 1999
	Snarkhurst Wood	Oxford Archaeological Unit 1996
	Southern and Northern Occupation Floors, Lower Halstow	Burchill 1925a;1925b
	West of Scalers Hill	Oxford Archaeological Unit 1998
Lincolnshire	Uffington Estate (UFF 91)	ADS (EHNMR-656013)

County	Site*	Reference(s)
Norfolk	Two Mile Bottom	Robbins 1998
Oxfordshire	Radley Vale of White Horse	ADS (NMR_NAT-INV-1396922)
Salop	Woodhouse Farm	Wymer 1977
Somerset	Avelines Hole	ADS (NMR_NAT-INV-194278)
	Goughs Cave	Jacobi 2004
	Lower Pitts Farm	ADS (SOMS-SMR-13965)
	Milsoms Corner	ADS (SOMSSMR-6312)
	Moor Lane, Backwell	ADS (EHNMR-1314897)
Suffolk	Mildenhall #1	ADS (NMR_NAT-INV-377541)
Suffolk	The Wangford Site - The Carr	ADS (EHNMR-647204)
Surrey	Abinger Manor	Leakey 1951
	Bourne Mill Spring Site, Farnham	Rankine and Clark 1939; Palmer 1971
	Kettlebury 103	Reyneir 1993; 2002
	North Park Farm	(Shaikhley 2005, Personal Communication)
Sussex	Angmering Decoy	Graves and Hammond 1993
	Iping Common	Keef et al. 1965
	'Site 5' at Stretto	ADS (WARWSMR-3000)
	Stretton On Fosse (Site 5)	ADS (EHNMR-631112)
West Sussex	Castle Park (West Durrington)	ADS (EHNMR-1333145); Archaeology South-East, forthcoming
West Sussex	Rock Common	Harding 2000
Wiltshire	Downton	Higgs 1959; Palmer and Dimbleby 1979
	Golden Ball Hill	Dennis and Hamilton 1997

*Note where no reference is available then the database record number is supplied.

Historically, the available dataset represents over 100 years of archaeological investigations in the region. The methodologies used to investigate these sites are extremely diverse. Overall the investigation of British hunter-gatherer sites has been one in which excavation has been the primary means of engaging with the archaeological record. In contrast to the Southern High Plains dataset, surface survey methods are rarely employed, as sites with in situ evidence for domestic fire technology tend to be well buried and therefore require subsurface excavation in order to be investigated. In areas not conducive to rapid burial, sites have been eradicated by the later activities of humans and natural processes.

Analysis of the number of sites identified per decade indicates a positive growth rate of 13% in the total number of sites being identified per decade. This trend would appear to indicate that it is very likely that significant numbers of Mesolithic sites with evidence for domestic fire technology are going to be identified in the coming decades.

Geographically, the distribution of thermal-feature sites is concentrated in the central and southern portion of the study area (Figure 9.9). It is very unlikely, however, that this distribution is the direct result of prehistoric population dynamics. Instead, the distribution is most likely to be a product of two inter-related factors. Firstly, the geomorphology of the study has resulted in rapid burial of well-stratified sites along the late Quaternary river valleys in the southeast. Concomitantly, higher rates of erosion in this area are also exposing sites more rapidly than on the more stable land surfaces that dominate the West Country. Secondly, economic development in the second half of the twentieth century has focused on the more prosperous southeast, which has therefore been the focus of more intensive contract-based excavation than is the case in other portions of the study area.

Concentrating on the central southern portion of the study area, the geomorphic factors underlying the spatial distribution are particularly apparent. Mesolithic thermal features are being periodically exposed along the cliff faces of the rapidly eroding muds that comprise the western section of the Isle of Wight. These sites are subsequently being claimed by the seas at an alarming rate. On the north side of the Island, the recently identified underwater site at Bouldnor Cliff, illustrates the potential of submerged valley systems for the investigation of Mesolithic landscapes. The identification of a localised thermal feature with organic preservation at this site is, therefore, highly significant (Momber 2000).

A simple examination of the estimated site sizes (by the excavator) reveals that only 15% of the sites investigated contain any information as to the ground area covered by either the modern or prehistoric site footprints. Furthermore the high standard deviation (SD = 6439m^2) indicates there is a massive range in the size of sites that do have size estimates. This result means that we have no real idea what a typical post-glacial hunter-gatherer site comprises in terms of spatial extent. This obviously has important implications for research planning purposes.

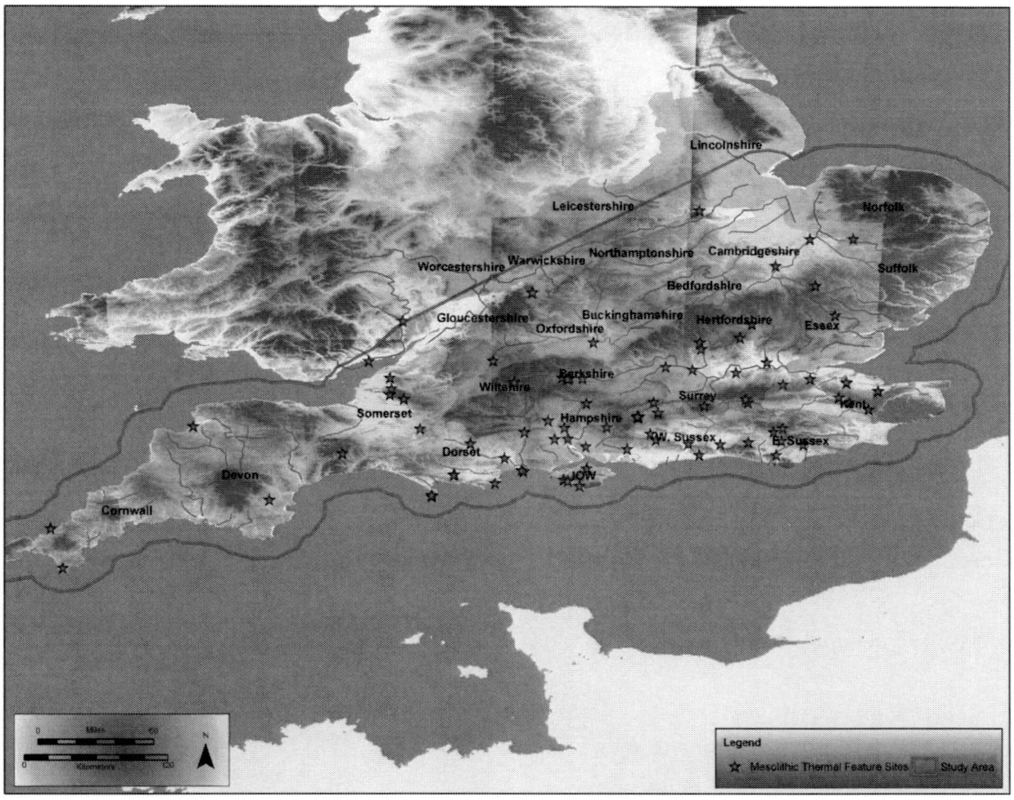

Figure 9.9. Distribution of hunter-gatherer sites with evidence for localised thermal features identified within the southern British test area.

Examination of the number of LTFs identified per site indicates that at nearly half of the identified sites, no estimate was given as to how many features were present. This stands in contrast to the rather detailed lists of lithic tools and even faunal assemblage data that were often present. Where LTFs were identified, it seems that single features are typical of Mesolithic sites and only rarely are more than ten features identified at any one site.

In order to qualitatively examine the variation in the physical evidence for identified LTFs, I applied the same hierarchical key classificatory structure as used on the Southern High Plains dataset. The results indicate that insufficient data exist for even a basic classification for over half of the sites included in the British study area (Figure 9.10). Of the sites that could be classified, there is a fairly even distribution of types across the full range of possible structures. The spread in types is encouraging and suggests that meaningful research directions can be framed to explore whether the variation is behaviourally significant. Although fewer in number, the spread in types identified in the British dataset contrasts with the results from the Southern High Plains where high numbers of features with low structural integrity were identified. This result suggests that although Mesolithic LTFs are harder to encounter, they have a high generally higher likelihood of structural preservation and, therefore, increased potential for meaningful interpretation.

Although the result of undoubtedly very different populations, the physical evidence for domestic fire technology utilised by hunter-gatherer groups occupying southern Britain in the early Holocene is not overtly dissimilar to that observed on the Southern High Plains of North America. As on the High Plains, evidence for domestic fire technology has been largely undervalued by the archaeological community as the technological 'poor relation' of the more accepted hunter-gatherer technologies, notably stones and bones. The reasons for this situation are numerous and the conception that a hearth or localised thermal feature is a passive element of site furniture has perhaps been the most detrimental.

Nevertheless, considerable research has already taken place and much more is predicted to be unearthed in the coming decades. Based on this evidence, there is clearly a research need to develop an archaeology of hunter-gatherer domestic fire technology within the Southern British study area. Not only would such an approach be informative of the fire technologies utilised by post-glacial populations, but it would also facilitate the construction of a practical framework for their investigation and interpretation. In this way, research agendas and priorities could be identified and met in terms of a common dialogue between archaeological professionals within both the public and private sectors.

Resituating Fire Technology Research in Hunter-Gatherer Studies

Fire technology is well represented at hunter-gatherer sites identified in the two discussed regions and there is little reason to suspect that either is atypical of the wider distribution for these features in the archaeological record. Stepping back from the two study areas, it is possible then to outline the implications of theoretically refocusing our interpretation on these features as culturally constructed technologies worthy of detailed recording, analysis, and interpretation.

As discussed in Chapter 5, sites and features are not physical artefacts in the sense that they could be picked up whole and transported back to a museum for display. Theoretically at least, a feature could be transported to a museum by means of the mechanical removal of a large block of sediment. It does not seem to be likely that a museum would want to display such a block and unusual situations aside (e.g. waterlogged deposits or rescue situations) there is little practical benefit to undertaking such an endeavour. In most cases, features are recorded in the field.

The simple dichotomy between object and feature recording is stressed here because it underscores the relative importance normally assigned to the study of each. Artefacts are typically studied by materials specialists who, with a preconceived common technological agenda, offer the opportunity for capturing replicable quantitative data (see Andrefsky 1998 for examples pertaining to lithic analysis). Arbitrary constructs such as sites and features, on the other hand, are at the mercy of whoever discovers them. This discoverer may or may not be an expert in their recording, but it is highly unlikely that they will have engaged in fieldwork with the common agenda for technological research demonstrated by the materials specialists. After all, fieldworkers have limited time and more than the feature to worry about. Specialists on the other hand have more control over the way they apportion their time and energy and are more likely to be senior members of the archaeological community.

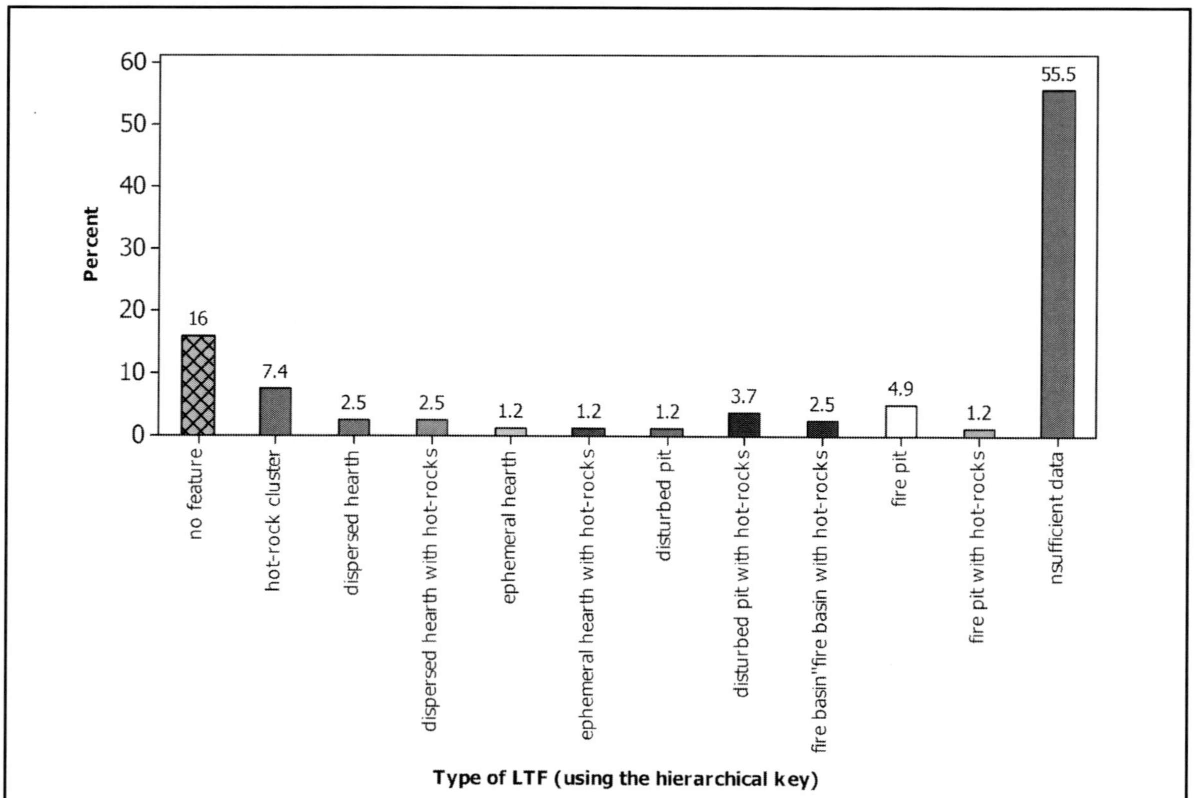

Figure 9.10. Range of types of LTF in the southern British study area after classification.

A disjunction is therefore apparent, which privileges the analysis and interpretation of portable material culture (studied by the specialist) at the expense of sites and features that are encountered by the fieldworker (for a similar argument see Lucas 2001). This is particularly a problem in hunter-gatherer archaeology where external factors (e.g. low site-encounter rates in the Old World and the methodological dominance of survey in North America) have resulted in disproportionate attention being placed on objects rather than features.

Emphasis placed on the investigation of domestic fire technology is not an advocation of the rejection of traditional perspectives. Instead, feature-based technologies (and sites) provide critical context for the portable material culture and in turn can benefit directly from their study. The identification of grinding stones at a site next to a large LTF yielding plant materials is an obvious example of this linkage. This has clearly been long realised by archaeologists but it seems that too often, interpretations look toward the ethnographic record for the quick explanation of site structure.

A commitment to integrating evidence for fire technology as a locus of human activity at the site level is required by the archaeological community. This must necessarily begin in the field if it is going to be successfully achieved. The low cultural/temporal resolution of the interpretative potential observed in the two examined geographic regions is a result of the level of recording available at the state level. The present project demonstrates a method by which a commitment to technological analysis can be holistically achieved. The resulting coarse-grained framework is the first necessary step onto which further sites and high-resolution datasets can be grafted. The classification of features by means of the hierarchical key facilitates the development of regional aims and objectives, which can be tailored to individual programmes of research by archaeologists of all theoretical persuasions.

Establishing Research Values

For instance, on the Southern High Plains a list of general research values for LTFs can be set out:

High Research Value

- Any LTF with at least some structural integrity (those that would be classified to the left of type-2 'hot-rock clusters' using the hierarchical key);
- Any LTF with associated faunal and botanical remains that might be useful in determining the function of the feature;
- Any LTF with potential for absolute dating assay.
- Any LTF with evidence for hunter-gatherer use of fire not associated with subsistence activities.
- Any LTF located in the interior of the Llano Estacado with special importance placed on sites associated with Playa basins.
- Any LTF located along the northwestern portion of the study area along the southern margin of the Canadian river valley.
- Any LTF associated with Paleoindian-type cultural material.
- Any LTF associated with the Early and Middle Archaic-type cultural material.
- Any LTF associated with Protohistoric-type cultural material.

Low Research Value

- Any LTF with no structural integrity (those that would be classified as type-2 'hot-rock clusters' or type-1 'no feature' using the hierarchical key).

Research values are of course relative, and should a new archaeometric technique for the absolute dating of LTFs be developed that can reliably be utilised on hearthstones, then the mass of hot-rock clusters identified in the dataset might be elevated in the priority list. Indeed, researchers working within particular areas of the region or within particular time periods may develop their own criteria for assessment. The important point is that differential research potential is recognised in the LTF dataset prior to fieldwork. With this in mind, the amount of energy devoted to recording different LTFs can be apportioned in fluid field situations, as is often the case during walking surveys.

The realisation that LTFs, when correctly identified, always constitute an *in situ* expression of hunter-gatherer technology is a significant outcome of this research. This claim cannot be consistently applied to any other aspects of hunter-gatherer technology that have the archaeological visibility of fire technology. Furthermore, the construction of a feature requires the careful selection and collection of different resources (fuel, hearthstones, food), which have the potential to be preserved and, therefore, identified by archaeologists. The use-life of individual features is, therefore, an ideal candidate to be studied from a *Chaîne opératoire* perspective (e.g. Thieme 2005).

The interpretative project has indicated that data generated from the investigation of LTFs might usefully be utilized by archaeologists for purposes other than the investigation of fire technologies by hunter-gatherer groups. Foremost amongst these is the realization that LTFs appear to be excellent proxy indicators for the expected type and intensity of site-formation processes. For instance, the observation that hot-rock clusters dominate the record of a particular area is useful for planning purposes and also for expectations as to the nature of the surficial record when conducting research. Clearly, these expectations require groundtruthing and are not a substitute for actually conducting fieldwork.

Also significant is the potential association between LTF location, prehistoric demographics, and land-use patterns. Unfortunately, the geospatial patterning of LTFs identified for both study areas has been demonstrated to be largely the result of taphonomic factors skewing the visibility of sites in particular areas and regions. Interpretation of the observed distribution is, therefore, necessarily coarse-grained. The potential undoubtedly exists, however, for the development of sophisticated diachronic geospatial model,s which can begin to factor in these biases. Continued commitment to a long-term regional perspective is clearly required to achieve this goal.

Conclusions

The interpretative project has been successful in identifying and simply classifying the material evidence for a significant number of sites with LTFs on the Southern High Plains of North America. The success in compiling two disparate datasets suggests that there is considerable potential for such approaches to structure the vast bodies of data that have been accumulating in various archives for the better part of a century. Classification of the data has contributed greatly to understanding the variation in fire technology expressed in the regional record both over time and across space. This process has also highlighted areas and time periods for which we still know very little. Fine-grained archaeological examples, ethnographic descriptions, and programmes of experimental research all complement the development of the wider project and ultimately a holistic approach to technological analysis. Ethnographic research, in particular, has indicated that fire technology is always a fundamental component of hunter-gatherer life (see Figure 2.1. for an overview of the range of technologies observed ethnographically). Subsistence-based interpretation, a necessary component of an archaeological approach, must be tempered with the realisation that fire technologies utilised by hunter-gatherers were often far more complex and heterogeneous than the archaeological record alone suggests.

Variability in the physical manifestation of these frequently encountered feature classes is, therefore, potentially behaviourally and culturally meaningful. Experimentation has proven that this variation is not always the result of cultural variation and is often a product of various site-formation processes that have conspired to alter the physical signature observed by the archaeologists. Nonetheless, by consistently targeting this feature class as a phenomenon worth studying and by situating analysis in a broad regional framework, then behaviourally meaningful inferences can be teased out.

Furthermore, it has been demonstrated that although the Southern High Plains is an ideal laboratory for examining hunter-gatherer fire technology, it is not dissimilar to any geographic region once inhabited by hunter-gatherer groups. A case study exploring the post-glacial archaeological record of Southern Britain confirms the general applicability of the wider interpretative project to other regions. Regardless of the area of investigation, the underlying objective of the project is to stress the importance of focussing analytical attention on fire features. Many calls for new directions in hunter-gatherer research have been made in the zeitgeist of a new millennium, but few novel and practically achievable approaches have been put forward. The present project is demonstrably one such approach.

CHAPTER 10 - CONCLUSIONS

Abstract

A research programme was developed which utilised four analytical methods to investigate hunter-gatherer fire technology on the Southern High Plains of North America. The development of a taxonomic key by which future fieldworkers can classify and interpret the fire features they record has been critical to the success of the project. Application of this key to the Southern High Plains dataset represents the first serious attempt to comprehensively query and evaluate feature data in the site files of Texas and New Mexico. Concurrent programmes of ethnographic, experimental and archaeological research focussed on important aspects of the regional dataset. The research demonstrates that LTFs are the most visible component of complex fire technologies and that the analysis of the physical structure and materials contained within these features is behaviourally significant. The results confirm that LTFs should no longer be arbitrarily interpreted as passive elements of site-furniture, and clearly demonstrate that a focus on fire features can progress significantly beyond basic descriptive recording. On the Southern High Plains two major findings of this research are that shifts in domestic fire technology were apparently the result of increasing resource intensification, strongly suggesting human populations living in the region were widening their diet breadth as a function of time, and that hunter-gatherer campsites appear to have been preferentially located in the ecologically diverse and resource-abundant periphery of the region. The development of the research project establishes a baseline and classificatory method by which future fieldworkers can integrate LTF data into regional models of fire technology, subsistence, and demography. The success of the project on the Southern High Plains and a preliminary study in Southern Britain suggest that it might be applied to other geographic regions in which the detailed examination of hunter-gatherer fire has yet to be prioritised. Significant conclusions are presented as a series of answers to the basic questions which the current research sought to address.

The production and use of fire, often as a vital component of complex technologies, is an important part of all modern human societies. Hunter-gatherer fire technology is no exception. Ethnographic evidence suggests that the use of fire is essentially embedded in all known hunter-gatherer lifeways. Physical differences in fire technology are clearly behaviourally significant and we are left with no reason to suppose that this was not the case in the past as well. Despite a general disinterest in the holistic investigation of domestic fire technology it is often routinely identified in the archaeological record

Examination of the archaeological record of the Southern High Plains of North America has demonstrated the potential for the regional analysis of domestic fire technology that complements the traditional focus on portable material culture, the "stones and bones" approach that has been the mainstay of archaeological analysis for more than a century. The design and implementation of a hierarchical key for the classification of LTFs has for the first, time, allowed two very different states' cultural resource databases to be efficiently queried and classified for the purposes of technological analysis. Results of this analysis form a baseline from which research hypotheses which target different aspects of domestic fire technology are tested through a cohesive programme of experimental, ethnographic, and fieldwork based research.

The results clearly illustrate that the 'campfires' once built by hunter-gatherers are a small (but archaeologically visible) component of broad and oftentimes sophisticated human uses of fire. Physical variation within these structures, their location within the wider landscape, and the methods and energy used in their construction and maintenance demonstrate that they are extremely useful tools with which to examine the technologies, subsistence strategies, and demographics of past human populations (Figure 8.33). This realisation challenges the simplistic 'campfire as a social gathering place' concept that continues to essentialise interpretations of past domestic fire technology, by inscribing it with a familiar emic status, which probably derives more from camping trips than archaeological evidence. A focus on small fire features now allows archaeologists to prioritize their investigation by conducting regional programmes of research that integrate feature data within holistic frameworks for investigation and interpretation.

Are all campfires the same?

For the most part, the function of LTFs within hunter-gatherer sites has historically been uncritically assumed, rather than actively explored by archaeologists (e.g. Sidell *et al.* 2002). This situation appears to be largely the result of the strength of ethnographic research that has dominated epistemological discussions concerning the dynamic features identified as LTFs in the archaeological record. This theoretical homogenization of interpretation has previously been observed in European contexts (Gamble and Porr 2005) and has resulted in an abstracted perspective of the function of LTFs, e.g. as somewhere to gather and tell stories. The result is arguably the perception that hunter-gatherer thermal features are more or less similar i.e. pieces of site furniture and are therefore not interesting or worth investigating on their own terms.

In contrast, a significant conclusion of the presented research is that LTFs are meaningfully constituted technologic features. The identification of the localised anthropogenic use of fire in the archaeological record always indicates at least a temporary cessation of movement. Conceptualised from a landscape perspective, the decision to build an LTF can be used as an important theoretical construct in understanding indigenous preferences for particular places. At a smaller scale, the processes involved in the construction, use-life, and abandonment of an LTF are the embodiment of the concept of human agency in the production of technology (cf. Dobres 2000). Furthermore, the extended use-life of these features, the ability to study them *in situ,* and their relationship to surrounding artefact scatters make them ideally suited to investigation through a technological *chaine opèratoire*. This is similar to but much more the case than Pauketat and Alt's (2005) "agency in a postmold" (posthole) because LTFs arguably have a much more fundamental relationship to situating human activities in relation to a single feature.

Can state level site records be effectively queried and classified?

The ability to electronically access and query state level archives (site files) has greatly enhanced our ability to examine large quantities of information over large geographic areas. Unfortunately when study areas cross state lines (as they will invariably do) then it becomes necessary to access and attempt to integrate information from one or more state level repository. Research undertaken for the present study would suggest that this is no small task and that whilst gaining access to the data is now relatively straightforward, making sense of it in a standard format currently takes a great deal of patience and manual sifting of the site records. The main obstacle to attempts to undertake regional scale analysis would seem to be a lack of standardisation in terms of what and how sites are recorded at a state level. The uneasy conclusion of the present study, therefore, is that technology has unfortunately revealed that we need to seriously consider the infrastructure of how we store, record, and disseminate information about archaeological sites. The archaeological community needs to look up from the trenches and assess what is important about the way we collect data before continuing to fill state level repositories with information that will be of little use to research or cultural heritage management in general.

How different are the results of a regional 'plus' approach to analysis?

The creation of a ten mile buffer around the periphery of the Southern High Plains turned out to have unexpected consequences in terms of the sheer number of archaeological sites that were identified within the buffer zone at the boundary of the region. In retrospect this result should perhaps not be surprising, as all the available evidence suggests that the boundaries of the region were the most ecologically diverse and dependable in terms of the critical resources such as water, fuel, and plant communities. It is interesting to note that sites within the interior of the region tend to follow the system of draws and it is tempting to think that these systems acted as arteries for the movement of people onto and off of the Southern High Plains in search of the abundant, but less predictable faunal resources that were found on the huge grassland.

The ten mile buffer is an arbitrary concept designed to test the relationship of the regional archaeological record in relation to areas adjacent to it. Obviously a decision to set the limits of the buffer at 20 miles, 2 miles, or not to use a buffer at all would have resulted in a radically different dataset. The application of a novel geographic perspective has arguably aided the current research generating a set of evidence to be analysed which is complimentary too but not typical of regional perspectives which tend to be geographically constrained to ecologically homogenous areas.

Was the hierarchical key successful in structuring the regional dataset?

If the identification of campsites is useful to hunter-gatherer research questions in general, then the development of a regional dataset, based on the presence or absence of LTFs and their organisation by means of a hierarchical key, facilitates an important step in establishing a qualitative structure with which to classify site data. This dataset makes the assumption that campsites can be identified from the archaeological evidence for thermal features and is therefore in error where thermal features were not constructed, or where the evidence for them has been removed by subsequent site-formation processes. Nevertheless, based on the available evidence, the presence or absence of LTFs does appear to be a useful conceptual method for qualifying the presence or absence of campsite activities.

The development of a hierarchical key for the classification of the regional record facilitates, in general, incorporation of the archaeological record already investigated within a region. The model, therefore maximises previous data recovery efforts and has the ability to rapidly generate a set of quantitative characteristics, which can be examined by fieldwork and actualistic investigation. Application of the latter facilitates prioritization of particular feature classes, geographic areas, and geomorphologic contexts, allowing archaeologists to be proactive rather than reactive in the creation of research designs. This is an important consideration when presented with the difficult economic/political constraints of modern research. The framework facilitates rapid assessment of what to and what not to investigate, and how to go about it.

In practical terms the hierarchical key required a significant investment both in terms of time to formulate and also to sift through the massive number of site records to determine the type of evidence present from a particular site. The identification of simple variables with which to classify the features required both background knowledge of the types of evidence available in the region and the information that was captured on the site survey forms that a held on file by the State Site Files. There is no doubt that more accurate classifications of individual features could be deduced by obtaining the site reports for all investigations within the area, however, this is not yet practically possible due to the vast amount of 'grey literature' (site reports) that would need to be sifted in order to obtain the necessary information. For this reason the development of an electronic 'grey literature' archive by the State Historic Preservation Offices (or a larger organisation such as the National Park Service) would be of tremendous benefit to extracting meaningful inferences from the huge number of archaeological surveys which take place annually.

The variables selected for inclusion in the classificatory structure did manage to discriminate patterns of variation within the combined regional dataset and are therefore considered largely successful. However, the huge number of features that could not be classified suggests that any future attempt to classify such features might be wise to concentrate on the poorly documented and preserved features that dominate the right hand section of the classificatory schema.

Is the focused examination of Localised Thermal Features useful?

Emphasis on LTFs complements portable material culture approaches in providing greater site-level context to the archaeological observation of human behaviours. A basic component of this research project was to place the examination of thermal features squarely in this site-specific context. A significant conclusion then is that the variability identified in technologic processes during the creation and use-life of LTFs directly relates to the spatial structuring of human behaviour in relation to the feature. This relationship is often not clear because active site-formation processes, subsequent occupation debris, and the conflated time-scales presented by the archaeological record act to cloud interpretative potential. Nevertheless, if the function of a specific feature can be identified, for instance, as an earth-oven facility, then not only can something be said of the type of food being processed, but also about the effort put into the construction of the facility and the length of potential occupation. Site-level observations should expect such facilities to be at the periphery of occupation areas as they provide little indirect benefits (heat and light). This situation should therefore, be reflected in limited material evidence for activities taking place in direct relationship to the identified feature. In our haste to put actors into the past, we should do so from an informed perspective and not in a manner transplanted directly from ethnographic observation. By foregrounding the role of LTFs at a site scale, this project allows for a contextualised interpretative relationship between humans and thermal features to be developed rather than one in which the site furniture forms a passive backdrop for human behaviour.

The construction of an overarching research framework

allowed for the holistic inclusion of archaeometric methods that otherwise tend to be seen as separate from the interpretation of the archaeological record beyond the site level. In this sense, the non-specialist may be able to observe that if lipid analysis produced useful results for a type-12 LTF from Garza County, then it may well be applicable to a type 12 under excavation in Borden County. The two results can be compared with the potential for deriving behaviourally meaningful inferences. The research framework draws analytical focus towards building useful datasets that can be related to regional prehistory. This methodological cookbook allows some of the mystery and atomism to be removed from the analysis of features and allows modern political modes of Cultural Resource Management (CRM)-based investigations positively to contribute to the active production of regional frameworks of knowledge. The dynamic nature of thermal-feature technology requires experimental archaeology be undertaken in order to examine and test the validity of physical processes and observations inferred from the archaeological record. Drawing from the collective regional record and site-specific examples, simple hypothesis testing can yield significant results for explaining LTFs across a range of scales. Experimental research conducted for the Southern High Plains dataset (Backhouse 2004; Backhouse *et al.* 2005b; Backhouse and Johnson 2007a, c) focused primarily on the active investigation of hot-rock technologies and site-formation processes. The results of these experiments, when situated within a regional perspective, help to restructure the focus of future research strategies by predicting the types of evidence to look for, where to expect it, and the physical characteristics of the types(s) of LTFs represented.

Experimental archaeology engages the archaeologists with largely unfamiliar technology. An indirect benefit is, therefore, the self-reflexive pedagogy of learning what is and what is not involved in working with different elements of fire-based technologies (e.g. Townend 2002). Obviously, there is no psychological inference inferred in this process; however, the material engagement does allow for processes to be experienced that are not normal during typical archaeological engagements with the material record. Examining LTFs from a practical experience of engagement certainly helps when conceptualising the time and energy requirements that would have been needed to be invested in different types of features. In addition, less formal observations of how far to be away from the fire when working with the flame, when to add fuel, or how hot-rocks will perform given the daily weather were all intuitively instructive. The realization that a significant range of physical processes useful in the examination of pyrotechnologic strategies can be explored through experimental research is a powerful argument for the continued development of the experimental research programme.

In archaeological terms, experimental archaeology always seeks to explore processes within relatively short time ranges. The realization is that *chaine opèratoire*-type technologic analyses are ideally suited for investigating dynamic features such as LTFs. Put another way, all we see in the archaeological record is the static remains (some hot-rocks and maybe some charcoal if we are lucky). In contrast, making an experimental hearth requires the collection of fuel, excavation of a pit, collection of food to cook, preparation of fuel and food, lighting of fuel, maintaining the fire once lit, etc.

Experimental studies can reveal important indirect processes that require explanation when considering the archaeological record. For example, the production and transmission of light and heat at a site scale in relation to artefact scatters or the post-depositional erosion processes acting to break up features. Louis Binford's ethnographic observations hint at the physical relationships between thermal features and human behaviour; however, his major interest was in human activities that generated observable archaeological signatures of artifacts (Binford 1978:333) and not the features as active components of site structure *per se*.

In a similar sense to the experimental work, ethnographic research can open up the range of possibilities for technology and symbolism commensurate with the creation of an LTF. Ethnographic research often illustrates that people's actions are often highly unstructured in relation to LTF use (Galanidou 2000). Symbolic acts such as needing to throw bones in a fire to appease spirits or being afraid of the dark are not measured by economic models but were very important to people and the subsequent formation of the archaeological record.

Perhaps the most obvious conclusion that can be drawn from the ethnographic record is that not all LTFs were used for cooking purposes. The observed variability in features (many with similar archaeological signatures) should be a cause of concern to archaeologists eager to fit LTFs into economic models of optimisation. The ethnographic record indicates that a huge range of human activities involves the use of fire and unfortunately, for archaeologists, the material remains of many of these activities leave very similar archaeological signatures. Great care, therefore, needs to be taken in assessing the context of a particular feature or features prior to placement within behavioural models, which when extended to a regional scale, can have profound impacts on our understanding of prehistoric lives.

Prehistoric Fire Technology on the Southern High Plains

The examination of the Southern High Plains archaeological record indicates that LTFs are routinely identified or inferred during all types of archaeological investigation. Analysis of the spatial distribution of sites containing these feature types indicates that they are most frequently recorded at the periphery of the region along the eroding Caprock escarpment (Figure 8.2). Where identified in the interior of the region, LTF sites tend to be located along the course of the incised drainage systems (known as draws). On the basis of only this evidence, the location of hunter-gatherer campsites appears to be directly related to important economic resources, such as the water contained within the draw systems and the greater diversity of fuel and plant and animal life found at the periphery of the region (e.g., Boyd 1995).

Application of the hierarchical key to the LTF dataset indicates that the situation is much more complex. The results demonstrate that the dataset is dominated by surface scatters of hearthstones that are most likely to be encountered during survey modes of research. The intensity of sites located at the edge of the region and the number of LTF features comprised of hearthstone clusters only are positively correlated. The periphery of the region is also the most dynamic in terms of erosional processes. It is perhaps not surprising then that the highest frequency of features should be found in these areas. In contrast, buried, structurally intact features are more likely to be encountered in the interior of the region within the late Quaternary valley fill of the incised draw systems (e.g. Johnson and Holliday 1995, 2004). Examination of the modern land ownership and land-use strategies in the interior of the region indicates that the prehistoric record is most likely to be highly obscured and in some cases destroyed by agricultural development during the twentieth century.

The quantitative dataset, therefore, indicates that LTF will probably be encountered in surveys undertaken at the periphery of the Southern High Plains. Specifically, the eroding eastern escarpment and western Mescalero escarpment appear to have a high potential for encountering LTFs, although again this is likely to be the result of twentieth-century land-use strategies rather than prehistoric occupation preferences. Profuse LTFs in contexts easily visible to survey crews should be expected in these areas, they are also the areas indicated by the hierarchical key as containing the lowest research potential due to the highly dynamic nature of this landscape setting. In contrast, the results of this study indicate that deeply buried sites with the highest research potential are relatively few in number and are most likely to be located within the interior of the area in the incised draw systems. This spatial model is obviously useful when considering the theoretical and practical considerations of conducting research on the Southern High Plains.

The results of a focus on localised thermal features appear to correlate well with Alston Thoms (2003, 2008, 2009) models for the development of North American cook stone technology. Trends in the type and frequency of the Southern High Plains evidence indicates a situation in which diet breadth increases and mobility decreases throughout the Holocene until a major disruption in the Protohistoric period (Figure 9.3). The Southern High Plains dataset suggests that Paleoindian and Early Archaic domestic fire technology was not intensive and did not result in significant archaeological signatures that are easily identified in the archaeological record. This result is consistent with traditional interpretations of foraging groups targeting faunal resources.

An exponential increase in evidence from Middle Archaic contexts onwards indicates resource intensification both in the types of thermal structures being built and the investment in domestic fire technology by social groups. The types of facilities documented strongly suggest vegetal processing was an increasing component of hunter-gatherer lifeways and a widening of diet breadth relating to lower group mobility as a function of population pressure, lack of faunal resources, or change in dietary preferences.

During the later Ceramic period thermal feature

technology becomes increasingly common and is prolific in the southwestern portion of the study area where semi-sedentary communities of people appear to have mapped onto the reliable plant resources that were available.

Protohistoric evidence for domestic fire technology is severely limited and this may be a function of a return to a more mobile lifestyle focused on faunal resources. It should be remembered, however, that this period is by far the shortest of the time periods considered and the lack of evidence may be as much a result of the limited amount of time for archaeological evidence to accumulate as well as the research agendas of archaeologists that tend to prioritize earlier periods.

The reality of imposing a qualitative structure for the assessment of LTFs enables cultural resource managers, planners, and state officials to more objectively assess the significance of a site without having been involved in the fieldwork. This is particularly important as the discovery of an LTF during the testing phase of a cultural resource project is generally taken to be significant as it implies the presence of *in situ* traces of past cultural activities. If cultural resource managers were supplied with a qualitative classification of the type of feature encountered, they could quickly place this discovery within a spatiotemporal regional framework, assess its significance, and devise research strategies to maximise research potential if subsequent field activities were warranted.

Potential for Extension

The presented framework for investigating LTFs is a useful scheme that opens up new avenues for research that are applicable to the examination of the archaeological record in any regional context. The first stage of the research model engages with poor, patchy datasets that have been accumulated over ca. 100 years of archaeological research. This is a situation that is true in most regions globally and is not restricted to the archaeological record of the Southern High Plains. The effectiveness of the model is that it acknowledges that hunter-gatherer archaeological sites are comparatively rare in the regional record. Therefore, it is imperative that we build research structures that incorporate what little data we have already collected, no matter how poor we perceive it to be. The presented model uses and then builds on previous research by generating a baseline with which to question our current state of knowledge. This process is appropriate to any region in which at least some prior research has been undertaken, no matter how poorly it may (or may not) have been recorded.

The emphasis of hunter-gatherer research on elements of small-scale discarded material culture (which can be removed to museums) appears to be a fairly universal phenomenon. Therefore, the analysis of LTFs is almost always deficient in most discussions of hunter-gatherer archaeology. This realization indicates that there is a high potential for extension of the current research programme to other regions.

Survey of the distribution of sites located in the Southern British research area indicate that hunter-gatherer sites containing LTFs are most likely to be encountered in the south and east of the area. These sites tend to be deeply buried and, therefore, are not typically observed during surface surveys. The buried context of hunter-gatherer sites in south-eastern English counties has resulted in fewer sites being identified, as compared to the Southern High Plains dataset. Qualitative analysis, however, suggests that these sites have a generally higher research potential due to the increased likelihood for encountering relatively intact features. Furthermore, a brief examination of the year of investigation indicates that an increasing trend in the identification of sites appears to be occurring. Ongoing coastal erosion, advances in archaeological techniques, and large-scale excavation in advance of major construction projects all appear to be responsible for this trend. The British dataset, therefore, has high potential for the development of a research structure that allows the disparate evidence for LTFs to be brought together and examined.

The relative rarity of hunter-gatherer sites identified within the British study area and the abundance of later prehistoric remains from sedentary societies have arguably contributed to the academic short-shrift afforded to the investigation of post-glacial LTFs. This unfortunate situation has resulted in the adoption of uncritical interpretative structures, which often serve to effectively decontextualise the importance of site data in favour of the recovery of artefacts of transportable material culture (stones and bones). The identification of LTFs at numerous well-excavated sites throughout the research area is, however, a cause for optimism, and clearly indicates the need for the development of a research project similar to that developed for the Southern High Plains.

Limitations

The research framework is scaled to highlight regional patterns of archaeological data. The arbitrary division of the landscape into physiographic regions based on physical geography data is not necessarily relevant to past cultural demographics (Fish and Kowalewski 1990).

The data used in the present study include only LTFs that have, for whatever reason, an identifiable archaeological signature. On the Southern High Plains, the use of hearthstones in hot-rock technology has probably skewed the archaeological record towards their distribution. This is due to the increased likelihood for preservation and visibility on the modern ground surface during surveys long after the feature that once contained them has been eradicated by the natural processes of deflation and eolian transport. In consequence, LTF technologies that did not use hot-rocks are most likely to be severely underrepresented in the regional record.

Suggestions for Future Research

Future research should continue to develop and assess the usefulness of the project's focus on fire technology to exploring the archaeological record of the Southern High Plains in North America. Experimental research should concentrate on examining basic and often assumed properties of physical structures and materials. This research should concentrate on the transformative nature of fire-material relationships that are often highly dynamic and result in the physical remains recovered in the archaeological record. Suggestions for further experimental research include:

Data Capture and Organisation

- The development of a database of fuel-type properties qualified in terms of energy output, intensity, longevity of burn, and performance characteristics with reference to different thermal-feature types.
- The development of a hot-rock utility index based on type of material, abundance in local setting, likeliness to fracture, energy retentive characteristics, and performance characteristics with reference to different thermal-feature types.
- The development of a large-scale radiocarbon database is critical to unifying regional attempts to characterise North American datasets.
- Site reports should be made available electronically and integrated with the sites records systems.

Ethnographic

- Compilation of the ethnographic evidence for Native American use of fire for the North American continent as a reference work intended for archaeological researchers.

Experimentation

- Building datasets on cooking various food resources using different techniques and structural facilities.
- Examination of labour requirements in building and maintaining different types of LTFs.
- Continued development of experiments that examine the role of site-formation processes in structuring the types of LTF identified in the archaeological record.

Archaeological Research

- The development of standardised procedures for recording localised thermal features.
- Development of the investigation of the relationship between artefact distributions surrounding different types of LTFs.
- Prioritisation of macro and micro botanical sampling procedures to determine the role of plant resources in prehistoric diets.
- Prioritisation of the collection of samples for absolute dating assay.

The time and energy required to compile, store, manage, and house regional datasets of hunter-gatherer thermal-feature technology is clearly justified in its dual potential to inform interpretative models of hunter-gatherer subsistence and also to inform research planning strategies concerning what areas to target, what samples to collect, and how to go about the practical concerns of investigation. It is suggested that research structures such as this one could be housed at regional higher-education centres and made explicit through web-based technologies available to be queried and supplemented by local heritage sector contractors.

Drawing together the rich history of archaeological research within a framework of knowledge-sharing and dissemination will undoubtedly facilitate a broader understanding of LTFs and the hunter-gatherer groups that constructed them. In the end, then the creation of an explicit 'pyroarchaeology' is moot, for what is required are active frameworks of knowledge-building with which to contextualise all aspects of prehistory and thus avoid the specificity of a past that is re-created around a single object or artefact class.

REFERENCES CITED

Abbott, J.T. and Frederick, C.D., 1992. Proton Magnetometer Investigations of Burned Rock Middens in West-Central Texas: Clues to Formation Processes. *Journal of Archaeological Science*, 17, 535-545.

Akins, N., 1988. Firecracked Rock. *In*: Marshall, M.P., ed. *The Excavation of the Cortez CO2 Pipeline Project Sites 1982-1983.* Albuquerque: Contract Archaeology, University of New Mexico, 173-178.

Albert, R.M., Bar-Yosef, O., Meignen, L., Weiner, S., 2002. Quantitative Phytolith Study of Hearths from the Natufian and Middle Paleolithic Levels of Hayonim Cave (Galilee, Israel). *Journal of Archaeological Science*, 29, 1-20.

Andrefsky, W., Jr. 2001. Emerging Directions in Debitage Analysis. *In*: Andrefsky Jr. W. ed. *Lithic Debitage: Context from Meaning*. Utah: The University of Utah Press, 2-14.

Anonymous, 1999. Channel Tunnel Rail Link Union Railways (South) Limited Archaeological Excavation at Sandway Road (Arc Swr99), Nr Sandway, Kent Environmental Statement Route Window 26: Final Interim Report. Report 45997c. Salisbury: Wessex Archaeology.

Armstrong, S.W., 1993. Alder Complex Kitchens: Experimental Replication of Paleoindian Cooking Facilities. *Archaeology in Montana*, 34(2), 1-66.

Ascher, R., 1961. Experimental Archaeology. *American Anthropologist*, 63(4), 793-816.

Atkinson, T.C., Briffa, K.R., Coope G.R., 1987. Seasonal Temperatures in Britain During the Last 22,000 Years Reconstructed Using Beetle Remains. *Nature,* 325, 587-93.

Backhouse, P.N., 2002. Excavation Program at 41LU1 Area 10 for the 2000 Field Season. *In*: E. Johnson, ed. *Holocene Investigations at the Lubbock Lake Landmark. The 1991 through 2000 Work.* Lubbock Lake Landmark Quaternary Research Center Series, 11, 183-194.

Backhouse, P.N., 2003. Archaeological Investigations at Tahoka Lake (41LY52) Area 14. *In*: E. Johnson, ed. *Explorations at Tahoka Lake, a Salina on the Llano Estacado.* Museum of Texas Tech University, 65-83.

Backhouse, P.N., 2004. Conceptualizing the Social Use of Space Around a Prehistoric Hearth Pit. Paper Presented at the 69th Society for American Archaeology Conference, Montral, Canada.

Backhouse, P.N. and Johnson, E., 2007a. Hearth Life: An Experimental Examination of Site-Formation Processes Acting on Upland Hunter-Gatherer Camp Site Assemblages on the Southern High Plains. *Plains Anthropologist*, 52(202),175-194.

Backhouse, P.N. and Johnson, E., 2007b. Exploring Prehistoric Lifeways in the Shadow of the Llano Estacado: Recent Research in the Upper Brazos. *In*: Wiseman, R. N., O'Laughlin, T.C., Snow, C.T., eds. Texas and Points West: Papers in Honor of John A. Hedrick and Carrol P. Hedrick. *The Archaeological Society of New Mexico Memoir,* 33, 19-38.

Backhouse, P.N. and Johnson, E., 2007c. Where Were the Hearths: An Experimental Investigation of the Archaeological Signature of Prehistoric Fire Technology in the Alluvial Gravels of the Southern Plains. *Journal of Archaeological Science*, 34, 1367-1378.

Backhouse, P.N. and Johnson, E., 2009. Discussion. *In*: Johnson, E., ed. Return to the Nash Site (41LU35) in the Yellowhouse System on the Southern High Plains. *Lubbock Lake Landmark Quaternary Research Center Series*, 16, 245-264.

Backhouse, P.N., Johnson, E., Holliday, V.T., 2005a. Intensive Survey and Initial Testing at Jones Station Overlooking Yellowhouse Canyon. *In*: Johnson, E. ed. Explorations along the Yellowhouse System and Playas of the Southeastern Llano Estacado. The 2002 and 2003 Work. Museum of Texas Tech University, *Lubbock Lake Landmark Quaternary Research Center Series*, 12, 421-490.

Backhouse, P.N., Johnson, E., Brackenreed-Johnston, A., Buchanan, B., 2005b. Experimental Hearths and the Thermal Alteration of Caliche on the Southern High Plains. *Georarchaeology*, 20(7), 695-716.

Backhouse, P. N., Johnson, E., Hewat, C., Brackenreed-Johnston, A., Holiday, V.T., 2001 Archaic Times in Yellowhouse Canyon. Paper presented at the 39th annual Plains Anthropological Conference, Lincoln, Nebraska.

Balme, J. and Beck, W.E., 2002. Starch and Charcoal: Useful Measures of Activity Areas in Archaeological Rockshelters. *Journal of Archaeological Science*, 29, 157-166.

Barton, R.N.E., 1992. *Hengistbury Head, Dorset, Volume 2: The Late Upper Palaeolithic and Early Mesolithic Sites.* Oxford: Oxford University Committee for Archaeology, Monograph 34.

Bartram, L.E., Kroll, E.M., Bunn, H.T., 1991. Variability in Camp Structure and Bone Refuse Patterning at Kua San Hunter-Gatherer Camps. *In*: Kroll, E.M., Price, T. D., eds. *The Interpretation of Archaeological Spatial Patterning.* New York: Plenum Press, 77-144.

Basford, H.V., 1980. *The Vectis Report a Survey of Isle of Wight Archaeology.* Newport: Isle of Wight County Council.

Bell, M., 1994. Field Survey and Excavation at Goldcliff, Gwent. *In*: Bell, M., ed. *Archaeology in the Severn Estuary 1994.* Lampter: Saint David's University College, 115-144.

Bellamy, P., 2000. Portland Land Off Reap Lane, Southwell. *Proceedings of the Dorset Natural History and Archaeological Society*, 122, 163-164.

Bellomo, R.V., 1994. Methods of Determining Early Hominid Behavioural Activities Associated with the Controlled Use of Fire at FxJj 20 Main, Koobi Fora, Kenya. *Journal of Human Evolution*, 27, 173-195.

Bender, B., 2001. Landscapes-On-The-Move. *Journal of Social Archaeology*, 1(1), 75-89.

Bennett, K.D., 1989. A Provisional Map of Forest Types for the British Isles 5,000 Years Ago. *Journal of Quaternary Science,* 4, 141-144.

Bevan, B., 1998. Geophysical Exploration for Archaeology: An Introduction to Geophysical Exploration. Lincoln: *Midwest Archaeological Center Special Report* 1, 1-107.

Biesaart, L.A., Roberson, W.R., Spotts L.C., 1985. Prehistoric Archaeological Sites in Texas, A Statistical Overview. *Office of the State Archaeologist Special Report 28.* Austin: Texas Historical Commission.

Binford, L.R., 1977. *For Theory Building in Archaeology: Essays on Faunal Remains, Aquatic Resources, Spatial Analysis, and Systemic Modelling.* New York: Academic Press.

Binford, L.R., 1978. Dimensional Analysis of Behaviour and Site Structure: Learning from an Eskimo Hunting Stand. *American Antiquity*, 43(3), 330-361.

Binford, L.R., 1980. Willow Smoke and Dogs' Tails: Hunter-Gatherer Settlement Systems and Archaeological Site Formation. *American Antiquity*, 45(1), 4-20.

Binford, L.R., 1983. *In Pursuit of the Past.* London: Thames and Hudson.

Binford, L.R., 1987. Researching Ambiguity: Frames of Reference and Site Structure. *In*: Kent, S., ed. *Method and Theory for Activity Area Research.* New York: Columbia University Press, 449-512.

Binford, L.R., 1992. Seeing the Present and Interpreting the Past - and Keeping Things Straight. *In*: Rossignol, J., Wandsnider L., eds. *Space, Time, and Archaeological Landscapes.* New York: Plenum Press, 43-59.

Bird, D.W. and O'Connell, J.F., 2006. Behavioral Ecology and Archaeology. *Journal of Archaeological Research*, 14, 143-188.

Bird, J.B., 1988. *Travels and Archaeology in South Chile.* Iowa City: University of Iowa Press.

Black, S.L. and Ellis L.W., 1997. Introduction: A Study of Four Burned Rock Midden Sites. *In*: Black, S.L., Ellis, L.W., Creel, D.G., Goode, G.T., eds. *Hot Rock Cooking on the Greater Edwards Plateau: Four Burned Rock Midden Sites in West Central Texas.* Volume 1, Studies in Archaeology 22, Texas Archaeological Research Laboratory, The University of Texas at Austin, 1-22.

Blackman, W.S., 1919. The Magical and Ceremonial Uses of Fire. *Folklore*, 27(4), 352-377.

Bleed, P., 2006. Sequences Have Length and Breadth and Both Matter: Some Thoughts on Addressing Cognition with Sequence Models. Paper presented at the 71st meeting of the Society for American Archaeology, San Juan, Puerto Rico.

Bleed, P. and Meier, M., 1980. An Objective Test of the Effects of Heat Treatment on Flakeable Stone. *American Antiquity*, 45(3), 502-507.

Boado, F.C., 2001. Problems, Functions and Conditions of Archaeological Knowledge. *Journal of Social Archaeology*, 1(1), 126-146.

Boaza, N.T., Ciochonb, R.L., Xuc, Q., Liuc, J., 2004. Mapping and Taphonomic Analysis of the Homo Erectus Loci at Locality 1 Zhoukoudian, China. *Journal of Human Evolution*, 46, 519-549.

Boyd, D., 1995. The Palo Duro Complex: Redefining the Early Ceramic Period in the Caprock Canyonlands. *Bulletin of the Texas Archaeological Society*, 66, 461-518.

Boyd, D., 1997. Caprock Canyonlands Archaeology: A Synthesis of The Late Prehistory and History of Lake Alan Henry and The Texas Panhandle-Plains Volume II. Austin: Prewitt and Associates, 110.

Boyd, D., 2001. Querechos and Teyas: Protohistoric Hunter and Gatherers in the Texas Panhandle-Plains, A.D. 1540-1700. *Bulletin of the Texas Archaeological Society*, 72, 5-22.

Boyd, D.K., Abbott, J.T., Bryan, W.A., Garvey, C.M., Tomka, S.A., Fields, R.C., 1987. *Phase II Investigations at Prehistoric and Rock Art Sites, Justiceburg Reservoir, Garza and Kent Counties, Texas*. Austin: Reports of Investigations 71, Prewitt and Associates, Inc.

Boyd, D.K., Peck, J., Tomka, S.A., Kibler, K.W., Freeman, M.D., 1994. *Data Recovery at Lake Alan Henry (Justiceburg Reservoir), Garza and Kent Counties, Texas: Phase III, Season 3*. Austin: Reports of Investigations 93, Prewitt and Associates, Inc.

Bradley, B., 2003. Seeing Things: Perception, Experience and the Constraints of Excavation. *Journal of Social Archaeology*, 3(2), 151-168.

Brethauer, D.P., 1977. *An Archaeological Survey of a Proposed 42" Oil Pipeline from Jal, New Mexico to Midland, Texas*. New Mexico State University, Las Cruces. Report of investigations on file at the Texas Historical Commission, Austin.

Bretz, J.H. and Horberg, L., 1949. Caliche in Southeastern New Mexico. *Journal of Geology*, 57, 491-511.

Brink, J.W. and Dawe, B., 2003. Hot Rocks as Scarce Resources: The Use, Re-Use and Abandonment of Heating Stones at Head-Smashed-In Buffalo Jump. *Plains Anthropologist*, 48(186), 85-104.

Brown, C., 1986. SPAS-LU-6. *In*: Runkles, F., Riggs, A., Cox, F., Brantley, B., eds. Selected Articles from the South Plains Archaeological Society Newsletters, South Plains Archaeological Society, 75-78.

Brown, C., 1999. Perry (41LU75) – A Lubbock County Playa Site. *Bulletin of the South Plains Archaeological Society*, 5, 1-56.

Brown, H., 2000. Wildland Burning by American Indians in Virginia. *Fire Management Today*, 60(3), 29-39.

Buchanan, B., 2002. Excavation Program at 41LU1 Area 10 for the 1993 through 1997 Field Seasons. *In*: Johnson, E., ed. *Holocene Investigations at the Lubbock Lake Landmark. The 1991 through 2000 Work*. Lubbock Lake Landmark Quaternary Research Center Series, 11, 139-158.

Buchanan, B., Litwinonek, L., Johnson, E., Holliday, V.T., Hicks, J. Kent, 1996. Renewed Investigations at Milnesand and Ted Williamson Paleoindian Sites, Southern High Plains. *Current Research in the Pleistocene*, 13, 8-10.

Buonasera, T., 2005. Fatty Acid Analysis of Prehistoric Burned Rocks: a Case Study From Central California. *Journal of Archaeological Science*, 32(6), 957-965.

Burchill, J., 1925a. The Shell Mound Industry of Denmark as Represented at Lower Halstow, Kent. *Proceedings of the Prehistoric Society of East Anglia*, 5, 217-227.

Burchill, J., 1925b. Further Report on the Epi-Palaeolithic Factory Site at Lower Halstow, Kent. *Proceedings of the Prehistoric Society of East Anglia*, 5, 288-296.

Burrow, S., 2003. *Catalogue of the Mesolithic and Neolithic Collections in the National Museums & Galleries of Wales*. Cardiff: National Museum of Wales.

Cabeza de Baca, F., 1994. *We Fed Them Cactus*. Second Edition. Albuquerque: University of New Mexico Press.

Cain, C. R., 2005. Using Burned Animal Bone to Look at Middle Stone Age Occupation and Behaviour. *Journal of Archaeological Science*, 32, 873-884.

Calkin, J.B., 1952. *Archaeological Discoveries in the Isle of Purbeck. Proceedings of the Dorset Natural History and Archaeological Society, 74,* 48-49.

Campling, N.R., 1999. An Earth Oven from British Columbia, Canada. *In*: Hodder, M.A., Barfield, L.H., eds. *Burnt Mounds Hot Stone Technology: papers from the second international burnt mound conference*. Sandwell, West Midlands, 93-96.

Canti, M.G. and Linford, N., 2000. The Effects of Fire on Archaeological Soils and Sediments: Temperature and Colour Relationships. *Proceedings of the Prehistoric Society,* 66, 385-395.

Carlson, P. H., 1998. *The Plains Indians*. College Station: Texas A&M University Press.

Carlson, P. H., 2005. *Deep Time and the Texas High Plains: History and Geology.* Lubbock: University of Texas Press.

Carr, C., 1984. "The Nature of Organization of Intrasite Archaeological Records and Spatial Analytic Approaches to their Investigation". *Advances in Archaeological Method and Theory,* 7, 103-222.

Carr, C., 1991. Left in the Dust: Contextual Information in Model-Focussed Archaeology. *In*: Kroll, E.M., Price, T. D., eds. *The Interpretation of Archaeological Spatial Patterning*, New York: Plenum Press, 221-255.

Carter, H.H., 1976. Fauna of an Area of Mesolithic Occupation in the Kennet Valley Considered in Relation to Contemporary Eating Habits. *Berkshire Archaeological Journal*, 68, 1-3.

Carter, R.J., 2001. New Evidence for Seasonal Human Presence at the Early Mesolithic Site of Thatcham, Berkshire, England. *Journal of Archaeological Science*, 28, 1055-1060.

Caspar, J. and De Bie, M., 1996. Preparing for the Hunt in the Late Paleolithic Camp at Rekem, Belgium. *Journal of Field Archaeology*, 23, 437-460.

Centennial Archaeology, 1994. LA107646. Site survey form on file at the New Mexico Laboratory of Archaeology, Santa Fe.

Church, R.R. and Lyman R.L., 2003. Small Fragments Make Small Difference in Efficiency When Rendering Grease From Fractured Artiodactyl Bones By Boiling. *Journal of Archaeological Science*, 30, 1077-1084.

Churchill, D.M., 1965. The Kitchen Midden Site at Westward Ho!, Devon, England: Ecology, Age, and Relation to Changes in Land and Sea Level. *Proceeding of the Prehistoric Society*, 31, 74-84.

Clarke, D., 1973. Archaeology: The Loss of Innocence. *Antiquity*, 47, 6-18.

Coles, J., 1973. *Archaeology by Experiment*. New York: Charles Scribner's Sons.

Collins, M. 1966. 41MT1. Unpublished site survey form on file at the Texas Archaeological Research Laboratory, Austin.

Collins, D., and Lorimer, D., 1989. *Excavations at the Mesolithic Site on West Heath, Hampstead 1976-1981: Investigations by Members of the Hendon and District Archaeological Society*. BAR British Series, 217. Oxford: Archaeopress.

Conneller, C. and Warren, G. 2006. *Mesolithic Britain and Ireland: New Approaches*. Stroud: Tempus.

Cook, S.F. and Heizer, R.F., 1965. Studies on the Chemical Analysis of Archaeological Sites. *University of California Publications in Anthropology, Volume 2*, Los Angeles: University of California Press.

Corley, J.A., 1965. A Proposed Extension of the Jornada Branch of the Mogollon. *Bulletin of the Lea County Archaeological Society*, 1, 30-36.

Cruse, B., 2002. 41MY18. Unpublished site survey form on file at the Texas Archaeological Research Laboratory, Austin.

Curtis, E.S., 1907. *The North American Indian : Being a Series of Volumes Picturing and Describing the Indians of the United States, and Alaska. Vol. 1. The Apache. The Jicarillas. The Navaho.* Cambridge Mass: The University Press. Digital version accessed [23/7/06] and available from Northwestern University Digital Library Collections. Online: http://curtis.library.northwestern.edu/index.html.

Curtis, E.S., 1930. *The North American Indian : Being a Series of Volumes Picturing and Describing the Indians of the United States, and Alaska. Vol. 30. The Indians of Oklahoma. The Wichita. The Southern Cheyenne. The Oto. The Comanche. The Peyote Cult* [online]. Cambridge Mass: The University Press. Available from: http://curtis.library.northwestern.edu/index.html [Accessed 23 July 2006]

Darvill, T., 1996. *Prehistoric Britain*. London: Routledge.

Darvill, T., 2002. *The Concise Oxford Dictionary of Archaeology*. Oxford: Oxford University Press.

Davidson, D.S., 1947. Fire-Making in Australia. *American Anthropologist*, 49(3), 426-437.

Davies, W., 2000. The Paleolithic and Post-Processualism; a Pragmatic Approach? *Archaeological Review from Cambridge*, 17(1), 5-18.

Dawson, P.C., 2002. Space Syntax Analysis of Central Inuit Snow Houses. *Journal of Anthropological Archaeology*, 21, 464-480.

Dennis, I. and Hamilton, M.A., 1997. *Exploratory Excavations and Survey of Mesolithic, Late Bronze and Ironage Sites at Golden Ball hill Near Alton Barnes Wiltshire: An Interim Report for 1997*. Cardiff Specialist Report. Cardiff: Cardiff University.

Dering, P., 1999. Earth-Oven Plant Processing in Archaic Period Economies an Example From a Semi-Arid Savannah in South-Central North America. *American Antiquity*, 64(4), 659-674.

Dillehay, T.D., 1974. Late Quaternary Bison Population Changes on the Southern High Plains. *Plains Anthropologist*, 19(6), 180-196.

Dobres, M.A., 1995. Gender and Prehistoric Technology: on the Social Agency of Technical Strategies. *World Archaeology*, 27(1), 25-49.

Dobres, M.A., 2000. *Technology and Social Agency*. Malden: Blackwell Publishers.

Drass, R., 1995. A Bison Processing Camp, 34CU209, in Western Okalahoma. *Oklahoma Archaeological Society Bulletin*, 44, 85-90.

Driver, H.E. and Massey, W.C., 1957. Comparative Studies of North American Indians. *Transactions of the American Philosophical Society*, 47(2), 165-456.

Du Bois, C., 1940. Wintu Ethnography. *University of California Publications in American Archaeology and Ethnology*, 36, 1-148. Berkley: University of California Press.

Dufraisse, A., 2006. Firewood Economy During the 4th Millennium BC at Lake Clairvaux, Jura, France. *Environmental Archaeology*, 11(1), 87-99.

Dunell, R.C., 1971. *Systematics in Prehistory*. New York: The Free Press.

Edmonds, M., 1990. Description, Understanding, and the *Chaine Opèratoire*. *Archaeological Review from Cambridge*, 9(1), 55-70.

Elbaum, R., Weiner, S., Albert R.M., Elbaum, M., 2003. Detection of Burning of Plant Materials in the Archaeological Record by Changes in the Refractive Indices of Siliceous Phytoliths. *Journal of Archaeological Science*, 30(2), 217-226.

Ellis, L.W., 1997. Hot Rock Technology. *In*: Black, S., Ellis, L.W., Creel, D.G., Goode, G.T., eds. Hot Rock Cooking on the Greater Edwards Plateau: Four Burned Rock Midden Sites in West Central Texas. *Texas Archaeological Research Laboratory, Studies in Archaeology, 22*, The University of Texas at Austin, 43-81.

Ellis, C.J., Allen, M.J., Gardiner, J., Harding, P., Ingrem, C., Powell, A., Scaife, R., 2003. An Early Mesolithic Seasonal Hunting Site in the Kennet Valley, Southern England. *Proceedings of the Prehistoric Society*, 69, 107-135.

Fagan, B., 1991. *Ancient North America: The Archaeology of a Continent.* London: Thames and Hudson.

Finch, D., 2004. Assessment of Grassland Ecosystem Conditions in the Southwestern United States: Volume 1. USDA Forest Service. Report RMRS-GTR-135, 1:1-167.

Fish, S.K. and Kowalewski, S.A., 1990. *The Archaeology of Regions: A Case for Full-Coverage Survey*. Washington D.C.: Smithsonian Institution Press.

Fisher, J.W. and Strickland, H.C., 1991. Dwellings and Fireplaces; Keys to Efe Pygmy Campsite Structure. *In*: Gamble C.S., Boismier, W.A., eds. *Ethnoarchaeological Approaches to Mobile Campsites. Hunter-Gatherer and Pastrolist Case Studies*, Ann Arbor, Michigan, Ethnoarchaeological Series 1, 215-236.

Flores, D., 1990. *Caprock Canyonlands: Journeys into the Heart of the Southern Plains*. Austin: University of Texas Press.

Frederick, C.D. and Abbott, J.T., 1990. Magnetic Prospection of Prehistoric Sites in an Alluvial Environment: Examples from NW and West-Central Texas. *Journal of Field Archaeology*, 19, 139-153.

Frink, L., Hoffman, B.W., Knudson, K.J., Price, T.D., 2004. Chemical Characterization of Arctic Soils: Activity Area Analysis in Contemporary Yup'ik Fish Camps Using ICP-AES. *Journal of Archaeological Science*, 31, 443-456.

Frison, G.C., 1983. Stone Circles, Stone-Filled Fire Pits, Grinding Stones and High Plains Archaeology. *Memoir 19, Plains Anthropologist*, 28(2), 81-91.

Froom, F.R., 1972. Some Mesolithic Sites in South-West Berkshire. *Berkshire Archaeological Journal*, 61, 11-22.

Froom, F.R., 1976. Wawcott III: A Stratified Mesolithic Succession. *Berkshire Archaeological Journal*, 66, 1-209.

Galanidou, N., 1997. 'Home is Where the Hearth is': The *Spatial Organization of the Upper Palaeolithic Rockshelter Occupations at Klithi and Kastristsa in Northwest Greece*. BAR International Series, S687. Oxford: Archaeopress.

Galanidou, N., 2000. Patterns in Caves: Foragers, Horticulturists, and the Use of Space. *Journal of Anthropological Archaeology*, 19, 243-275.

Gamble, C. and Porr, M., 2005. *The Hominid Individual in Context; Archaeological Investigations of Lower and Middle Palaeolithic Landscapes, Locales and Artefacts*. London: Routledge.

Gamble, C. S. and Boismier, W.A., 1991. *Ethnoarchaeological Approaches to Mobile Campsites*. Ethnoarchaeological Series 1, International Monographs in Prehistory, Ann Arbor.

Gheorghiu, D., (ed.) 2002. Introduction: Towards Pyroarchaeology. *In*: Gheorghiu, D., ed. *Fire in Archaeology*, BAR International Series, S1089. Oxford: Archaeopress, 1-6.

Gidley, M., 2001. Ways of Seeing the Curtis Project on the Plains. *In*: Curtis, E.S., The Plains Indian Photographs of Edward S. Curtis. Lincoln: University of Nebraska Press, 39-66.

Gifford-Gonzalez, D.P., 1989. Ethnographic Analogues for Interpreting Modified Bones: Some Cases from East Africa. *In*: Bonnichsen R., Sorg, M.H. eds. *Bone Modification*. Peopling of the Americas Publications. Orono: Center for the Study of the First Americans, Institute for Quaternary Studies, University of Maine, 179-246.

Gifford-Gonzalez, D.P., Damrosch, D.B., Damrosch, D.R., Pryor, J., Thunen, R.L., 1985. The Third Dimension in Site Structure: An Experiment in Trampling and Vertical Dispersal. *American Antiquity*, 50(4), 803-818.

Gilmore, M.R., 1917. Ancient Indian Fireplaces in South Dakota Bad-Lands: Fact and Fantasy. *American Anthropologist*, 19(4), 583-585.

Gose, W.A., 2000. Paleomagnetic Studies of Burned Rocks. *Journal of Archaeological Science*, 27, 409-421.

Gose, W.A., Collins, K.S., Collins, M.B., Muckle R.J., 2004. Paleomagnetic Studies from the Moore-Hancock Farmstead, Austin, Texas. *Journal of Field Archaeology*, 21(1), 125-131.

Gott, B., 2002. Fire-Making in Tasmania: Absence of Evidence Is Not Evidence of Absence. *Current Anthropology*, 43(4), 649-656.

Gould, R.A., (ed.) 1978. *Explorations in Ethnoarchaeology*. Albuquerque: University of New Mexico Press.

Gowlett, J.A.J., 1987. The Archaeology of Radiocarbon Accelerator Dating. *Journal of World Prehistory*, 2 (2), 127-170.

Graves, P. and Hammond, P., 1993. A Mesolithic Site at Angmering Decoy, West Sussex. *Sussex Archaeology Collections*, 131, 195–198.

Gravina, B., 2004. Agency, Technology, and the 'Muddle in the Middle': The Case of the Middle Palaeolithic. *In*: Gardner, A., ed. *Agency Uncovered: Archaeological Perspectives on Social Agency, Power, and Being Human*. London: UCL Press, 65-78.

Grayson, D.K. and Meltzer, D.J., 2003. A Requiem for North American Overkill. *Journal of Archaeological Science*, 30, 585-593.

Green, F., 1991. Mesolithic Structures in the Test Valley: Bowman's Farm. *PAST*, 1-2.

Greene, K., 2004. Archaeology and Technology. *In*: Bintliff, J., ed. *A Companion to Archaeology*. Oxford: Blackwell Publishing, 155-173.

Grøn, O., 2003. Mesolithic Dwelling Places in South Scandinavia: Their Definition and Social Interpretation. *Antiquity*, 77(298), 685-708.

Grøn, O. and Kuznetsov, O., 2003. Ethno-archaeology Among Evenkian Forest Hunters. Preliminary Results and a Different Approach to Reality!. *In*: Larsson, L., Kindgren, H., Knutsson, K., Leoffler, D., Åkerlund, A., eds. *Mesolithic on the Move: Papers Presented at the Sixth International Conference on the Mesolithic in Europe, Stockholm 2000*. Oxford: Oxbow Books, 216-229.

Guenther, M., 1996. Diversity and Flexibility: the Case of the Bushmen of Southern Africa. *In*: Kent, S., ed. *Cultural Diversity Among Twentith Century Foragers: An African Perspective*. Cambridge: Cambridge University Press, 65-86.

Gustavson, T.C., 1986. Geomorphology and Quaternary Stratigraphy of the Rolling Plains, Texas Panhandle. *Guidebook* 22. Bureau of Economic Geology, The University of Texas at Austin.

Gustavson, T.C. and Simpkins, W. W., 1989. Geomorphic Processes and Rate of Retreat Affecting the Caprock Escarpment, Texas Panhandle. Bureau of Economic Geology, 180.

Gunnerson, J.H., 1992. Protohistoric Apaches in Northeastern New Mexico. *In*: Johnson, E., ed. Cultural Encounters and Episodic Droughts: The Protohistoric Period on the Southern Plains. Guidebook to the Ethnohitsory and Selected Protohistoric Sites of the Southern High Plains. Museum of Texas Tech University, *Lubbock Lake Landmark Quaternary Research Center Series*, 3, 91-102.

Hagmeier, E.M. and Stults, C.D., 1964. A Numerical Analysis of the Distributional Patterns of North American Mammals. *Systematic Zoology*, 13(3), 125-155.

Hall, R.L., 1997. *An Archaeology of the Soul: North American Indian Belief and Ritual*. Chicago: University of Illinois Press.

Hall, S.A., 2002. Field Guide to the Geoarchaeology of the Mescalero Sands, Southeastern New Mexico. Report Prepared for the New Mexico State Historic Preservation Division and the New Mexico Bureau of Land Management. Red Rock Geological Enterprises, Santa Fe.

Hamilton, M.A., 1998. *Broken Bison Bones: A Faunal Analysis of 41LU1, Lubbock Lake Landmark, Texas, USA*. Thesis (BSc), Institute of Archaeology, University College London.

Hans, D., 2002. An In Situ Fireplace from the Mesolithic Period at "Ronases Skov" in the Gamborg Fjord. *Nachrichtenblatt Arbeitskreis Unterwasserarchäologie*, 9, 29-34.

Harding, P., 2000. A Mesolithic Site at Rock Common, Washington. *West Sussex. Sussex Archaeological Collections,* 138, 29-48.

Hawkins, R.A., 1998. Coming Full Circle: Plowzone Assemblages and the Interpretation of Fort Ancient Settlement Structure. *In*: Sulivan III, A.P., ed. *Surface Archaeology*. Albuquerque: University of New Mexico Press, 91-109.

Healy, F., Heaton, M., Lobb, S.J., 1992. Excavations of a Mesolithic Site at Thatcham, Berkshire. *Proceedings of the Prehistoric Society*, 58, 41-76.

Hester, J.J., 1972. *Blackwater Locality No. 1 A Stratified, Early Man Site in Eastern New Mexico*. Rancho de Taos: Fort Burgwin Research Center, Southern Methodist. University.

Hester, T.R., 1999. Artifacts, Archaeology and Cabeza De Vaca in Southern Texas and Northeastern Mexico. *In*: Windows to the Unknown: Cabeza de Vaca's Journey to the Southwest. An Interdisciplinary Symposium Hosted by the Center for the Study of the Southwest at Southwest Texas State University, San Marcos.

Higgs, E., 1959. Excavation at a Mesolithic site at Downton, Near Salisbury, Wiltshire. *Proceedings of the Prehistoric Society*, 25, 209-212.

Hill, M.E. Jr., 2002. The Milnesand Site: Site Formation Study of a Paleoindian Bison Bonebed in Eastern New Mexico. *Plains Anthropologist*, 47(183), 323-337.

Hill, M.E. Jr., 2006. Regional Differences in Great Plains Paleoindian Occupational Intensity and Duration. Paper Presented at the 71st Annual meeting of the Society for American Archaeology, San Juan, Puerto Rico.

Hodder, I., 1992. *Theory and Practice in Archaeology*. London: Routledge.

Hodge, F.W., 1912. Handbook of American Indians North of Mexico. *Smithsonian Institution Bureau of American Ethnology*, Bulletin 30. Government Printing Office, Washington.

Hoffman, J. L., 1981. The Refitting of Chipped-Stone Artifacts as an Analytical and Interpretative Tool. *Current Anthropology*, 22(6), 691-693.

Hoffman, J. L., 1989. Prehistoric Cultural History. Hunters and Gatherers in the Southern Great Plains. *In*: Hoffman, J., Brooks, R.L., Hays, J.S., Owsley, D., Jantz, R.L., Marks, M.K., Manhein, M.H., eds. *From Clovis to Comanchero: Archaeological Overview of the Southern Great Plains*. Fayetteville, Arkansas Archaeological Survey Research Series 35, 25-60.

Holdaway, S.J., Fanning, P.C., Jones, M., Shiner, J., Witter, D.C., Nicholls, G., 2002. Variability in the Chronology of Late Holocene Aboriginal Occupation on the Arid Margin of Southeastern Australia. *Journal of Archaeological Science*, 29, 351-363.

Holland, T.D., 1984. A Pilot Study in the Thermal Properties of Buffalo Chips. *Plains Anthropologist*, 29(104):161-165.

Holliday, V.T., 1987. Re-examination of Late-Pleistocene Boreal Forest Reconstructions for the Southern High Plains. *Quaternary Research*, 23, 388-402.

Holliday, V.T., 1988a. Genesis of a Late-Holocene Soil Chronosequence at the Lubbock Lake Archaeological Site. *Annals of the Association of American Geographers,* 78(4), 594-610.

Holliday, V.T., 1988b. Late Pleistocene and Holocene Stratigraphy Southern High Plains of Texas. *Geological Society of America Centennial Field Guide*, 69, 293-297.

Holliday, V.T., 1989. Middle Holocene Drought on the Southern High Plains. *Quaternary Research,* 31, 74-82.

Holliday, V.T., 1990. Soils and Landscape Evolution of Eolian Plains: The Southern High Plains of Texas and New Mexico. *Geomorphology*, 3, 489-515.

Holliday, V.T., 1995. Stratigraphy and Paleoenvironments of Late Quaternary Valley Fills on the Southern High Plains. *Geological Society of America Memoir*, 186, 1-136.

Holliday, V.T., 1997. *Paleoindian Geoarchaeology of the Southern High Plains*. Austin: University of Texas Press.

Holliday, V.T., 2000. Folsom Drought and Episodic Drying on the Southern High Plains from 10,900-10,200 14C yr B.P. *Quaternary Research*, 53,1-12.

Holliday, V.T., Johnson, E., Haas, H., Stuckenrath, R., 1985. Radiocarbon Ages from the Lubbock Lake Site: 1981-1984. *Plains Anthropologist*, 110(1), 277-291.

Holliday, V.T. and Welty, C.M. 1981. Lithic Tool Resources of the Eastern Llano Estacado. *Bulletin of the Texas Archeological Society*, 52, 201-214.

Holmes, W.H., 1903. Aboriginal Pottery of the Eastern United States. *Bureau of American Ethnology (Washington, D.C.). Annual Report of the Bureau of American Ethnology to the Secretary of the Smithsonian institution,* 20,1898-1899.

Hood, B.C., 1981. Miscellaneous Archaeological Investigations Along The North-Central Labrador Coast: 1980. *In:* Sproull Thomson, J., Ransom, B., eds. *Archaeology in Newfoundland & Labrador*. Research Report 1. St. Johns: Historic Resources Division, Department of Tourism and Culture, 58-63.

Horsley, E.M., 1979. *The New Hutchinson 20th Century Encyclopaedia*, 4th revised impression. London: Hutchinson and Co.

Hough, W., 1890. Aboriginal Fire-Making. *American Anthropologist*, 3(4), 359-372.

Hughes, J.T., 1989. Prehistoric Cultural Developments on the Texas High Plains. *Bulletin of the Texas Archaeological Society*, 60, 1-55.

Hughes, M.F. and ApSimon, A.M., 1978. A Mesolithic Flint Working Site on the South Coast Motorway (M27) Near Fort Wallington, Fareham, Hampshire, 1972. *Proceedings of the Hampshire Field Club & Archaeological Society,* 34, 25-35.

Huizenga, C., Hui, Z., Arens, E., 2001. A Model of Human Physiology and Comfort for Assessing Complex Thermal Environments. *Building and the Environment*, 36, 691-699.

Hunt, C.B., 1974. *Natural Regions of the United States and Canada*. San Francisco: W.H. Freeman and Co.

Hurst, S., Johnson, E., Martinez McCoy, Z., Cunningham, D., 2010. The Lithology of Ogallala Gravels and Hunter-Gatherer Procurement Strategies Along the Southern High Plains Eastern Escarpment of Texas, USA. *Geoarchaeology:An International Journal*, 25(1):96-121.

Jacobi, R., 2004. The Late Upper Palaeolithic Lithic Collection from Gough's Cave, Cheddar, Somerset and Human Use of the Cave. *Proceedings of the Prehistoric Society*, 70, 1-92.

Johnson II, C.A. and Stafford, T., 1976. Report of the 1975 Archaeological Survey Investigation of the Canyon Lakes Project ,Yellowhouse Canyon, Lubbock, Texas. Report submitted to the Texas Antiquities Committee (Austin) and the City of Lubbock.

Johnson, E., 1977. Animal Food Resources of Paleoindians. *The Museum Journal*, 17, 65-77.

Johnson, E., (ed.) 1987a. *Lubbock Lake. Late Quaternary Studies on the Southern High Plains*. College Station: Texas A&M University Press.

Johnson, E., 1987b. Cultural Activities and Interactions. *In*: Johnson, E., ed. *Lubbock Lake. Late Quaternary Studies on the Southern High Plains.* College Station: Texas A&M University Press, 120-158.

Johnson, E., (ed.) 1989. Fenceline Corridor Survey and Testing Program. Lubbock Lake Landmark Quaternary Research Center Series. Museum of Texas Tech University, *Lubbock Lake Landmark Quaternary Research Center Series*, 1, 1-293.

Johnson, E., (ed.) 1990. Fifty Years of Discovery: The Lubbock Lake Landmark, Guidebook to the Quaternary History of the Llano Estacado. Lubbock Lake Landmark Quaternary Research Center Series. Museum of Texas Tech University, *Lubbock Lake Landmark Quaternary Research Center Series*, 2, 1-155.

Johnson, E., 1991. Late Pleistocene Cultural Occupation of the Southern High Plains. *In*: Bonnichsen, R., Turnmire, K.L., eds. *Clovis Origins and Adaptations*. Corvallis, Oregon State University, 215-326.

Johnson, E., (ed.) 1993. Late Holocene Investigations at the Lubbock Lake Landmark. Volume 1 The 1988 Work. Lubbock Lake Landmark Quaternary Research Center Series. Museum of Texas Tech University, *Lubbock Lake Landmark Quaternary Research Center Series*, 5, 1-243.

Johnson, E., (ed.) 1995. Late Holocene Investigations at the Lubbock Lake Landmark. Volume 2 The 1989 and 1990 Work. Lubbock Lake Landmark Quaternary Research Center Series. Museum of Texas Tech University, *Lubbock Lake Landmark Quaternary Research Center Series*, 8, 1-447.

Johnson, E., (ed.) 2002. Holocene Investigations at the Lubbock Lake Landmark. The 1991 through 2000 Work. Lubbock Lake Landmark Quaternary Research Center Series. Museum of Texas Tech University, *Lubbock Lake Landmark Quaternary Research Center Series*, 11, 1-324.

Johnson, E., (ed.) 2005. Explorations along The Yellowhouse System and Playas of the Southeastern Llano Estacado. The 2002 and 2003 Work. Museum of Texas Tech University, *Lubbock Lake Landmark Quaternary Research Center Series*, 12.

Johnson, E., 2007. Grassland Ecosystems of the Llano Estacado. *In:* Sosebee, R.E., Wester, D.B., Britton, C.M., McArthur, E. D., Kitchen, S.G., eds. *Shrubland dynamics --fire and water, 2004 August 10-12, Lubbock, TX. Proceedings*. Forest Service, Rocky Mountain Research Station, RMRS-P-47, 11-23.

Johnson, E., 2008. Landscapes and Peoples of the Llano Estacado. *In*: Scheiber, L.L., Clark, B., eds. *Archaeological Landscapes on the High Plains*. Boulder: University of Colorado Press, 115-156.

Johnson, E. and Backhouse P.N., 2001. Amphitheatre Site (41LU118) Testing Results. Interim Report 2, Submitted to the Texas Historical Commission, Austin, 1-15.

Johnson, E. and Hartwell W.T., 1989. Analysis of Survey and Testing Generated Materials. *In*: Johnson, E., ed. Lubbock Lake Landmark: 1987 Fenceline Corridor Survey and Testing Program, *Lubbock Lake Landmark Quaternary Research Center Series*, 1, 83-234.

Johnson, E. and Holliday, V. T., 1981. Late Paleo-Indian Activity at the Lubbock Lake Site. *Plains Anthropologist*, 26(93), 172-193.

Johnson, E. and Holliday, V. T., 1986. The Archaic Record at Lubbock Lake. *In*: Baugh, T.G., ed. Current Trends in Southern Plains Archaeology. *Plans Anthropologist Memoir*, 31, 7-54.

Johnson, E. and Holliday, V. T., 1989. Lubbock Lake: Late Quaternary Cultural and Environmental Change on the Southern High Plains, U.S.A. *Journal of Quaternary Science*, 4(2), 145-165.

Johnson, E. and Holliday, V. T., 1990. (eds.). Fifty Years of Discovery: The Lubbock Lake Landmark, Guidebook to the Quaternary History of the Llano Estacado. Museum of Texas Tech University, *Lubbock Lake Landmark Quaternary Research Center Series*, 2.

Johnson, E. and Holliday, V. T., 1995. Archaeology and Late Quaternary Environments of the Southern High Plains of Texas. *Bulletin of the Texas Archeological Society*, 66, 519-540.

Johnson, E. and Holliday, V. T., 2004. Archaeology and Late Quaternary Environments of the Southern High Plains. *In*: Perttula, T. K. ed. *The Prehistory of Texas*. College Station: Texas A&M University Press, 283-295.

Johnson, E., Holliday, V.T., Kaczor, M.J., Stuckenrath, R., 1977. The Garza Occupation at the Lubbock Lake Site. *Bulletin of the Texas Archaeological Society*, 48, 83-109.

Johnson, E., Holliday, V.T., Warnica, J.M., Williamson, T., 1986. The Milnesand and Ted Williamson Paleoindian Sites, East-Central New Mexico. *Current Research in the Pleistocene*, 3, 9-11.

Johnson, L. Jr., 1986. A Plague of Phases. *Bulletin of the Texas Archeological Society*, 57, 1-25.

Johnson, M., 1999. *Archaeological Theory an Introduction*. Oxford: Blackwell Publishers.

Jones, K.T., 1993. The Archaeological Structure of a Short-Term Camp. *In*: Hudson, J., ed. *From Bones to Behavior: Ethnoarchaeological and Experimental Contributions to the Interpretation of Faunal Remains*. Occasional Paper 21. Carbondale: Southern Illinois University, Center for Archaeological Investigations, 101-114.

Jones, G. and Munson, G., 2005. Geophysical Survey as an Approach to the Ephemeral Campsite Problem: Case Studies from the Northern Plains. *Plains Anthropologist*, 50(193), 31-43.

Karlin, C. and Julien, M., 1994. Prehistoric Technology: A Cognitive Science?. *In*: Renfrew C., Zubrow, E.B.W., eds. *The Ancient Mind*. Cambridge: Cambridge University Press, 152-164.

Katz, P.R. and Katz, S.R, 1974. Archaeological Investigations in Lower Tule Canyon Briscoe County, Texas. Texas Tech University, Archaeological Survey Report 16.

Katz, S.R. and Katz, P.R., 1985. The Prehistory of the Carlsbad Basin, Southeastern New Mexico. Technical Report of Prehistoric Archaeological Investigations in the Brantley Project Locality. San Antonio: Texas, Unpublished Report Prepared by Incarnate Word College for the Bureau of Reclamation, Contract No. 3-CS-57-01690.

Katz, S.R. and Katz, P.R., 2000. Prehistory of the Pecos Country, Southeastern New Mexico. Report submitted to the New Mexico Historic Preservation Division, Office of Cultural Affairs, Santa Fe, in fulfillment of Historic Preservation Division Project 35-99-14264. 19.

Keef, P.A.M., Wymer, J. J., Dimbleby, G.W., 1965. A Mesolithic Site on Iping Common, Sussex, England. *Proceedings of the Prehistoric Society*, 31, 85-92.

Kelly, I. 1932. Ethnography of the Surprise Valley Paiute. *University of California Publications in American Archaeology and Ethnology*, 31, 67-210.

Kelly, R.L., 1995. *The Foraging Spectrum: Diversity in Hunter-Gatherer Lifeways*. London: Smithsonian Institution Press.

Kent, S., 1990. A Cross-cultural Study of Segmentation, Architecture and the Use of Space. *In*: Kent, S., ed. *Domestic Architecture and the Use of Space: An Interdisciplinary Cross-cultural Study*. Cambridge: Cambridge University Press, 127-152.

Kent, S., 1993. Variability in Faunal Assemblages: The Influence of Hunting Skill, Sharing, Dogs, and Mode of Cooking on Faunal Remains at a Sedentary Kalahari Community. *Journal of Anthropological Archaeology*, 12, 323-385.

Knudson, K.J., Frink, L., Hoffman, B.W., Price, T.D., 2004. Chemical Characterization of Arctic Soils: Activity Area Analysis in Contemporary Yup'ik Fish Camps Using ICP-AES. *Journal of Archaeological Science*, 31, 443-456.

Kotter, S.M., and Cruse, B., 1983. 41OL178. Site Survey Form on File at the Texas Archaeological Research Laboratory, Austin.

Kreiger, A.D., 2002. *We Came Naked and Barefoot : The Journey of Cabeza de Vaca Across North America*. Austin: University of Texas Press.

Kroll, E.M. and Price, T.D., (eds.) 1991. *The Interpretation of Archaeological Spatial Patterning*. New York: Plenum Press.

Kruger, P., 1957. Specification Survey for Radiocarbon Dating. *American Antiquity*, 23(2), 182.

LaBelle, J.M., 2005. Hunter-Gatherer Foraging Variability During the Early Holocene of the Central Plains of North America. Thesis (Phd). Southern Methodist University.

La Flesche, F., 1921. The Osage Tribe Rite of the Chiefs; Sayings of the Ancient Men. *Bureau of American Ethnology (Washington, D.C.). Annual report of the Bureau of American Ethnology to the Secretary of the Smithsonian Institution,* 36:1914-1915, 37-597.

La Vere, D., 1998. *Life Among the Texas Indians: The WPA Narratives.* College Station: Texas A&M University Press.

Lambeck, K., 1995. Late Devensian and Holocene Shorelines of the British Isles and North Sea from Models of Glacio-hydro-isostatic Rebound. *Journal of the Geological Society*, 152, 437-448.

Laubin, R. and Laubin, G., 1977. *The Indian Tipi: Its History, Construction, and Use* (second edition). Norman: University of Oklahoma Press.

Leakey, L.S.B., 1951. Preliminary Excavations of Mesolithic Site At Abinger Common, Surrey. *Research Papers of the Surrey Archaeological Society*, 3,1-44.

Lehmer, D.J., 1948. The Jornada Branch of the Mogollon. *University of Arizona Bulletin,* 19(2). Social Science Bulletin 17.

Lemonier, P., 1993. *Technological Choices*. London: Routledge.

Leonard, W.R., 2002. Food for Thought: Dietary Change was a Driving Force in Human Evolution. *Scientific American*, December 2002, 64-71.

Leroi-Gourhan, A. and Brézillon, M., 1966. 'L'habitation Magdalènienne.' *Gallia Prèhistoire*, 9, 263-385.

Lewis, J., 1991. A Late Glacial and Early Postglacial Site at Three Ways Wharf, Uxbridge, England: Interim Report. *In*: Barton, R.N.E., Roberts, A.J., Roe, D., eds. *The Late Glacial in North-West Europe*. BAR British Series, 77, 246-255.

Lewis, H.T., 2003. An Anthropological Critique. *In*: Stewart, O. C., *Forgotten Fires, Native Americans and the Transient Wilderness*. Norman: University of Oklahoma Press.

Lewis, P.J., Gutierrez, M., Johnson, E., 2000. Ondarta Zibethicus (Arvicolinae, Rodentia) Dental Microwear Patterns as a Potential Tool for Paleoenvironmental Reconstruction. *Journal of Archaeological Science*, 27(9), 789-798.

Lian, O.B. and Brooks, G.R., 2004. Optical Dating Studies of Mud-dominated Alluvium and Buried Hearth-like Features from Red River Valley, Southern Manitoba, Canada. *The Holocene*, 14(4), 570-578.

Lintz, C., 1989. Experimental Thermal Discoloration and Heat Conductivity Studies of Caliche from Eastern New Mexico. *Geoarchaeology,* 4(4), 319-346.

Lovick, Steven K., 1983. Fire-Cracked Rock as Tools: Wear-Pattern Analysis. *Plains Anthropologist*, 28(99), 41-52.

Lowell, J.C., 1999. The Fires of Grasshopper: Enlightening Transformations in Subsistence Practices Through Fire-Feature Analysis. *Journal of Anthropological Archaeology*, 18, 441-470.

Lowenthal, D. and Price, H.C., 1964. The English Landscape. *Geographical Review*, 54(3), 309-346.

Lucas, G., 2001. Destruction and the Rhetoric of Excavation. *Norwegian Archaeological Review*, 34(1), 35-46.

Martin, H., 1793. An Account of Some of the Principal Dies Employed by the North-American Indians, Extracted from a Paper, Communicated by the Late Mr. Hugh Martin. *Transactions of the American Philosophical Society*, 3, 222-225.

McAllister, M., 2000. Archaeopolitics. *The S.A.A. Bulletin*, 18(1), 3.

McDowell-Loudan, E.E., 1983. FIRE - CRACKED ROCK : Preliminary Experiments To Determine Its Nature and Significance in Archaeological Contexts. *Chesopien*, 21(1), 21-29.

Meese, D. A., Gow, A.J., Grootes, P., Mayewski, P.A., Ram, M., Stuiver, M., Taylor, K.C., Waddington E.D., Zielinski, G.A., 1994. The Accumulation Record from the GISP2 Core as an Indicator of Climate Change Throughout the Holocene. *Science*, 266, 1680-1682.

Melody, M.E., 1989. *The Apache*. New York: Chelsea House Publishers.

Meltzer, D., 1991. Altithermal Archaeology and Paleoecology at Mustang Springs, on the Southern High Plains of Texas. *American Antiquity*, 56(2), 236-267.

Meltzer, D., 1993. *Search for the First Americans*. Washington: Smithsonian Books.

Meltzer, D., 1995. Modelling The Prehistoric Response to Altithermal Climates on the Southern High Plains. *In*: Johnson, E., ed. *Ancient Peoples and Landscapes*. Lubbock: Museum of Texas Tech University, 349-368.

Meltzer, D., 1999. Human Responses to Middle Holocene (Altithermal) Climates on the North American Great Plains. *Quaternary Research*, 52, 404–416.

Mercado-Allinger, P.A., 1984. Archaeological Investigations Caprock Canyons State Park. Briscoe County Texas, 41BI452. Texas Parks and Wildlife Department, Austin Texas.

Metcalfe, D. and Heath, K.M., 1990. Microrefuse and Site Structure: The Hearths and Floors of the Heartbreak Hotel. *American Antiquity*, 55(4), 781-796.

Miller, M. and Kenmotsu, N., 2004. Prehistory of the Jornada Mogollon and Eastern Trans-Pecos Regions of West Texas. *In*: Perttula, T.K., ed. *The Prehistory of Texas*. College Station: Texas A&M University Press, 205-265.

Miller Morris, J., 1997. *El Llano Estacado: Exploration and Imagination on the High Plains of Texas and New Mexico, 1536-1860*. Austin: Texas State Historical Association.

Mithen, S., 2000. *Hunter-Gatherer Landscape Archaeology, The Southern Hebrides Mesolithic Project 1988-98. Volume 1: Project Development, Palaeoenvironmental Studies and Archaeological Fieldwork on Islay*. Cambridge: Macdonald Institute for Archaeological Research.

Momber, G., 2000. Drowned and Deserted: a Submerged Prehistoric Landscape in the Solent, England. *International Journal of Nautical Archaeology*, 29(1), 86-99.

Morris, J.M., 1997. *El Llano Estacado: Exploration and Imagination on the High Plains of Texas and New Mexico*. Austin: Texas State Historical Association.

Muir, R., 1998. Geography and the History of Landscape: Half a Century of Development as Recorded in the Geographical Journal. *The Geographical Journal*, 164(2), 148-154.

Museum of London Archaeology Service, 1998. Beechbrook Wood (ARC BBW 98), An Archaeological Evaluation, Final Report, Prepared for Union Railways Limited, contract 194/870P4, 1-29.

Neck, R.W., 1987. Changing Holocene Snail Faunas and Environments along the Eastern Caprock Escarpment of Texas. *Quaternary Research*, 27, 312-322.

Neck, R.W., 1995. Extirpation of Quaternary Molluscs on the Southern High Plains. *In*: Johnson, E., ed. *Ancient Peoples and Landscapes*. Lubbock: Museum of Texas Tech University, 283-287.

Neumann, N., 2004. Earliest Fire Sheds Light on Hominids. Nature News Service [online]. New York, Nature Publishing Group. Available from: http://www.nature.com/nsu/nsu/pf/040426/040426-16.html. [Accessed 1 May 2004].

Newcomb, W.W. Jr., 1963. *The Indians of Texas*. Austin: University of Texas Press.

Nickels, D.L., 2001. 41CB47. Site survey form on file at the Texas Archaeological Research Laboratory, Austin.

Nickels, D.L., 2002. A Cultural Resource Inventory of the L7 Ranch, Crosby County, Texas. Archaeological Studies Report No.1. San Marcos: Center for Archaeological Studies, Southwest Texas State University.

O'Connell, J.F., Hawkes, K., Blurton Jones, N., 1991. Distribution of Refuse-Producing Activities at Hazda Residential Base Camps: Implications for Analyses of Site Structure. *In*: Kroll, E.M., Price, T. D., eds. *The Interpretation of Archaeological Spatial Patterning*. New York: Plenum Press, 61-76.

Odell, G.H., 2001. Stone Tool Research at the End of the Millennium: Classification, Function, and Behaviour. *Journal of Archaeological Research*, 9(1), 45-100.

Odell, G.H. and Cowan, F., 1988. Experiments with Spears and Arrows on Targets. *Journal of Field Archaeology*, 13(2), 195-212.

O'Malley, M. and Jacobi, R.M., 1978. The Excavation of a Mesolithic Occupation Site at Broom Hill, Braishfield, Hampshire. *Rescue Archaeology in Hampshire*, 4, 16-39.

Opler, E.M., 1941. *An Apache Life-Way: The Economical, Social, and Religious Institutions of the Chiricuahua Indians.* The University of Chicago Publications in Anthropology, Ethnological Series. Chicago: University of Chicago Press.

Osterkamp, W.R., Fenton, M.M., Gustavson T.C., Hadley, R.F., Holliday, V.T., Morison, R.B., Toy T.J., 1987. Great Plains. *In*: Graf, W.L., ed. *Geomorphic Systems of North America*: Boulder, Colorado: Geologic Society of America, Centennial Special Volume 2, 163-210.

Otte, M., 2002. Fire as an Evolutionary Factor. *In*: Gheorghiu, D., ed. *Fire in Archaeology*. BAR International Series, S1089. Oxford: Archaeopress, 7-10.

Oxford Archaeological Unit, 1996. Land South of Snarkhurst Wood, Eyhorne Street, Hollingbourne, Kent (SNK 95), Archaeological Evaluation Report, Prepared for Union Railways Limited, contract 192/84, 1-43.

Oxford Archaeological Unit, 1998. West of Scalers Hill, Cobham, Kent (ARC WSH 98), Archaeological Evaluation Report, Prepared for Union Railways Limited, contract 194/870, 1-20.

Pagoulatos, P., 2006. Experimental Burned Rock Studies on the Edwards Plateau: A View From Camp Bullis Texas. *North American Archaeologist*, 26(3), 289-329.

Palmer, S., 1971. Second Report on Excavations at Portland Site 1, 1967 to 1968. *Proceedings Dorset Natural History and Archaeological Society*, 92, 168-180.

Palmer, S., 1976. The Mesolithic Habitation Site at Culverwell, Portland, Dorset. Interim Note. *Proceedings of the Prehistoric Society*, 42, 324-327.

Palmer, S., 1999. *Culverwell Mesolithic Habitation Site, Isle of Portland, Dorset, Excavation Report and Research Studies*. BAR British Series, 287. Oxford: Archaeopress.

Palmer, S. and Dimbleby, G., 1979. *A Mesolithic Habitation Site on Winfrith Heath, Dorset. Proceedings of the Dorset Natural History and Archaeological Society, 101-27-50.*

Pauketat, T. and Alt, S.M., 2005. Agency in a Postmold? Physicality and the Archaeology of Culture-Making. *Journal of Archaeological method and theory*, 12(3), 213-235.

Perles, C., 1981. Hearth and Home in the Old Stone Age. *Natural History*, 90(10), 38-40.

Perry, D., 1997. *The Archaeology of Hunter-Gatherers: Plant Use in the Dutch Mesolithic*. Thesis (Phd). New York University.

Perry, D., 1999. Vegetative Tissues from Mesolithic Sites in the Northern Netherlands. *Current Anthropology*, 40(2), 231-237.

Petraglia, M.D., 2002. The Heated and the Broken: Thermally Altered Stone, Human Behavior, and Archaeological Site Formation. *North American Archaeologist*, 23(3), 241-270.

Pfaffenberger, B., 1988. Fetished Objects and Humanised Nature: Towards an Anthropology of Technology. *MAN*, 23, 236-252.

Pfaffenberger, B., 1992. Social Anthropology of Technology. *Annual Review of Anthropology*, 21, 491-516.

Pigeot, N., 1990. Technical and Social Actors, Flintknapping Specialists and Apprentices at Magdalenian Etiolles. *Archaeological Review from Cambridge*, 9(1), 126-141.

Potter, D., 2006. Destruction of Archaeological Sites in Texas: Results of the THC Stewards' Resurvey Project. *Current Archaeology in Texas*, 8(1),1-8.

Powell, R., 1999. *Hampreston, a New Mesolithic site in Dorset. Proceedings of the Dorset Natural History and Archaeological Society,* 121, 162.

Prewitt, E.R., 1995. Distributions of Typed Projectile Points in Texas. *Bulletin of the Texas Archeological Society,* 66, 83-173.

Proctor, R.N., 2003. Three Roots of Human Recency: Molecular Anthropology, the Refigured Acheulean, and the UNESCO Response to Auschwitz. *Current Anthropology*, 44(2), 213-239.

Proudfoot, B., 1976. The Analysis and Interpretation of Soil Phosphorus in Archaeological Contexts. *In*: Davidson, D.A., Shackley, M.L., eds. *Geoarchaeology: Earth Science and the Past.* London: Duckworth, 93-113.

Purdy, B.A., 1975. Fractures for the Archaeologist. *In*: Swanson, E., ed. *Lithic Technology Making and Using Stone Tools*. Chicago: World Anthropology, 133-141.

Pyne, S. J., 1991. *Burning Bush: A Fire History of Australia.* New York: Henry Holt and Company.

Pyne, S. J., 2001. *Fire, a Brief History.* London: British Museum Press.

Quigg, J.M., Malainey, M.E., Przybylski, R., Monks, G., 2001. No Bones About It: Using Lipid Analysis of Burned Rock and Groundstone Residues to Examine Late Archaic Subsistence Practices in South Texas. *Plains Anthropologist*, 46(177), 283-303.

Rackham, O., 1976. *Trees and Woodland in the British Landscape.* London: Dent.

Radovanović , I., 1996. *The Iron Gates Mesolithic.* International Monographs in Prehistory 11. Michigan: Ann Arbor.

Randerson, J., 2004. Charred Remains May be Earliest Human Fires. *Science*, 304, 725.

Rankine, W.F., 1952. A Mesolithic Chipping Floor at the Warren, Oakhanger, Selborne, Hants. *Proceedings of the Prehistoric Society*, 18, 21-35.

Rankine, W.F., 1953. Mesolithic Research in East Hampshire. The Hampshire Greensand. *Proceeding of the Hampshire Field Club & Archaeological Society,* 18(2), 170.

Rankine, W.F. and Clark, J.G.D., 1939. Excavations at Farnham, Surrey (1937-38); the Horsham Culture and Question of Mesolithic Dwellings. *Proceedings of the Prehistoric Society*, 5, 61-118.

Rankine, W.F., Rankine, W.M., Dimbleby, G.W., 1960. Further Excavations at a Mesolithic Site at Oakhanger, Selborne, Hants. *Proceedings of the Prehistoric Society*, 26, 242-262.

Rapp, G. Jr., Balescu, S., Lamothe, M., 1999. The Identification of Fire Cracked-Rock Using Luminescence of Alkali Feldspars. *American Antiquity,* 64(1), 71-78.

Rathje, W.L., 1978. Archaeological Ethnography... Because Sometimes it is Better to Give than to Receive. *In*: Gould, R.A., ed. *Explorations in Ethnoarchaeology.* Albuquerque: University of New Mexico Press, 49-75.

Reeves, C.C. Jr., 1976. *Caliche: Origin, Classification, Morphology and Uses.* Lubbock: Estacado Books.

Reyneir, M.J., 1993. A Statistical Analysis of Ten Early Mesolithic Sites in SE England. *Proceedings of the Lithic Studies Society*, 4, 199-205.

Reyneir, M.J., 2002. Kettlebury 103: A Mesolithic 'Horsham' Type Stone Assemblage from Hankley Common, Elstead. *Surrey Archaeological Collections*, 89, 211-231.

Rincon, P., 2004. Early Human Fire Mastery Revealed. BBC News [online]. London, Broadcasting House. Available from: http://www.bbc.co.uk/go/pr/fr/-/hi/sci/tech/3670017.stm. [Accessed 5 February 2004].

Roberts, S.J., Smith, C.I., Millard, A., Collins, M.J., 2002. The Taphonomy of Cooked Bone: Characterizing Boiling and its Physico-chemical Effects. *Archaeometry*, 44(3), 485-494.

Robin, C. and Rothschild, N.A., 2002. Archaeological Ethnographies, Social Dynamics of Outdoor Space. *Journal of Social Archaeology*, 2(2), 159-172.

Robins, P., 1998. Mesolithic Sites at Two Mile Bottom Near Thetford, Norfolk. *In*: Ashton, N., Healy, F., Pettitt, P., eds. *Stone Age Archaeology: Essays in honour of John Wymer*. Oxbow Monograph 102. Lithic Studies Occasional Paper 6, 205-210.

Rolland, N., 2004. Was the Emergence of Home Bases and Domestic Fire a Punctuated Event? A Review of the Middle Pleistocene Record in Eurasia. *Asian Perspectives*, 43(2), 248-280.

Rollings, W.H., 1989. *The Comanche*. New York: Chelsea House Publishers.

Rosenfeld, A., 1964. Excavations in the Torbryan Caves, Devonshire. *Proceedings of the Devon Archaeological Society*, 22, 3-26.

Roskams, S., 2001. *Excavation*. Cambridge: Cambridge University Press.

Russell, F., 1898. Myths of the Jicarilla Apache. *Journal of American Folklore*, 11(43), 253-271.

Saraydar, S.C. and Shimada, I., 1973. Experimental Archaeology: A New Outlook. *American Antiquity*, 38(3), 344-350.

Sabin, T. and Holliday, V.T., 1995. Morphological and Spatial Relationships of Playas and Lunettes on the Southern High Plains. *Annals of the Association of American Geographers*, 25, 286-305.

Schiffer, M., 1983. Towards the Identification of Formation Processes. *American Antiquity*, 48, 675-706.

Schiffer, M., Skibo, J.M., Boelke, T.C., Neupert, M.A., Aronson, M., 1994. New Perspectives on Experimental Archaeology: Surface Treatments and Thermal Response of the Clay Cooking Pot. *American Antiquity*, 59(2), 197-217.

Schlanger, S., 1992. Recognizing Persistent Places in Anasazi Settlement Systems. *In*: Rossignol, J., Wandsnider, L., eds. *Space, Time, and Archaeological Landscapes*. New York: Plenum Press, 91-112.

Schroeder, E.A. and Holliday, V.T., 1994. Physiographic Setting. *In*: Johnson, E., ed. *Archaeological Survey Along the Mobil ESTE CO2 Pipeline Corridor from Denver City to Clairemont, Texas*. Lubbock Lake Landmark Quaternary Research Center Series, 6, 17-37.

Score, D. and Mithen, S., 2000. The Experimental Roasting of Hazelnuts. *In*: Mithen, S., ed. *Hunter-Gatherer Landscape Archaeology, The Southern Hebrides Mesolithic Project 1988-98, Volume 2: Archaeological Fieldwork on Colonsay, Computer Modelling, Experimental Archaeology, and Final Interpretations*. Cambridge: Macdonald Institute for Archaeological Research, 507-512.

Sebastian, L. and Larralde, S., 1989. *Living on the Land: 11,000 Years of Human Adaptation in Southeastern New Mexico. An Overview of Cultural Resources in the Roswell District, Bureau of Land Management*. Bureau of Land Management, Cultural Resources Series, 6, 1-174.

Sellards, E.H., 1955. Fossil Bison and Associated Artifacts from Milnesand, New Mexico. *American Antiquity*, 20(4), 336-344.

Selwood, S.M. and Holliday, V.T., 2002. Excavation Program at 41LU1 Area 10 for the 1999 Field Season. *In*: Johnson, E., ed. *Holocene Investigations at the Lubbock Lake Landmark. The 1991 through 2000 Work*. Lubbock Lake Landmark Quaternary Research Center Series, 11, 167-181.

Sergant, J., Crombe, P., Perdaen, Y., 2006. The 'Invisible' Hearths: a Contribution to the Discernment of Mesolithic Non-Structured Surface Hearths. *Journal of Archaeological Science*, 33(7), 999-1007.

Shea, J.J., 2006. The Origins of Lithic Projectile Point Technology: Evidence from Africa, the Levant, and Europe. *Journal of Archaeological Science*, 33(6), 823-846.

Sheldon, A.E., 1905. Ancient Indian Fireplaces of the South Dakota Bad-Lands. *American Anthropologist*, 7(1), 44-48.

Shelley, P.H., 1994. A Geoarchaeological and Technological Evaluation of the Archaic Archaeology of the Llano Estacado and Adjacent Areas of New Mexico. *In*: Shelley, P., ed. *Archaic Hunter-gatherer Archaeology in the Southwest*. Eastern New Mexico University Contributions in Anthropology, 13(1), 372-404.

Sheridan, R., Sheridan, D., Hassell, P., 1963. Rescue Excavation of a Mesolithic Site at Greenham Dairy Farm, Newbury. *Transactions of the Newbury and District Field Club*, 11, 69-73.

Shockey, D., 1997. An Experimental Inquiry into the Thermal Characteristics of Square and Round Hearths. *Oklahoma Anthropological Society Bulletin*, 46, 25-33.

Shott, M., 2005. Two Cultures: Thought and Practice in British and North American archaeology. *World Archaeology*, 37(1), 1-10.

Sidell, J., Cotton, J., Rayner, L., Wheeler, L., 2002. The Prehistory and Topography of Southwark and Lambeth. *MoLAS Monograph 14*, London: Museum of London Archaeology Service.

Simms, S.R. and Heath, K.M., 1990. Site Structure of the Orbit Inn: An Application of Ethnoarchaeology. *American Antiquity*, 55(4), 797-813.

Smith, A.G., Whittle, A., Cloutman, E.W., Morgan, L.A., 1989. Mesolithic and Neolithic Activity and Environmental Impact on the South-east Fen-edge in Cambridgeshire. *Proceedings of the Prehistoric Society*, 55, 207-249.

Smith, C., 1992. *Late Stone Age Hunters of the British Isles*. London: Routledge.

Smith, V.L., 1993. Humakind in Prehistory: Ecology, Economy, and Institutions. *In*: Anderson, T.L., Simmons, R.T., eds. *The Political Economy of Customs and Culture*, Rowman & Littlefield Publishers, Inc, 157-184.

Smith, C.S. and McNess, L.M., 1999. Facilities and Hunter-Gatherer Long-Term Land Use Patterns: An Example from Southwest Wyoming. *American Antiquity*, 6(1), 117-136.

Smith, C.S., William, M., Johansen, K.A., 2001. Sego Lilies and Prehistoric Foragers: Return Rates, Pit Ovens, and Carbohydrates. *Journal of Archaeological Science*, 28, 169-183.

Smith, E.A. and Winterhalder, B., (eds.) 1992. *Evolutionary Ecology and Human Behaviour*. New York: Aldine de Gruyter.

Solecki, R.S., 1995. The Cultural Significance of the Fire Hearths in the Middle Paleolithic of Shanidar Cave, Iraq. *In*: Johnson, E., ed. *Ancient Peoples and Landscapes*. Lubbock: Museum of Texas Tech University.

Speth, J.D., 2000. Boiling Vs. Baking and Roasting: a Taphonomic Approach to the Recognition of Cooking Techniques in Small Mammals. *In*: Rowley-Conwy, P., ed. *Animal Bones, Human Societies*. Oxford: Oxbow Books, 89-105.

Spencer, R. F., Jennings, J.D., Dibble, C.D., Johnson, E., King, A.R., Stern, T., Stewart, K.M., Stewart, O.C., Wallis, W.J., 1977. *The Native Americans: Ethnology and Backgrounds of the North American Indians*. New York: Harper and Row.

Spier, L., 1930. Klamath Ethnography. *University of California Publications in American Archaeology and Ethnology*, 30. Berkley: University of California Press.

Spikins, P., 1999. *Mesolithic Northern England: Environment, Population and Settlement*. BAR British Series, 283. Oxford: Archaeopress.

Spikins, P., 2000. Go Forth and Multiply? Gradual Population Growth Re-assessed – A Case Study from Mesolithic Northern England. *Archaeological Review from Cambridge,* 17, (1) 99-121.

Stafford, C.R. and Hajic, E.R., 1992. Landscape Scale: Geoenvironmental Approaches to Prehistoric Settlement Strategies. *In*: Rossignol, J., Wandsnider, L., eds. *Space, Time, and Archaeological Landscapes*. New York: Plenum Press, 137-161.

Stafford, T., 1981. Alluvial Geology and Archaeological Potential of the Texas Southern High Plains. *American Antiquity*, 48, 548-565.

Stahl, A.B., 1989. Plant Food Processing: Implications for Dietary Quality. *In*: Harris, D.R., Hillman, G.C., eds. *Foraging and Farming: The Evolution of Plant Exploitation*, 171-196. London: Unwin Hyman.

Stapert, D. and Street, M., 1997. High Resolution or Optimum Resolution? Spatial Analysis of the *Federmesser* site at Andenrach, Germany. *World Archaeology*, 29(2), 172-194.

State of New Mexico Office of Cultural Affairs Historic Preservation Division, 1993. User's Guide. New Mexico Cultural Resource Information System. Guidelines for Submitting Archaeological Records. Santa Fe: Archaeological Records Management Section.

Stevenson, M., 1985. The Formation of Artifact Assemblages at Workshop/Habitation Sites: Models from Peace Point Northern Alberta. *American Antiquity,* 50(1), 63-81.

Stevenson, M., 1991. Beyond the Formation of Hearth-Associated Artifact Assemblages. *In*: Kroll, E.M., Price, T. D., eds. *The Interpretation of Archaeological Spatial Patterning*. New York: Plenum Press, 269-300.

Steward, J.H., 1933. Ethnography of the Owens Valley Paiute. *University of California Publications in American Archaeology and Ethnology*, 33, 233-350. University of California Press, Berkley.

Stewart, O.C., 2002. *Forgotten Fires: Native Americans and the Transient Wilderness*. Norman: University of Oklahoma Press.

Stiner, M.C. and Kuhn, S.L., 1995. Differential Burning, Recrystallization, and Fragmentation of Archaeological Bone. *Journal of Archaeological Science*, 22, 223-227.

Straus, L.G., 2006. Of Stones and Bones: Interpreting Site Function in the Upper Paleolithic and Mesolithic of Western Europe. *Journal of Anthropological Archaeology*, 25, 500-509.

Stringer, C. and Andrews P., 2005. *The Complete World of Human Evolution*. London: Thames & Hudson.

Stout, J.E., 2001. Dust and Environment in the Southern High Plains of North America. *Journal of Arid Environments*, 47, 425-441.

Sulivan III, A.P. and Rozen, K.C., 1985. Debitage Analysis and Archaeological Interpretation. *American Antiquity*, 50(4), 755-779.

Sulivan III, A.P., Cook, R.A., Purtill, M.P., Uphus, P.M., 2001. Economic and Land-Use Implications of Prehistoric Fire-Cracked-Rock Piles, Northern Arizona. *Journal of Field Archaeology*, 28, 367-382.

Summers, P.G., 1941. A Mesolithic Site, Near Iwerne Minster, Dorset. *Proceedings of the Prehistoric Society*, 7, 145-146.

Surovell, T.A. and Brantigham, J.P., 2007. A Note on the Use of Temporal Frequency Distributions in Studies of Prehistoric Demography. *Journal of Archaeological Science,* 34, 1868-1877.

Swanton, J.R., 1928. Social Organization and Social Usages of the Indians of the Creek Confederacy. *Bureau of American Ethnology (Washington, D.C.). Annual Report of the Bureau of American Ethnology to the Secretary of the Smithsonian Institution,* 42,1924-1925, 1-472.

Sweeney, J.C. and O'Hare, G.P., 1992. Geographical Variation in Precipitation Yields and Circulation Types in Britain and Ireland. *Transactions of the Institute of British Geographers*, 17(4), 448-463.

Telles, J.A. 2000. Biographical Connections: Experiences as Sources of Legitimate Knowledge in Qualitative Research. *Qualitative Studies in Education*, 13(3), 251-262.

Texas Historical Commission, 2006. A Property Owners Guide to Archaeological Sites. Austin, Texas: Texas Historical Commission.

Théry-Parisot, I., 2002. Fuel Management (Bone and Wood) During the Lower Aurignacian in the Pataud Rock Shelter (Lower Paleolithic, Les Eyzies de Tayac, Dordogne, France). Contribution of Experimentation. *Journal of Archaeological Science*, 29, 1415-1421.

Thieme, H., 2005. The Lower Palaeolithic Art of Hunting: The Case of Schoningen 13 11-4, Lower Saxony, Germany. *In*: Gamble C., Porr, M., eds. *The Hominid Individual in Context Archaeological Investigations of Lower and Middle Palaeolithic Landscapes, Locales and Artefacts*. London: Routledge, 115-132.

Thomas, J., 2000. *Interpretive Archaeology: A Reader*. Leicester: Leicester University Press.

Thompson, J. L., 1987. Modern, Historic, and Fossil Flora. *In*: Johnson, E., ed. *Lubbock Lake. Late Quaternary Studies on the Southern High Plains.* College Station: Texas A&M University Press, 26-35.

Thoms, A.V., 2003. Cook-Stone Technology in North-America: Evolutionary Changes in Domestic Fire Structures During the Holocene. *In*: Frère-Sautot, M., ed. *Le Feu Domestique et Ses Structures au Néothlique et Âges des Métaux*. Monique Mergoil, Montagnac, France : Âctes du Colloque de Bourg-en-Bresse et Breune, Prèhistories, 9, 87-96.

Thoms, A.V., 2008. The Fires Stones Carry: Ethnographic Records and Archaeological Expectations for Hot-Rock Cookery in Western North America. *Journal of Anthropological Archaeology*, 27, 443-460.

Thoms, A.V., 2009. Rocks of Ages: Propagation of Hot-Rock Cookery in Western North America. *Journal of Archaeological Science*, 36(3), 573-591.

Timmins, P.A., 1997. The Little Shaver Site: Site Structure and Excavation Methodology on a Small Unploughed Site. *In*: Pilon J., Perkins, R., eds. *Home is where the Hearth is: The Contribution of Small Sites to our Understanding of Ontario's Past*. Ontario: Proceedings of the 1996 Ontario Archaeological Society Symposium, 72-88.

Torrence, R., 2001. Hunter-Gatherer Technology: Macro-and Microscale Approaches. *In*: Panter-Brick, C., Layton, R.H., Rowley-Conwy, P., eds. H*unter-Gatherers An Interdisciplinary Perspective*. Cambridge: Cambridge University Press, 73-98.

Townend, S., 2002. Interpreting People Interpreting Things: A Heideggerian Approach to 'Experimental Reconstruction'. *Papers from the Institute of Archaeology*, 14, 72-93.

Trigger, B., 1986. Prospects for a World Archaeology. *World Archaeology*, 18(1), 1-20.

Trigger, B., 1989. *A History of Archaeological Thought*. Cambridge: Cambridge University Press.

Tringham, R., 1978. Experimentation, Ethnoarchaeology, and the Leapfrogs in Archaeological Methodology. *In*: Gould, R.A., ed. *Explorations in Ethnoarchaeology*. Albuquerque: University of New Mexico Press.

Tringham, R., Cooper, G., Odell, G., Voytek, B., Whitman, A., 1974. Experimentation in the Formation of Edge Damage: A New Approach to Lithic Analysis. *Journal of Field Archaeology*, 1(1/2), 171-196.

Tucker, G.C., and Winters, R., 1990. 41AD48. Site Survey Form on File at the Texas Archaeological Research Laboratory, Austin.

Vallin, L., Masson, B., Caspar, J., 2001. Taphonomy at Hermies, France: A Mousterian Knapping Site in a Loessic Context. *Journal of Field Archaeology*, 28(3/4), 419-436.

Vallverdú, J., Allué, E., Bischoff, J.L., Caceres, I., Carbonell, E., Cebrià, A., Garcia-Antón, D., Huguet, R., Ibáñez, N., Martínez, K., Pastó, I., Rosell, J., Saladié, P., Vaquero, M., 2005. Short Human Occupations in the Middle Palaeolithic Level I of the Abric Romaní Rock-Shelter (Capellades, Barcelona, Spain). *Journal of Human Evolution*, 48, 157-174.

Vanpool, T.L. and Vanpool, C.S., (eds.) 2003. *Essential Tensions in Archaeological Method and Theory*. Salt Lake City: University of Utah Press.

Vaquero, M. and Pastó, I., 2001. The Definition of Spatial Units in Middle Palaeolithic Sites: The Hearth-Related Assemblages. *Journal of Archaeological Science*, 28, 1209-1220.

Wade, M., 2002. Go-between: The Roles of Native American Women and Alvar Nenez Cabeza de Vaca in Southern Texas in the 16th Century. *Journal of American Folklore*, 122(445), 332-342.

Wade, M., 2003. *The Native Americans of the Texas Edwards Plateau 1582-1799*. Austin: University of Texas Press.

Wandsnider, L., 1996. Describing and Comparing Archaeological Spatial Structures. *Journal of Archaeological Method and Theory*, 3(4), 319-384.

Wandsnider, L., 1997. The Roasted and the Boiled: Food Composition and Heat Treatment with Special Emphasis on Pit-Hearth Cooking. *Journal of Anthropological Archaeology,* 16,1-48.

Warnica, J.M. and Williamson, T., 1968. The Milnesand Site. Revisited. *American Antiquity*, 33(1), 16-24.

Wendland, W.M., 1995. Environmental History at Lubbock Lake and the Episodic Nature of Holocene Climates. *In*: Johnson, E., ed. *Ancient Peoples and Landscapes*. Lubbock: Museum of Texas Tech University, 341-348.

Wendorf, F., (ed.) 1961. Paleoecology of the Llano Estacado. *Report No. 1. of the Fort Burgwin Research Center*. Santa Fe: The Museum of New Mexico Press.

Westbroek, P., Collins, M.J., Jansen, J.H., Talbot, L.M., 1993. World Archaeology and Global Change: Did Our Ancestors Ignite the Ice Age?. *World Archaeology*, 25(1), 122-133.

Whalen, M.E., 1994. Moving Out of the Archaic on the Edge of the Southwest. *American Antiquity*, 59(4), 622-638.

Whallon, R. Jr., 1973. Spatial Analysis of Occupation Floors I: Application of Dimensional Variance. *American Antiquity,* 38(3), 266-278.

Whallon, R. Jr., 1974. Spatial Analysis of Occupation Floors II: The Application of Nearest Neighbour Analysis. *American Antiquity*, 39(1), 16-34.

White, J.R., 1980. A Closer Look at Clusters. *American Antiquity*, 45(1), 68-74.

Whitelaw, T.M., 1994. Order Without Architecture: Functional, Social and Symbolic Dimensions in Hunter-Gatherer Settlement Organization. *In*: Parker Pearson, M., Richards, C., eds. *Architecture and Order: Approaches to Social Space*. London: Routledge, 217-43.

Williams, G., 1990. Burnt Mounds in South-West Wales. *In*: Buckley, V., ed. *Burnt Offerings: International Contributions to Burnt Mound Archaeology*. Dublin: Wordwell, 129-140.

Williams, G.W., 2000. Introduction to Aboriginal Fire Use in North America. *Fire Management Today*, 60(3), 8-10.

Winship, G.P., 1922. *The Journey of Coronado, 1540-1542, From The City of Mexico to the Grand Canon of the Colorado and the Buffalo Plains of Texas, Kansas and Nebraska, as Told by Himself and His Followers*. New York: Allerton Book Co.

Wishart, D.J., (ed.) 2004. *Encyclopedia of the Great Plains*. Lincoln: University of Nebraska Press.

Winterhalder, B., 2001. The Behavioural Ecology of Hunter-Gatherers. *In*: Panter-Brick, C., Layton, R.H., Rowley-Conwy, P., eds. *Hunter-Gatherers: An Interdisciplinary Perspective*. Cambridge: Cambridge University Press, 12-38.

Wobst, M. H., 1978. The Archaeo-Ethnology of Hunter-Gatherers or the Tyranny of the Ethnographic Record in Archaeology. *American Antiquity*, 43(2), 303-309.

Wood, W.W., Stokes, S., Rich, J., 2002. Implications of Water Supply for Indigenous Americans during Holocene Aridity Phases on the Southern High Plains, USA. *Quaternary Research*, 38, 139-148.

Wrangham, R.W. and Conklin-Brittain, N., 2003. 'Cooking as a Biological Trait'. *Comparative Biochemistry and Physiology*, Part A, 136, 35-46.

Wrangham, R.W., Jones, J.H., Laden, G., Pilbeam, D., Conklin-Brittain, N., 1999. The Raw and the Stolen: Cooking and the Ecology of Human Origins. *Current Anthropology*, 40(5), 567-594.

Wright, P., 2003. Preservation or Destruction of Plant Remains by Carbonization? *Journal of Archaeological Science*, 30, 577-583.

Wright, H.A. and Bailey, A.W., 1982. *Fire Ecology, United States and Southern Canada*. New York: Wiley-Interscience.

Wylie, A., 2002. *Thinking from Things, Essays in the Philosophy of Archaeology*. Berkley: University of California Press.

Wymer, J.J., 1977. *Gazetteer of Mesolithic sites in England and Wales with a Gazetteer of Upper Palaeolithic Sites in England and Wales*. Council for British Archaeology Research Reports 20, CBA, London.

Yellen, J., 1977. *Archaeological Approaches to the Present: Models for Reconstructing the Past*. New York: Academic Press.

Yvorra, P., 2003. Management of Space in a Paleolithic Rock Shelter: Defining Activity Areas by Spatial Analysis. *Antiquity*, 77(296), 336-344.

Zurel, R. L., 1979. Brief Comments Regarding the Nature of Fire Cracked Rock on Aboriginal Sites in the Great Lakes Area. Technical Report Series 3, Working Papers in Archaeology, Oakland University, Michigan.